T0275697

Alternative Energy in Power Electronics

Alternative Energy in Power Electronics

Edited by

Muhammad H. Rashid
Fellow IET (UK), Life Fellow IEEE (USA)
Department of Electrical and Computer Engineering
University of West Florida
USA

AMSTERDAM • BOSTON • HEIDELBERG • LONDON
NEW YORK • OXFORD • PARIS • SAN DIEGO
SAN FRANCISCO • SINGAPORE • SYDNEY • TOKYO

Butterworth-Heinemann is an imprint of Elsevier

Butterworth-Heinemann is an imprint of Elsevier
225 Wyman Street, Waltham, MA 02451, USA
The Boulevard, Langford Lane, Kidlington, Oxford, OX5 1GB, UK

Library of Congress Cataloging-in-Publication Data
Alternative energy in power electronics / [edited by] Muhammad H. Rashid.
   pages cm
   ISBN 978-0-12-416714-8 (pbk.)
 1. Power electronics. I. Rashid, Muhammad H. (Muhammad Harunur), 1945- editor.
 TK7881.15.A48 2015
 621.31'7–dc23
                           2014035517

British Library Cataloguing in Publication Data
A catalogue record for this book is available from the British Library

ISBN: 978-0-12-416714-8

For information on all Butterworth-Heinemann publications
visit our web site at http://store.elsevier.com

Working together
to grow libraries in
developing countries

www.elsevier.com • www.bookaid.org

Contents

3.   Photovoltaic System Conversion   155

Lana El Chaar

Contributors

Noradlina Abdullah Transmission and Distribution, TNB Research, Malaysia

Benjamin Blunier Universite de Technologie de Belfort-Montbeliard, Belfort Cedex, France

Juan M. Carrasco Department of Electronic Engineering, Engineering School, Seville University, Spain

Pablo A. Cassani Power Electronics and Energy Research (PEER) Group, P. D. Ziogas Power Electronics Laboratory, Department of Electrical and Computer Engineering, Concordia University, Montreal, QC, Canada

Lana El Chaar Electrical Engineering Department, The Petroleum Institute, Abu Dhabi, UAE

H. Dehbonei Department of Electrical and Computer Engineering, Curtin University of Technology, Perth, Western Australia, Australia

Eduardo Galván Department of Electronic Engineering, Engineering School, Seville University, Spain

Ir. Zahrul Faizi bin Hussien Transmission and Distribution, TNB Research, Malaysia

S. M. Islam Department of Electrical and Computer Engineering, Curtin University of Technology, Perth, Western Australia, Australia

Alireza Khaligh Electrical and Computer Engineering Department, University of Maryland at College Park, College Park, MD, USA

Srdjan Lukic Department of Electrical and Computer Engineering, North Carolina State University, Raleigh, NC, USA

S.K. Mazumder, Sr. Department of Electrical and Computer Engineering, Laboratory for Energy and Switching-Electronics Systems (LESES), University of Illinois, Chicago, IL, USA

C. V. Nayar Department of Electrical and Computer Engineering, Curtin University of Technology, Perth, Western Australia, Australia

Omer C. Onar Power Electronics and Electric Machinery Group, National Transportation Research Center, Oak Ridge National Laboratory, Oak Ridge, TN, USA

Ramón Portillo Department of Electronic Engineering, Engineering School, Seville University, Spain

Azlan Abdul Rahim Transmission and Distribution, TNB Research, Malaysia

H. Sharma Research Institute for Sustainable Energy, Murdoch University, Perth, Western Australia, Australia

K. Tan Department of Electrical and Computer Engineering, Curtin University of Technology, Perth, Western Australia, Australia

Sheldon S. Williamson Power Electronics and Energy Research (PEER) Group, P. D. Ziogas Power Electronics Laboratory, Department of Electrical and Computer Engineering, Concordia University, Montreal, QC, Canada

Preface

Electric power generated from renewable energy sources is getting increasing attention and supports for new initiatives and developments in order to meet the increased energy demands around the world. The availability of computer–based advanced control techniques along with the advancement in the high-power processing capabilities is opening new opportunity for the development, applications and management of energy and electric power. Power Electronics is integral part of the power processing and delivery from the energy sources to the utility supply and the electricity consumers.

The demand for energy, particularly in electrical forms, is ever-increasing in order to improve the standard of living. Power electronics helps with the efficient use of electricity, thereby reducing power consumption. Semiconductor devices are used as switches for power conversion or processing, as are solid state electronics for efficient control of the amount of power and energy flow. Higher efficiency and lower losses are sought for devices used in a range of applications, from microwave ovens to high-voltage dc transmission. New devices and power electronic systems are now evolving for even more effective control of power and energy.

Power electronics has already found an important place in modern technology and has revolutionized control of power and energy. As the voltage and current ratings and switching characteristics of power semiconductor devices keep improving, the range of applications continue to expand in areas, such as lamp controls, power supplies to motion control, factory automation, transportation, energy storage, multi-megawatt industrial drives, and electric power transmission and distribution. The greater efficiency and tighter control features of power electronics are becoming attractive for applications in motion control by replacing the earlier electromechanical and electronic systems. Applications in power transmission and renewable energy include high-voltage dc (VHDC) converter stations, flexible ac transmission system (FACTS), static var compensators, and energy storage. In power distribution, these include dc-to-ac conversion, dynamic filters, frequency conversion, and custom power system.

Audience:
The purpose of *Alternative Energy in Power Electronics* is a derivative of the best-selling *Power Electronics Handbook*, Third Edition. The purpose of *Alternative Energy in Power Electronics* is to provide a reference that is both concise and useful for engineering students and practicing professionals. It

is designed to cover topics that relate to renewable energy processing and delivery. It is designed as advanced textbooks and professional references. The contributors are leading authorities in their areas of expertise. All were chosen because of their intimate knowledge of their subjects, and their contributions make this a comprehensive state-of-the-art guide to the expanding filed energy of energy.

Muhammad H. Rashid, Editor-in-Chief

Any comments and suggestions regarding this book are welcome. They should be sent to

Dr. Muhammad H. Rashid
Professor
Department of Electrical and Computer Engineering
University of West Florida
11000 University Parkway Pensacola.
FL 32514-5754, USA
E-mail: mrashidfl@gmail.com
Web: http://uwf.edu/mrashid

Chapter 1

Power Electronics for Renewable Energy Sources

C. V. Nayar, S. M. Islam, H. Dehbonei, and K. Tan
Department of Electrical and Computer Engineering, Curtin University of Technology, Perth, Western Australia, Australia

H. Sharma
Research Institute for Sustainable Energy, Murdoch University, Perth, Western Australia, Australia

Chapter Outline

1.1 INTRODUCTION

The Kyoto agreement on global reduction of greenhouse gas emissions has prompted renewed interest in renewable energy systems worldwide. Many renewable energy technologies today are well developed, reliable, and cost competitive with the conventional fuel generators. The cost of renewable energy technologies is on a falling trend and is expected to fall further as demand and production increases. There are many renewable energy sources (RES) such as biomass, solar, wind, mini hydro and tidal power. However, solar and wind energy systems make use of advanced power electronics technologies and, therefore the focus in this chapter will be on solar photovoltaic and wind power.

One of the advantages offered by (RES) is their potential to provide sustainable electricity in areas not served by the conventional power grid. The growing market for renewable energy technologies has resulted in a rapid growth in the need of power electronics. Most of the renewable energy technologies produce DC power and hence power electronics and control equipment are required to convert the DC into AC power.

Inverters are used to convert DC to AC. There are two types of inverters: (a) stand-alone or (b) grid-connected. Both types have several similarities but are different in terms of control functions. A stand-alone inverter is used in off-grid applications with battery storage. With back-up diesel generators (such as photovoltaic (PV)/diesel/hybrid power systems), the inverters may have additional control functions such as operating in parallel with diesel generators and bi-directional operation (battery charging and inverting). Grid interactive inverters must follow the voltage and frequency characteristics of the utility generated power presented on the distribution line. For both types of inverters, the conversion efficiency is a very important consideration. Details of stand-alone and grid-connected inverters for PV and wind applications are discussed in this chapter.

Section 1.2 covers stand-alone PV system applications such as battery charging and water pumping for remote areas. This section also discusses power electronic converters suitable for PV-diesel hybrid systems and grid-connected PV for rooftop and large-scale applications. Of all the renewable energy options, the wind turbine technology is maturing very fast. A marked rise in installed wind power capacity has been noticed worldwide in the last decade. Per unit generation cost of wind power is now quite comparable with the conventional generation. Wind turbine generators are used in stand-alone battery charging applications, in combination with fossil fuel generators as part of hybrid systems and as grid-connected systems. As a result of advancements in blade design, generators, power electronics, and control systems, it has been possible to increase dramatically the availability of large-scale wind power. Many wind generators now incorporate speed control mechanisms like blade pitch control or use converters/inverters to regulate power output from variable speed wind turbines. In Section 1.3, electrical and power conditioning aspects of wind energy conversion systems were included.

1.2 POWER ELECTRONICS FOR PHOTOVOLTAIC POWER SYSTEMS

1.2.1 Basics of Photovoltaics

The density of power radiated from the sun (referred as "solar energy constant") at the outer atmosphere is $1.373 \, kW/m^2$. Part of this energy is absorbed and scattered by the earth's atmosphere. The final incident sunlight on earth's surface has a peak density of $1 \, kW/m^2$ at noon in the tropics. The technology of photovoltaics (PV) is essentially concerned with the conversion of this energy into usable electrical form. Basic element of a PV system is the solar cell. Solar cells can convert the energy of sunlight directly into electricity. Consumer appliances used to provide services such as lighting, water pumping, refrigeration, telecommunication, television, etc. can be run from PV electricity.

Solar cells rely on a quantum-mechanical process known as the "photovoltaic effect" to produce electricity. A typical solar cell consists of a p–n junction formed in a semiconductor material similar to a diode. Figure 1.1 shows a schematic diagram of the cross section through a crystalline solar cell [1]. It consists of a 0.2–0.3 mm thick monocrystalline or polycrystalline silicon wafer having two layers with different electrical properties formed by "doping" it with other impurities (e.g. boron and phosphorous). An electric field is established at the junction between the negatively doped (using phosphorous atoms) and the positively doped (using boron atoms) silicon layers. If light is incident on the solar cell, the energy from the light (photons) creates free charge carriers, which are separated by the electrical field. An electrical voltage is generated at the external contacts, so that current can flow when a load is connected. The photocurrent (I_{ph}), which is internally generated in the solar cell, is proportional to the radiation intensity.

A simplified equivalent circuit of a solar cell consists of a current source in parallel with a diode as shown in Fig. 1.2a. A variable resistor is connected to the solar cell generator as a load. When the terminals are short-circuited, the output voltage and also the voltage across the diode is zero. The entire photocurrent (I_{ph}) generated by the solar radiation then flows to the output. The solar cell current has its maximum (I_{sc}). If the load resistance is increased, which results in an increasing voltage across the p–n junction of the diode, a portion of the current flows through the diode and the output current decreases by the same amount. When the load resistor is open-circuited, the output current is zero and the entire photocurrent flows through the diode. The relationship between current and voltage may be determined from the diode characteristic equation

$$I = I_{ph} - I_0(e^{qV/kT} - 1) = I_{ph} - I_d \tag{1.1}$$

where q is the electron charge, k is the Boltzmann constant, I_{ph} is photocurrent, I_0 is the reverse saturation current, I_d is diode current, and T is the solar cell operating temperature ($^\circ$K). The current vs voltage (I–V) of a solar cell is thus equivalent to an "inverted" diode characteristic curve shown in Fig. 1.2b.

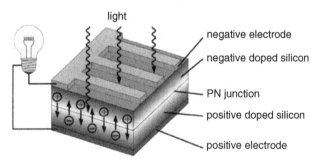

FIGURE 1.1 Principle of the operation of a solar cell [2].

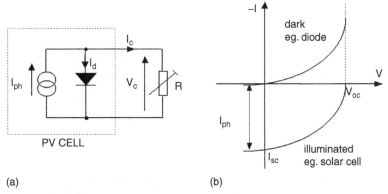

(a)   (b)

FIGURE 1.2   Simplified equivalent circuit for a solar cell.

A number of semiconductor materials are suitable for the manufacturing of solar cells. The most common types using silicon semiconductor material (Si) are:

- Monocrystalline Si cells.
- Polycrystalline Si cells.
- Amorphous Si cells.

A solar cell can be operated at any point along its characteristic current–voltage curve, as shown in Fig. 1.3. Two important points on this curve are the open-circuit voltage (V_{oc}) and short-circuit current (I_{sc}). The open-circuit voltage is the maximum voltage at zero current, while short-circuit current is the maximum current at zero voltage. For a silicon solar cell under standard test conditions, V_{oc} is typically 0.6–0.7 V, and I_{sc} is typically 20–40 mA for every square centimeter of the cell area. To a good approximation, I_{sc} is proportional to the illumination level, whereas V_{oc} is proportional to the logarithm of the illumination level.

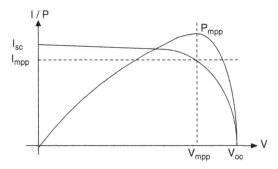

FIGURE 1.3   Current vs voltage (*I–V*) and current power (*P–V*) characteristics for a solar cell.

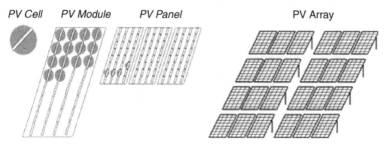

FIGURE 1.4 PV generator terms.

A plot of power (P) against voltage (V) for this device (Fig. 1.3) shows that there is a unique point on the I–V curve at which the solar cell will generate maximum power. This is known as the maximum power point (V_{mp}, I_{mp}). To maximize the power output, steps are usually taken during fabrication, the three basic cell parameters: open-circuit voltage, short-circuit current, and fill factor (FF) – a term describing how "square" the I–V curve is, given by

$$\text{Fill Factor} = (V_{mp} \times I_{mp})/(V_{oc} \times I_{sc}) \qquad (1.2)$$

For a silicon solar cell, FF is typically 0.6–0.8. Because silicon solar cells typically produce only about 0.5 V, a number of cells are connected in series in a PV module. A panel is a collection of modules physically and electrically grouped together on a support structure. An array is a collection of panels (see Fig. 1.4).

The effect of temperature on the performance of silicon solar module is illustrated in Fig. 1.5. Note that I_{sc} slightly increases linearly with temperature, but, V_{oc} and the maximum power, P_m decrease with temperature [1].

Figure 1.6 shows the variation of PV current and voltages at different insolation levels. From Figs. 1.5 and 1.6, it can be seen that the I–V characteristics of solar cells at a given insolation and temperature consist of a constant voltage segment and a constant current segment [3]. The current is limited, as the cell is short-circuited. The maximum power condition occurs at the knee of the characteristic curve where the two segments meet.

1.2.2 Types of PV Power Systems

Photovoltaic power systems can be classified as:

- Stand-alone PV systems.
- Hybrid PV systems.
- Grid-connected PV systems.

Stand-alone PV systems, shown in Fig. 1.7, are used in remote areas with no access to a utility grid. Conventional power systems used in remote areas often based on manually controlled diesel generators operating continuously

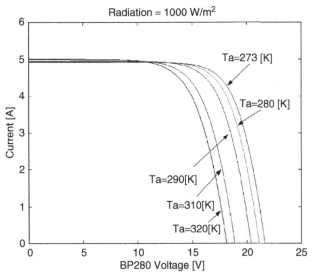

FIGURE 1.5 Effects of temperature on silicon solar cells.

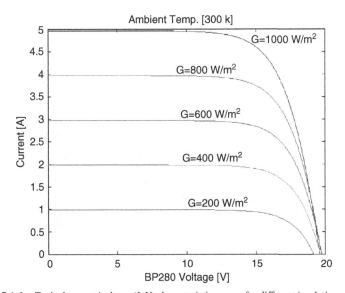

FIGURE 1.6 Typical current/voltage (*I–V*) characteristic curves for different insolation.

or for a few hours. Extended operation of diesel generators at low load levels significantly increases maintenance costs and reduces their useful life. Renewable energy sources such as PV can be added to remote area power systems using diesel and other fossil fuel powered generators to provide 24-hour power economically and efficiently. Such systems are called "hybrid energy systems."

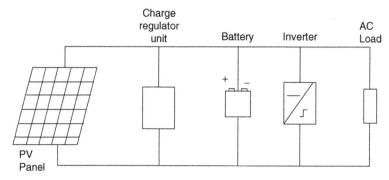

FIGURE 1.7 Stand-alone PV system.

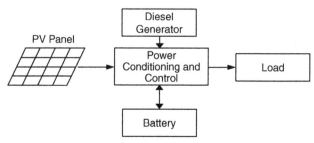

FIGURE 1.8 PV-diesel hybrid system.

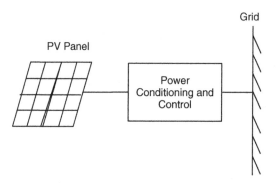

FIGURE 1.9 Grid-connected PV system.

Figure 1.8 shows a schematic of a PV-diesel hybrid system. In grid-connected PV systems shown in Fig. 1.9, PV panels are connected to a grid through inverters without battery storage. These systems can be classified as small systems like the residential rooftop systems or large grid-connected systems. The grid-interactive inverters must be synchronized with the grid in terms of voltage and frequency.

1.2.3   Stand-alone PV Systems

The two main stand-alone PV applications are:

- Battery charging.
- Solar water pumping.

1.2.3.1   Battery Charging

1.2.3.1.1   Batteries for PV Systems

Stand-alone PV energy system requires storage to meet the energy demand during periods of low solar irradiation and nighttime. Several types of batteries are available such as the lead acid, nickel–cadmium, lithium, zinc bromide, zinc chloride, sodium sulfur, nickel–hydrogen, redox, and vanadium batteries. The provision of cost-effective electrical energy storage remains one of the major challenges for the development of improved PV power systems. Typically, lead-acid batteries are used to guarantee several hours to a few days of energy storage. Their reasonable cost and general availability has resulted in the widespread application of lead-acid batteries for remote area power supplies despite their limited lifetime compared to other system components. Lead-acid batteries can be deep or shallow cycling gelled batteries, batteries with captive or liquid electrolyte, sealed and non-sealed batteries etc. [4]. Sealed batteries are valve regulated to permit evolution of excess hydrogen gas (although catalytic converters are used to convert as much evolved hydrogen and oxygen back to water as possible). Sealed batteries need less maintenance. The following factors are considered in the selection of batteries for PV applications [1]:

- Deep discharge (70–80% depth of discharge).
- Low charging/discharging current.
- Long duration charge (slow) and discharge (long duty cycle).
- Irregular and varying charge/discharge.
- Low self discharge.
- Long life time.
- Less maintenance requirement.
- High energy storage efficiency.
- Low cost.

Battery manufacturers specify the nominal number of complete charge and discharge cycles as a function of the depth-of-discharge (DOD), as shown in Fig. 1.10. While this information can be used reliably to predict the lifetime of lead-acid batteries in conventional applications, such as uninterruptable power supplies or electric vehicles, it usually results in an overestimation of the useful life of the battery bank in renewable energy systems.

Two of the main factors that have been identified as limiting criteria for the cycle life of batteries in PV power systems are incomplete charging and prolonged operation at a low state-of-charge (SOC). The objective of improved

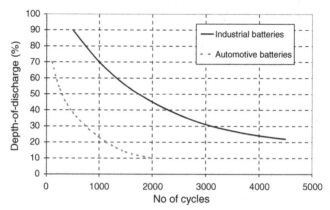

FIGURE 1.10 Nominal number of battery cycles vs DOD.

battery control strategies is to extend the lifetime of lead-acid batteries to achieve a typical number of cycles shown in Fig. 1.10. If this is achieved, an optimum solution for the required storage capacity and the maximum DOD of the battery can be found by referring to manufacturer's information. Increasing the capacity will reduce the typical DOD and therefore prolong the battery lifetime. Conversely, it may be more economic to replace a smaller battery bank more frequently.

1.2.3.1.2 PV Charge Controllers

Blocking diodes in series with PV modules are used to prevent the batteries from being discharged through the PV cells at night when there is no sun available to generate energy. These blocking diodes also protect the battery from short circuits. In a solar power system consisting of more than one string connected in parallel, if a short circuit occurs in one of the strings, the blocking diode prevents the other PV strings to discharge through the short-circuited string.

The battery storage in a PV system should be properly controlled to avoid catastrophic operating conditions like overcharging or frequent deep discharging. Storage batteries account for most PV system failures and contribute significantly to both the initial and the eventual replacement costs. Charge controllers regulate the charge transfer and prevent the battery from being excessively charged and discharged. Three types of charge controllers are commonly used:

- Series charge regulators.
- Shunt charge regulators.
- DC–DC converters.

1.2.3.1.2.1 A. Series Charge Regulators: The basic circuit for the series regulators is given in Fig. 1.11. In the series charge controller, the switch S_1 disconnects the PV generator when a predefined battery voltage is achieved.

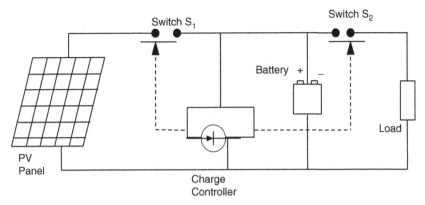

FIGURE 1.11   Series charge regulator.

When the voltage reduces below the discharge limit, the load is disconnected from the battery to avoid deep discharge beyond the limit. The main problem associated with this type of controller is the losses associated with the switches. This extra power loss has to come from the PV power and this can be quite significant. Bipolar transistors, metal oxide semi conductor field effect transistors (MOSFETs), or relays are used as the switches.

1.2.3.1.2.2 B. Shunt Charge Regulators: In this type, as illustrated in Fig. 1.12, when the battery is fully charged the PV generator is short-circuited using an electronic switch (S_1). Unlike series controllers, this method works more efficiently even when the battery is completely discharged as the short-circuit switch need not be activated until the battery is fully discharged [1].

The blocking diode prevents short-circuiting of the battery. Shunt-charge regulators are used for the small PV applications (less than 20 A).

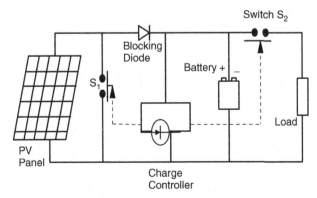

FIGURE 1.12   Shunt charge regulator.

Deep discharge protection is used to protect the battery against the deep discharge. When the battery voltage reaches below the minimum set point for deep discharge limit, switch S_2 disconnects the load. Simple series and shunt regulators allow only relatively coarse adjustment of the current flow and seldom meet the exact requirements of PV systems.

1.2.3.1.2.3 C. DC–DC Converter Type Charge Regulators: Switch mode DC-to-DC converters are used to match the output of a PV generator to a variable load. There are various types of DC–DC converters such as:

- Buck (step-down) converter.
- Boost (step-up) converter.
- Buck–boost (step-down/up) converter.

Figures 1.13–1.15 show simplified diagrams of these three basic types converters. The basic concepts are an electronic switch, an inductor to store energy, and a "flywheel" diode, which carries the current during that part of switching cycle when the switch is off. The DC–DC converters allow the charge current to be reduced continuously in such a way that the resulting battery voltage is maintained at a specified value.

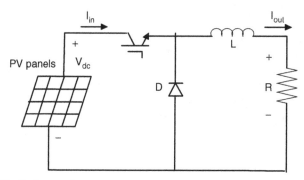

FIGURE 1.13 Buck converter.

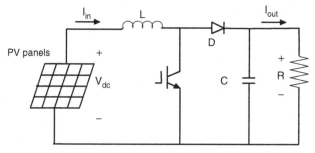

FIGURE 1.14 Boost converter.

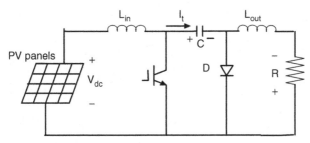

FIGURE 1.15   Boost–buck converter.

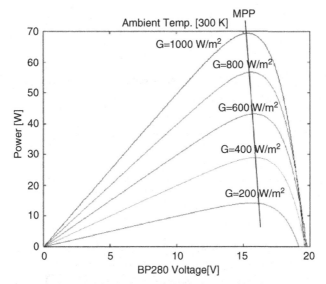

FIGURE 1.16   Typical power/voltage characteristics for increased insolation.

1.2.3.1.3   Maximum Power Point Tracking (MPPT)

A controller that tracks the maximum power point locus of the PV array is known as the MPPT. In Fig. 1.17, the PV power output is plotted against the voltage for insolation levels from 200 to 1000 W/m² [5]. The points of maximum array power form a curve termed as the maximum power locus. Due to high cost of solar cells, it is necessary to operate the PV array at its maximum power point (MPP). For overall optimal operation of the system, the load line must match the PV array's MPP locus.

Referring to Fig. 1.16, the load characteristics can be either curve OA or curve OB depending upon the nature of the load and it's current and voltage requirements. If load OA is considered and the load is directly coupled to the solar array, the array will operate at point A1, delivering only power P1. The maximum array power available at the given insolation is P2. In order to use

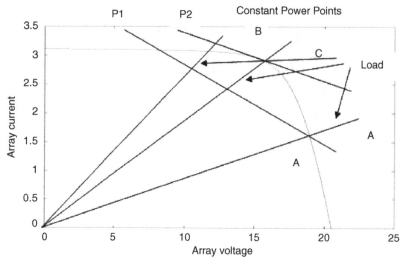

FIGURE 1.17   PV array and load characteristics.

PV array power P2, a power conditioner coupled between array and the load is needed.

There are generally two ways of operating PV modules at maximum power point. These ways take advantage of analog and/or digital hardware control to track the MPP of PV arrays.

1.2.3.1.4   Analog Control

There are many analog control mechanisms proposed in different articles. For instance, fractional short-circuit current (I_{SC}) [6–9], fractional open-circuit voltage (V_{OP}) [6, 7, 10–13], and ripple correlation control (RCC) [14–17].

Fractional open-circuit voltage (V_{OP}) is one of the simple analogue control method. It is based on the assumption that the maximum power point voltage, V_{MPP}, is a linear function of the open-circuit voltage, V_{OC}. For example $V_{MPP} = kV_{OC}$ where $k \approx 0.76$. This assumption is reasonably accurate even for large variations in the cell short-circuit current and temperature. This type of MPPT is probably the most common type. A variation to this method involves periodically open-circuiting the cell string and measuring the open-circuit voltage. The appropriate value of V_{MPP} can then be obtained with a simple voltage divider.

1.2.3.1.5   Digital Control

There are many digital control mechanisms that were proposed in different articles. For instance, perturbation and observation (P&O) or hill climbing [18–23], fuzzy logic [24–28], neural network [18, 29–31], and incremental conductance (IncCond) [32–35].

The P&O or hill climbing control involves around varying the input voltage around the optimum value by giving it a small increment or decrement alternately. The effect on the output power is then assessed and a further small correction is made to the input voltage. Therefore, this type of control is called a hill climbing control. The power output of the PV array is sampled at an every definite sampling period and compared with the previous value. In the event, when power is increased then the solar array voltage is stepped in the same direction as the previous sample time, but if the power is reduced then the array voltage is stepped in the opposite way and try to operate the PV array at its optimum/maximum power point.

To operate the PV array at the MPP, perturb and adjust method can be used at regular intervals. Current drawn is sampled every few seconds and the resulting power output of the solar cells is monitored at regular intervals. When an increased current results in a higher power, it is further increased until power output starts to reduce. But if the increased PV current results in lesser amount of power than in the previous sample, then the current is reduced until the MPP is reached.

1.2.3.2 Inverters for Stand-alone PV Systems

Inverters convert power from DC to AC while rectifiers convert it from AC to DC. Many inverters are bi-directional, i.e. they are able to operate in both inverting and rectifying modes. In many stand-alone PV installations, alternating current is needed to operate 230 V (or 110 V), 50 Hz (or 60 Hz) appliances. Generally stand-alone inverters operate at 12, 24, 48, 96, 120, or 240 V DC depending upon the power level. Ideally, an inverter for a stand-alone PV system should have the following features:

- Sinusoidal output voltage.
- Voltage and frequency within the allowable limits.
- Cable to handle large variation in input voltage.
- Output voltage regulation.
- High efficiency at light loads.
- Less harmonic generation by the inverter to avoid damage to electronic appliances like television, additional losses, and heating of appliances.
- Photovoltaic inverters must be able to withstand overloading for short term to take care of higher starting currents from pumps, refrigerators, etc.
- Adequate protection arrangement for over/under-voltage and frequency, short circuit etc.
- Surge capacity.
- Low idling and no load losses.
- Low battery voltage disconnect.
- Low audio and radio frequency (RF) noise.

Several different semiconductor devices such as metal oxide semiconductor field effect transistor (MOSFETs) and insulated gate bipolar transistors (IGBTs) are used in the power stage of inverters. Typically MOSFETs are used in units up to 5 kVA and 96 V DC. They have the advantage of low switching losses at higher frequencies. Because the on-state voltage drop is 2 V DC, IGBTs are generally used only above 96 V DC systems.

Voltage source inverters are usually used in stand-alone applications. They can be single phase or three phase. There are three switching techniques commonly used: square wave, quasi-square wave, and pulse width modulation. Square-wave or modified square-wave inverters can supply power tools, resistive heaters, or incandescent lights, which do not require a high quality sine wave for reliable and efficient operation. However, many household appliances require low distortion sinusoidal waveforms. The use of true sine-wave inverters is recommended for remote area power systems. Pulse width modulated (PWM) switching is generally used for obtaining sinusoidal output from the inverters.

A general layout of a single-phase system, both half bridge and full bridge, is shown in Fig. 1.18. In Fig. 1.18a, single-phase half bridge is with two switches, S_1 and S_2, the capacitors C_1 and C_2 are connected in series across the DC source. The junction between the capacitors is at the mid-potential. Voltage across each capacitor is $V_{dc}/2$. Switches S_1 and S_2 can be switched on/off periodically to produce AC voltage. Filter (L_f and C_f) is used to reduce high-switch frequency components and to produce sinusoidal output from the inverter. The output of inverter is connected to load through a transformer. Figure 1.18b shows the similar arrangement for full-bridge configuration with four switches. For the same input source voltage, the full-bridge output is twice and the switches carry less current for the same load power.

The power circuit of a three phase four-wire inverter is shown in Fig. 1.19. The output of the inverter is connected to load via three-phase transformer (delta/Y). The star point of the transformer secondary gives the neutral connection. Three phase or single phase can be connected to this system. Alternatively, a center tap DC source can be used to supply the converter and the mid-point can be used as the neutral.

Figure 1.20 shows the inverter efficiency for a typical inverter used in remote area power systems. It is important to consider that the system load is

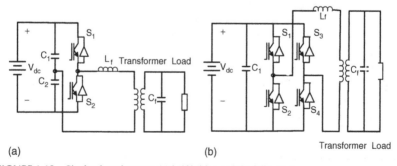

(a)   (b)

FIGURE 1.18   Single-phase inverter: (a) half bridge and (b) full bridge.

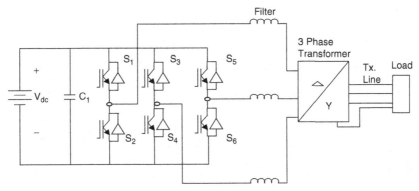

FIGURE 1.19   A stand-alone three-phase four wire inverter.

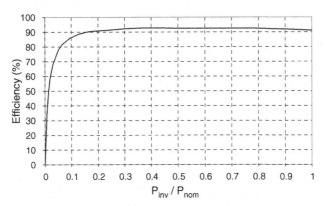

FIGURE 1.20   Typical inverter efficiency curve.

typically well below the nominal inverter capacity P_{nom}, which results in low conversion efficiencies at loads below 10% of the rated inverter output power. Optimum overall system operation is achieved if the total energy dissipated in the inverter is minimized. The high conversion efficiency at low power levels of recently developed inverters for grid-connected PV systems shows that there is a significant potential for further improvements in efficiency.

Bi-directional inverters convert DC power to AC power (inverter) or AC power to DC power (rectifier) and are becoming very popular in remote area power systems [4, 5]. The principle of a stand-alone single-phase bi-directional inverter used in a PV/battery/diesel hybrid system can be explained by referring Fig. 1.21. A charge controller is used to interface the PV array and the battery. The inverter has a full-bridge configuration realized using four power electronic switches (MOSFET or IGBTs) S_1–S_4. In this scheme, the diagonally opposite switches (S_1, S_4) and (S_2, S_3) are switched using a sinusoidally PWM gate pulses. The inverter produces sinusoidal output voltage. The inductors X_1, X_2,

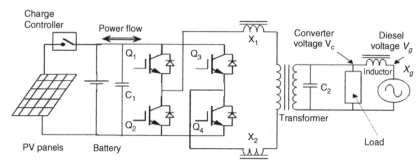

FIGURE 1.21 Bi-directional inverter system.

and the AC output capacitor C_2 filter out the high-switch frequency components from the output waveform. Most inverter topologies use a low frequency (50 or 60 Hz) transformer to step up the inverter output voltage. In this scheme, the diesel generator and the converter are connected in parallel to supply the load. The voltage sources, diesel and inverter, are separated by the link inductor X_m. The bi-directional power flow between inverter and the diesel generator can be established.

The power flow through the link inductor, X_m, is

$$S_m = V_m I_m^*$$ (1.3)

$$P_m = (V_m V_c \sin \delta)/X_m$$ (1.4)

$$Q_m = (V_m/X_m)(V_m - V_c \cos \delta)$$ (1.5)

$$\delta = \sin^{-1}[(X_m P_m)/(V_m V_c)]$$ (1.6)

where δ is the phase angle between the two voltages. From Eq. (1.4), it can be seen that the power supplied by the inverter from the batteries (inverter mode) or supplied to the batteries (charging mode) can be controlled by controlling the phase angle δ. The PWM pulses separately control the amplitude of the converter voltage, V_c, while the phase angle with respect to the diesel voltage is varied for power flow.

1.2.3.3 Solar Water Pumping

In many remote and rural areas, hand pumps or diesel driven pumps are used for water supply. Diesel pumps consume fossil fuel, affects environment, needs more maintenance, and are less reliable. Photovoltaic powered water pumps have received considerable attention recently due to major developments in the field of solar cell materials and power electronic systems technology.

1.2.3.3.1 Types of Pumps

Two types of pumps are commonly used for the water pumping applications: positive and centrifugal displacement. Both centrifugal and positive displacement pumps can be further classified into those with motors that are (a) surface mounted and those which are (b) submerged into the water ("submersible").

Displacement pumps have water output directly proportional to the speed of the pump, but, almost independent of head. These pumps are used for solar water pumping from deep wells or bores. They may be piston type pumps, or use diaphragm driven by a cam, rotary screw type, or use progressive cavity system. The pumping rate of these pumps is directly related to the speed and hence constant torque is desired.

Centrifugal pumps are used for low-head applications especially if they are directly interfaced with the solar panels. Centrifugal pumps are designed for fixed-head applications and the pressure difference generated increases in relation to the speed of pump. These pumps are rotating impeller type, which throws the water radially against a casing, so shaped that the momentum of the water is converted into useful pressure for lifting [4]. The centrifugal pumps have relatively high efficiency but it reduces at lower speeds, which can be a problem for the solar water pumping system at the time of low light levels. The single-stage centrifugal pump has just one impeller whereas most borehole pumps are multistage types where the outlet from one impeller goes into the center of another and each one keeps increasing the pressure difference.

From Fig. 1.22, it is quite obvious that the load line is located relatively faraway from P_{max} line. It has been reported that the daily utilization efficiency for a DC motor drive is 87% for a centrifugal pump compared to 57% for a constant torque characteristics load. Hence, centrifugal pumps are more compatible with PV arrays. The system operating point is determined by the intersection of the I–V characteristics of the PV array and the motor as shown in Fig. 1.22. The torque-speed slope is normally large due to the armature resistance being small. At the instant of starting, the speed and the back emf are zero. Hence

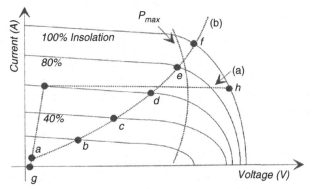

FIGURE 1.22 I–V characteristics of a PV array and two mechanical loads: (a) constant torque and (b) centrifugal pump.

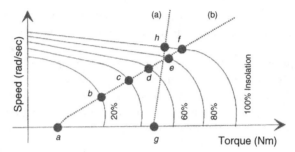

FIGURE 1.23 Speed torque characteristics of a DC motor and two mechanical loads: (a) helical rotor and (b) centrifugal pump.

the motor starting current is approximately the short-circuit current of the PV array. By matching the load to the PV source through MPPT, the starting torque increases.

The matching of a DC motor depends upon the type of load being used. For instance, a centrifugal pump is characterized by having the load torque proportional to the square of speed. The operating characteristics of the system (i.e. PV source, permanent magnet (PM) DC motor and load) are at the intersection of the motor and load characteristics as shown in Fig. 1.23 (i.e. points a, b, c, d, e, and f for centrifugal pump). From Fig. 1.23, the system utilizing the centrifugal pump as its load tends to start at low solar irradiation (point a) level. However, for the systems with an almost constant torque characteristics in Fig. 1.22, the start is at almost 50% of one sun (full insolation) which results in short period of operation.

1.2.3.3.2 Types of Motors

There are various types of motors available for the PV water pumping applications:

- DC motors.
- AC motors.

DC motors are preferred where direct coupling to PV panels is desired whereas AC motors are coupled to the solar panels through inverters. AC motors in general are cheaper than the DC motors and are more reliable but the DC motors are more efficient. The DC motors used for solar pumping applications are:

- Permanent magnet DC motors with brushes.
- Permanent DC magnet motors without brushes.

In DC motors with the brushes, the brushes are used to deliver power to the commutator and need frequent replacement due to wear and tear. These motors are not suitable for submersible applications unless long transmission shafts are used. Brush-less DC permanent magnet motors have been developed for submersible applications.

The AC motors are of the induction motor type, which is cheaper than DC motors and available, worldwide. However, they need inverters to change DC input from PV to AC power. A comparison of the different types of motors used for PV water pumping is given in Table 1.1.

1.2.3.3.3   Power Conditioning Units for PV Water Pumping

Most PV pump manufacturers include power conditioning units (PCU) which are used for operating the PV panels close to their MPP over a range of load conditions and varying insolation levels and also for power conversion. DC or AC motor-pump units can be used for PV water pumping. In its simplest form, a solar water pumping system comprises of PV array, PCU, and DC water-pump unit as shown in Fig. 1.24.

In case of lower light levels, high currents can be generated through power conditioning to help in starting the motor-pump units especially for reciprocating positive displacement type pumps with constant torque characteristics, requiring constant current throughout the operating region. In positive displacement type pumps, the torque generated by the pumps depends on the pumping head, friction, and pipe diameter etc. and needs certain level of current to produce the necessary torque. Some systems use electronic controllers to assist starting and operation of the motor under low solar radiation. This is particularly important when using positive displacement pumps. The solar panels generate DC voltage and current. The solar water pumping systems usually has DC or AC pumps. For DC pumps, the PV output can be directly connected to the pump through MPPT or a DC–DC converter can also be used for interfacing for controlled DC output from PV panels. To feed the AC motors, a suitable interfacing is required for the power conditioning. These PV inverters for the stand-alone applications are very expensive. The aim of power conditioning equipment is to supply the controlled voltage/current output from the converters/inverters to the motor-pump unit.

These power-conditioning units are also used for operating the PV panels close to their maximum efficiency for fluctuating solar conditions. The speed of the pump is governed by the available driving voltage. Current lower than the acceptable limit will stop the pumping. When the light level increases, the operating point will shift from the MPP leading to the reduction of efficiency. For centrifugal pumps, there is an increase in current at increased speed and the matching of $I–V$ characteristics is closer for wide range of light intensity levels. For centrifugal pumps, the torque is proportional to the square of speed and the torque produced by the motors is proportional to the current. Due to decrease in PV current output, the torque from the motor and consequently the speed of the pump is reduced resulting in decrease in back emf and the required voltage of the motor. Maximum power point tracker can be used for controlling the voltage/current outputs from the PV inverters to operate the PV close to

TABLE 1.1 Comparison of the Different Types of Motor Used for PV Water Pumping

Types of motor	Advantages	Disadvantages	Main features
Brushed DC	Simple and efficient for PV applications. No complex control circuits is required as the motor starts without high current surge. These motors will run slowly but do not overheat with reduced voltage.	Brushes need to be replaced periodically (typical replacement interval is 2000–4000 hr or 2 years).	Requires MPPT for optimum performance. Available only in small motor sizes. Increasing current (by paralleling PV modules) increases the torque. Increasing voltage (by series PV modules) increases the speed.
Brush-less DC	Efficient. Less maintenance is required.	Electronic computation adds to extra cost, complexity, and increased risk of failure/malfunction. In most cases, oil cooled, can't be submerged as deep as water cooled AC units.	Growing trend among PV pump manufacturers to use brush-less DC motors, primarily for centrifugal type submersible pumps.
AC induction motors	No brushes to replace. Can use existing AC motor/pump technology which is cheaper and easily available worldwide. These motors can handle larger pumping requirements.	Needs an inverter to convert DC output from PV to AC adding additional cost and complexity. Less efficient than DC motor-pump units. Prone to overheating if current is not adequate to start the motor or if the voltage is too low.	Available for single or three supply. Inverters are designed to regulate frequency to maximize power to the motor in response to changing insolation levels.

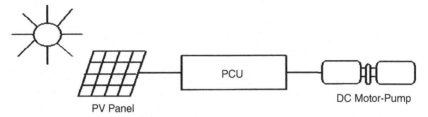

FIGURE 1.24 Block diagram for DC motor driven pumping scheme.

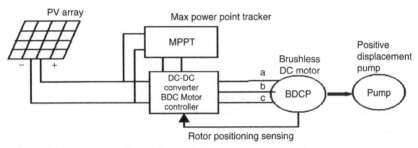

FIGURE 1.25 Block diagram for BDCM for PV application.

maximum operating point for the smooth operation of motor-pump units. The DC–DC converter can be used for keeping the PV panels output voltage constant and help in operating the solar arrays close to MPP. In the beginning, high starting current is required to produce high starting torque. The PV panels cannot supply this high starting current without adequate power conditioning equipment like DC–DC converter or by using a starting capacitor. The DC–DC converter can generate the high starting currents by regulating the excess PV array voltage. DC–DC converter can be boost or buck converter.

Brush-less DC motor (BDCM) and helical rotor pumps can also be used for PV water pumping [36]. Brush-less DC motors are a self synchronous type of motor characteristics by trapezoidal waveforms for back emf and air flux density. They can operate off a low voltage DC supply which is switched through an inverter to create a rotating stator field. The current generation of BDCMs use rare earth magnets on the rotor to give high air gap flux densities and are well suited to solar application. The block diagram of such an arrangement is shown in Fig. 1.25 which consists of PV panels, DC–DC converter, MPPT, and BDCM.

The PV inverters are used to convert the DC output of the solar arrays to the AC quantity so as to run the AC motors driven pumps. These PV inverters can be variable frequency type, which can be controlled to operate the motors over wide range of loads. The PV inverters may involve impedance matching to match the electrical characteristics of the load and array. The motor-pump unit and PV panels operate at their maximum efficiencies. Maximum power point

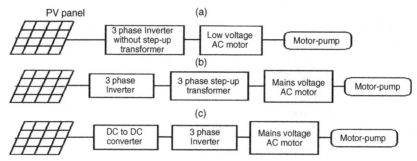

FIGURE 1.26   Block diagrams for various AC motor driven pumping schemes.

tracker is also used in the power conditioning. To keep the voltage stable for the inverters, the DC–DC converter can be used. The inverter/converter has a capability of injecting high-switch frequency components, which can lead to the overheating and the losses. So care shall be taken for this. The PV arrays are usually connected in series, parallel, or a combination of series parallel, configurations. The function of power electronic interface, as mentioned before, is to convert the DC power from the array to the required voltage and frequency to drive the AC motors. The motor-pump system load should be such that the array operates close to it's MPP at all solar insolation levels. There are mainly three types solar powered water pumping systems as shown in Fig. 1.26.

The first system shown in Fig. 1.26a is an imported commercially available unit, which uses a specially wound low voltage induction motor driven submersible pump. Such a low voltage motor permits the PV array voltage to be converted to AC without using a step-up transformer. The second system, shown in Fig. 1.26b makes use of a conventional "off-the-shelf" 415 V, 50 Hz, induction motor [6]. This scheme needs a step-up transformer to raise inverter output voltage to high voltage. Third scheme as shown in Fig. 1.26c comprises of a DC–DC converter, an inverter that switches at high frequency, and a mains voltage motor driven pump. To get the optimum discharge (Q), at a given insolation level, the efficiency of the DC–DC converter and the inverter should be high. So the purpose should be to optimize the output from PV array, motor, and the pump. The principle used here is to vary the duty cycle of a DC–DC converter so that the output voltage is maximum. The DC–DC converter is used to boost the solar array voltage to eliminate the need for a step-up transformer and operate the array at the MPP. The three-phase inverter used in the interface is designed to operate in a variable frequency mode over the range of 20–50 Hz, which is the practical limit for most 50 Hz induction motor applications. Block diagram for frequency control is given in Fig. 1.27.

This inverter would be suitable for driving permanent magnet motors by incorporating additional circuitry for position sensing of the motor's shaft. Also the inverter could be modified, if required, to produce higher output frequencies

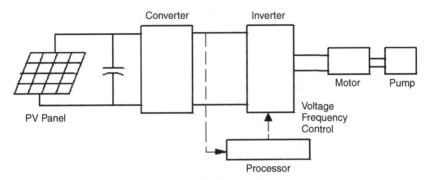

FIGURE 1.27   Block diagram for voltage/frequency control.

for high-speed permanent magnet motors. The inverter has a three-phase full-bridge configuration implemented by MOSFET power transistors.

1.2.4   Hybrid Energy Systems

The combination of RES, such as PV arrays or wind turbines, with engine-driven generators and battery storage, is widely recognized as a viable alternative to conventional remote area power supplies (RAPS). These systems are generally classified as hybrid energy systems (HES). They are used increasingly for electrification in remote areas where the cost of grid extension is prohibitive and the price for fuel increases drastically with the remoteness of the location. For many applications, the combination of renewable and conventional energy sources compares favorably with fossil fuel-based RAPS systems, both in regard to their cost and technical performance. Because these systems employ two or more different sources of energy, they enjoy a very high degree of reliability as compared to single-source systems such as a stand-alone diesel generator or a stand-alone PV or wind system. Applications of hybrid energy systems range from small power supplies for remote households, providing electricity for lighting and other essential electrical appliances, to village electrification for remote communities has been reported [37].

Hybrid energy systems generate AC electricity by combining RES such as PV array with an inverter, which can operate alternately or in parallel with a conventional engine-driven generator. They can be classified according to their configuration as [38]:

- Series hybrid energy systems.
- Switched hybrid energy systems.
- Parallel hybrid energy systems.

The parallel hybrid systems can be further divided to DC or AC coupling. An overview of the three most common system topologies is presented by

Bower [39]. In the following comparison of typical PV-diesel system configurations are described.

1.2.4.1 Series Configuration

In the conventional series hybrid systems shown in Fig. 1.28, all power generators feed DC power into a battery. Each component has therefore to be equipped with an individual charge controller and in the case of a diesel generator with a rectifier.

To ensure reliable operation of series hybrid energy systems both the diesel generator and the inverter have to be sized to meet peak loads. This results in a typical system operation where a large fraction of the generated energy is passed through the battery bank, therefore resulting in increased cycling of the battery bank and reduced system efficiency. AC power delivered to the load is converted from DC to regulated AC by an inverter or a motor generator unit. The power generated by the diesel generator is first rectified and subsequently converted back to AC before being supplied to the load, which incurs significant conversion losses.

The actual load demand determines the amount of electrical power delivered by the PV array, wind generator, the battery bank, or the diesel generator. The solar and wind charger prevents overcharging of the battery bank from the PV generator when the PV power exceeds the load demand and the batteries are fully charged. It may include MPPT to improve the utilization of the available PV energy, although the energy gain is marginal for a well-sized system. The system can be operated in manual or automatic mode, with the addition of appropriate battery voltage sensing and start/stop control of the engine-driven generator.

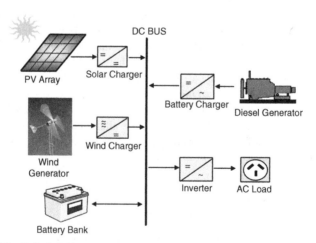

FIGURE 1.28 Series hybrid energy system.

Advantages:

- The engine-driven generator can be sized to be optimally loaded while supplying the load and charging the battery bank, until a battery SOC of 70–80% is reached.
- No switching of AC power between the different energy sources is required, which simplifies the electrical output interface.
- The power supplied to the load is not interrupted when the diesel generator is started.
- The inverter can generate a sine-wave, modified square-wave, or square-wave depending on the application.

Disadvantages:

- The inverter cannot operate in parallel with the engine-driven generator, therefore the inverter must be sized to supply the peak load of the system.
- The battery bank is cycled frequently, which shortens its lifetime.
- The cycling profile requires a large battery bank to limit the depth-of-discharge (DOD).
- The overall system efficiency is low, since the diesel cannot supply power directly to the load.
- Inverter failure results in complete loss of power to the load, unless the load can be supplied directly from the diesel generator for emergency purposes.

1.2.4.2   Switched Configuration

Despite its operational limitations, the switched configuration remains one of the most common installations in some developing countries. It allows operation with either the engine-driven generator or the inverter as the AC source, yet no parallel operation of the main generation sources is possible. The diesel generator and the RES can charge the battery bank. The main advantage compared with the series system is that the load can be supplied directly by the engine-driven generator, which results in a higher overall conversion efficiency. Typically, the diesel generator power will exceed the load demand, with excess energy being used to recharge the battery bank. During periods of low electricity demand the diesel generator is switched off and the load is supplied from the PV array together with stored energy. Switched hybrid energy systems can be operated in manual mode, although the increased complexity of the system makes it highly desirable to include an automatic controller, which can be implemented with the addition of appropriate battery voltage sensing and start/stop control of the engine-driven generator (Fig. 1.29).

Advantages:

- The inverter can generate a sine-wave, modified square-wave, or square-wave, depending on the particular application.

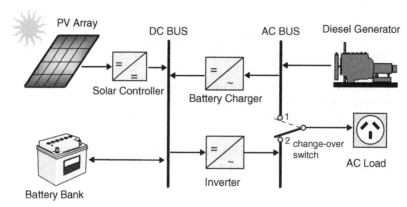

FIGURE 1.29 Switched PV-diesel hybrid energy system.

- The diesel generator can supply the load directly, therefore improving the system efficiency and reducing the fuel consumption.

Disadvantages:

- Power to the load is interrupted momentarily when the AC power sources are transferred.
- The engine-driven alternator and inverter are typically designed to supply the peak load, which reduces their efficiency at part load operation.

1.2.4.3 Parallel Configuration

The parallel hybrid system can be further classified as DC and AC couplings as shown in Fig. 1.30. In both schemes, a bi-directional inverter is used to link between the battery and an AC source (typically the output of a diesel generator). The bi-directional inverter can charge the battery bank (rectifier operation) when excess energy is available from the diesel generator or by the renewable sources, as well as act as a DC–AC converter (inverter operation). The bi-directional inverter may also provide "peak shaving" as part of a control strategy when the diesel engine is overloaded. In Fig. 1.30a, the renewable energy sources (RES) such as photovoltaic and wind are coupled on the DC side. DC integration of RES results in "custom" system solutions for individual supply cases requiring high costs for engineering, hardware, repair, and maintenance. Furthermore, power system expandability for covering needs of growing energy and power demand is also difficult. A better approach would be to integrate the RES on the AC side rather than on the DC side as shown in Fig. 1.30b.

Parallel hybrid energy systems are characterized by two significant improvements over the series and switched system configuration.

The inverter plus the diesel generator capacity rather than their individual component ratings limit the maximum load that can be supplied. Typically, this

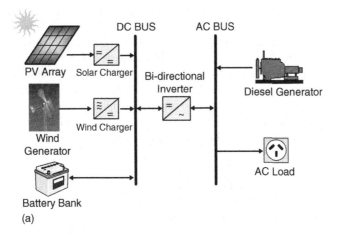

(a)

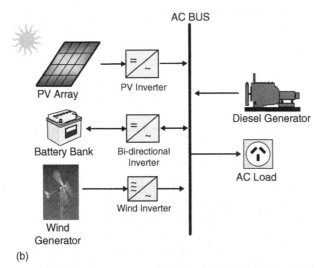

(b)

FIGURE 1.30   Parallel PV-diesel hybrid energy system: (a) DC decoupling and (b) AC coupling.

will lead to a doubling of the system capacity. The capability to synchronize the inverter with the diesel generator allows greater flexibility to optimize the operation of the system. Future systems should be sized with a reduced peak capacity of the diesel generator, which results in a higher fraction of directly used energy and hence higher system efficiencies.

By using the same power electronic devices for both inverter and rectifier operation, the number of system components is minimized. Additionally, wiring and system installation costs are reduced through the integration of all power-conditioning devices in one central power unit. This highly integrated system concept has advantages over a more modular approach to system design, but it may prevent convenient system upgrades when the load demand increases.

The parallel configuration offers a number of potential advantages over other system configurations. These objectives can only be met if the interactive operation of the individual components is controlled by an "intelligent" hybrid energy management system. Although today's generation of parallel systems include system controllers of varying complexity and sophistication, they do not optimize the performance of the complete system. Typically, both the diesel generator and the inverter are sized to supply anticipated peak loads. As a result most parallel hybrid energy systems do not utilize their capability of parallel, synchronized operation of multiple power sources.

Advantages:

- The system load can be met in an optimal way.
- Diesel generator efficiency can be maximized.
- Diesel generator maintenance can be minimized.
- A reduction in the rated capacities of the diesel generator, battery bank, inverter, and renewable resources is feasible, while also meeting the peak loads.

Disadvantages:

- Automatic control is essential for the reliable operation of the system.
- The inverter has to be a true sine-wave inverter with the ability to synchronize with a secondary AC source.
- System operation is less transparent to the untrained user of the system.

1.2.4.4 Control of Hybrid Energy Systems

The design process of hybrid energy systems requires the selection of the most suitable combination of energy sources, power-conditioning devices, and energy storage system together with the implementation of an efficient energy dispatch strategy. System simulation software is an essential tool to analyze and compare possible system combinations. The objective of the control strategy is to achieve optimal operational performance at the system level. Inefficient operation of the diesel generator and "dumping" of excess energy is common for many RAPS, operating in the field. Component maintenance and replacement contributes significantly to the lifecycle cost of systems. These aspects of system operation are clearly related to the selected control strategy and have to be considered in the system design phase.

Advanced system control strategies seek to reduce the number of cycles and the DOD for the battery bank, run the diesel generator in its most efficient operating range, maximize the utilization of the renewable resource, and ensure high reliability of the system. Due to the varying nature of the load demand, the fluctuating power supplied by the photovoltaic generator, and the resulting variation of battery SOC, the hybrid energy system controller has to respond to continuously changing operating conditions. Figure 1.31 shows different

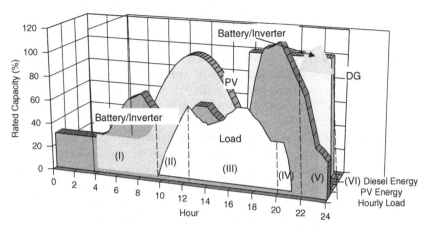

FIGURE 1.31   Operating modes for a PV single-diesel hybrid energy system.

operating modes for a PV single-diesel system using a typical diesel dispatch strategy.

Mode (I): The base load, which is typically experienced at nighttime and during the early morning hours, is supplied by energy stored in the batteries. Photovoltaic power is not available and the diesel generator is not started.

Mode (II): PV power is supplemented by stored energy to meet the medium load demand.

Mode (III): Excess energy is available from the PV generator, which is stored in the battery. The medium load demand is supplied from the PV generator.

Mode (IV): The diesel generator is started and operated at its nominal power to meet the high evening load. Excess energy available from the diesel generator is used to recharge the batteries.

Mode (V): The diesel generator power is insufficient to meet the peak load demand. Additional power is supplied from the batteries by synchronizing the inverter AC output voltage with the alternator waveform.

Mode (VI): The diesel generator power exceeds the load demand, but it is kept operational until the batteries are recharged to a high SOC level.

In principle, most efficient operation is achieved if the generated power is supplied directly to the load from all energy sources, which also reduces cycling of the battery bank. However, since diesel generator operation at light loads is inherently inefficient, it is common practice to operate the engine-driven generator at its nominal power rating and to recharge the batteries from the excess energy. The selection of the most efficient control strategy depends on fuel, maintenance and component replacement cost, the system configuration, environmental conditions, as well as constraints imposed on the operation of the hybrid energy system.

1.2.5   Grid-connected PV Systems

The utility interactive inverters not only conditions the power output of the PV arrays but ensures that the PV system output is fully synchronized with the utility power. These systems can be battery less or with battery backup. Systems with battery storage (or flywheel) provide additional power supply reliability. The grid connection of PV systems is gathering momentum because of various rebate and incentive schemes. This system allows the consumer to feed its own load utilizing the available solar energy and the surplus energy can be injected into the grid under the energy by back scheme to reduce the payback period. Grid-connected PV systems can become a part of the utility system. The contribution of solar power depends upon the size of system and the load curve of the house. When the PV system is integrated with the utility grid, a two-way power flow is established. The utility grid will absorb excess PV power and will feed the house during nighttime and at instants while the PV power is inadequate. The utility companies are encouraging this scheme in many parts of the world. The grid-connected system can be classified as:

- Rooftop application of grid-connected PV system.
- Utility scale large system.

For small household PV applications, a roof mounted PV array can be the best option. Solar cells provide an environmentally clean way of producing electricity, and rooftops have always been the ideal place to put them. With a PV array on the rooftop, the solar generated power can supply residential load. The rooftop PV systems can help in reducing the peak summer load to the benefit of utility companies by feeding the household lighting, cooling, and other domestic loads. The battery storage can further improve the reliability of the system at the time of low insolation level, nighttime, or cloudy days. But the battery storage has some inherent problems like maintenance and higher cost.

For roof-integrated applications, the solar arrays can be either mounted on the roof or directly integrated into the roof. If the roof integration does not allow for an air channel behind the PV modules for ventilation purpose, then it can increase the cell temperature during the operation consequently leading to some energy losses. The disadvantage with the rooftop application is that the PV array orientation is dictated by the roof. In case, when the roof orientation differs from the optimal orientation required for the cells, then efficiency of the entire system would be suboptimal.

Utility interest in PV has centered on the large grid-connected PV systems. In Germany, USA, Spain, and in several other parts of the world, some large PV scale plants have been installed. The utilities are more inclined with large scale, centralized power supply. The PV systems can be centralized or distributed systems.

Grid-connected PV systems must observe the islanding situation, when the utility supply fails. In case of islanding, the PV generators should be disconnected from mains. PV generators can continue to meet only the local load, if the PV output matches the load. If the grid is re-connected during islanding, transient overcurrents can flow through the PV system inverters and the protective equipments like circuit breakers may be damaged. The islanding control can be achieved through inverters or via the distribution network. Inverter controls can be designed on the basis of detection of grid voltage, measurement of impedance, frequency variation, or increase in harmonics. Protection shall be designed for the islanding, short circuits, over/under-voltages/currents, grounding, and lightening, etc.

The importance of the power generated by the PV system depends upon the time of the day specially when the utility is experiencing the peak load. The PV plants are well suited to summer peaking but it depends upon the climatic condition of the site. PV systems being investigated for use as peaking stations would be competitive for load management. The PV users can defer their load by adopting load management to get the maximum benefit out of the grid-connected PV plants and feeding more power into the grid at the time of peak load.

The assigned capacity credit is based on the statistical probability with which the grid can meet peak demand [4]. The capacity factor during the peaks is very similar to that of conventional plants and similar capacity credit can be given for the PV generation except at the times when the PV plants are generating very less power unless adequate storage is provided. With the installation of PV plants, the need of extra transmission lines, transformers can be delayed or avoided. The distributed PV plants can also contribute in providing reactive power support to the grid and reduce burden on VAR compensators.

1.2.5.1   Inverters for Grid-connected Applications

Power conditioner is the key link between the PV array and mains in the grid-connected PV system. It acts as an interface that converts DC current produced by the solar cells into utility grade AC current. The PV system behavior relies heavily on the power-conditioning unit. The inverters shall produce good quality sine-wave output. The inverter must follow the frequency and voltage of the grid and the inverter has to extract maximum power from the solar cells with the help of MPPT and the inverter input stage varies the input voltage until the MPP on the $I–V$ curve is found. The inverter shall monitor all the phases of the grid. The inverter output shall be controlled in terms of voltage and frequency variation. A typical grid-connected inverter may use a PWM scheme and operates in the range of 2–20 kHz.

1.2.5.2   Inverter Classifications

The inverters used for the grid interfacing are broadly classified as:

- Voltage source inverters (VSI).
- Current source inverters (CSI).

Whereas the inverters based on the control schemes can be classified as:

- Current controlled (CC).
- Voltage controlled (VC).

The source is not necessarily characterized by the energy source for the system. It is a characteristic of the topology of the inverter. It is possible to change from one source type to another source type by the addition of passive components. In the voltage source inverter (VSI), the DC side is made to appear to the inverter as a voltage source. The VSIs have a capacitor in parallel across the input whereas the CSIs have an inductor is series with the DC input. In the CSI, the DC source appears as a current source to the inverter. Solar arrays are fairly good approximation to a current source. Most PV inverters are voltage source even though the PV is a current source. Current source inverters are generally used for large motor drives though there have been some PV inverters built using a current source topology. The VSI is more popular with the PWM VSI dominating the sine-wave inverter topologies.

Figure 1.32a shows a single-phase full-bridge bi-directional VSI with (a) voltage control and phase-shift (δ) control – voltage-controlled voltage source inverter (VCVSI). The active power transfer from the PV panels is accomplished by controlling the phase angle δ between the converter voltage and the grid voltage. The converter voltage follows the grid voltage. Figure 1.32b shows the same VSI operated as a current controlled (CCVSI). The objective of this scheme is to control active and reactive components of the current fed into the grid using PWM techniques.

1.2.5.3 Inverter Types

Different types are being in use for the grid-connected PV applications such as:

- Line-commutated inverter.
- Self-commutated inverter.
- Inverter with high-frequency transformer.

1.2.5.3.1 Line-commutated Inverter

The line-commutated inverters are generally used for the electric motor applications. The power stage is equipped with thyristors. The maximum power tracking control is required in the control algorithm for solar application. The basic diagram for a single-phase line-commutated inverter is shown in the Fig. 1.33 [3].

The driver circuit has to be changed to shift the firing angle from the rectifier operation ($0 < \phi < 90$) to inverter operation ($90 < \phi < 180$). Six-pulse or 12-pulse inverter are used for the grid interfacing but 12-pulse inverters produce less harmonics. The thyristor type inverters require a low impedance grid interface connection for commutation purpose. If the maximum power

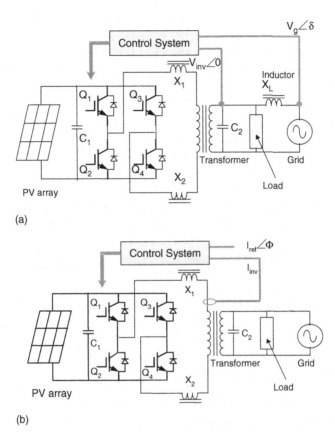

(a)

(b)

FIGURE 1.32   Voltage source inverter: (a) voltage control and (b) current control.

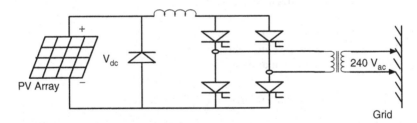

FIGURE 1.33   Line-commuted single-phase inverter.

available from the grid connection is less than twice the rated PV inverter power, then the line-commutated inverter should not be used [3]. The line-commutated inverters are cheaper but inhibits poor power quality. The harmonics injected into the grid can be large unless taken care of by employing adequate filters. These line-commutated inverters also have poor power factor, poor power quality, and need additional control to improve the power factor. Transformer can be used

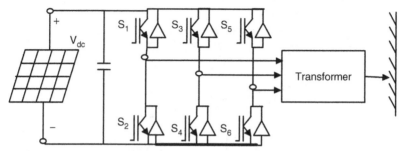

FIGURE 1.34   Self-commutated inverter with PWM switching.

to provide the electrical isolation. To suppress the harmonics generated by these inverters, tuned filters are employed and reactive power compensation is required to improve the lagging power factor.

1.2.5.3.2   Self-commutated Inverter

A switch mode inverter using pulse width modulated (PWM) switching control, can be used for the grid connection of PV systems. The basic block diagram for this type of inverter is shown in the Fig. 1.34. The inverter bridges may consist of bipolar transistors, MOSFET transistors, IGBT's, or gate turn-off thyristor's (GTO's), depending upon the type of application. GTO's are used for the higher power applications, whereas IGBT's can be switched at higher frequencies i.e. 16 kHz, and are generally used for many grid-connected PV applications. Most of the present day inverters are self-commutated sine-wave inverters.

Based on the switching control, the voltage source inverters can be further classified based on the switching control as:

- PWM (pulse width modulated) inverters.
- Square-wave inverters.
- Single-phase inverters with voltage cancellations.
- Programmed harmonic elimination switching.
- Current controlled modulation.

1.2.5.3.3   Inverter with High-frequency Transformer

The 50 Hz transformer for a standard PV inverter with PWM switching scheme can be very heavy and costly. While using frequencies more than 20 kHz, a ferrite core transformer can be a better option [3]. A circuit diagram of a grid-connected PV system using high frequency transformer is shown in the Fig. 1.35.

The capacitor on the input side of high frequency inverter acts as the filter. The high frequency inverter with PWM is used to produce a high frequency AC across the primary winding of the high frequency transformer. The secondary voltage of this transformer is rectified using high frequency rectifier. The DC

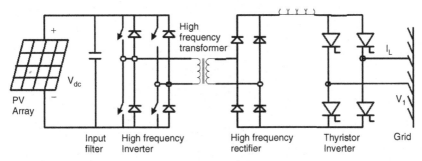

FIGURE 1.35   PV inverter with high frequency transformer.

voltage is interfaced with a thyristor inverter through low-pass inductor filter and hence connected to the grid. The line current is required to be sinusoidal and in phase with the line voltage. To achieve this, the line voltage (V_1) is measured to establish the reference waveform for the line current I_L^*. This reference current I_L^* multiplied by the transformer ratio gives the reference current at the output of high frequency inverter. The inverter output can be controlled using current control technique [40]. These inverters can be with low frequency transformer isolation or high frequency transformer isolation. The low frequency (50/60 Hz) transformer of a standard inverter with PWM is a very heavy and bulky component. For residential grid interactive rooftop inverters below 3 kW rating, high frequency transformer isolation is often preferred.

1.2.5.3.4   Other PV Inverter Topologies

In this section, some of the inverter topologies discussed in various research papers have been discussed.

A. Multilevel Converters Multilevel converters can be used with large PV systems where multiple PV panels can be configured to create voltage steps. These multilevel voltage-source converters can synthesize the AC output terminal voltage from different level of DC voltages and can produce staircase waveforms. This scheme involves less complexity, and needs less filtering. One of the schemes (half-bridge diode-clamped three level inverter [41]) is given in Fig. 1.36. There is no transformer in this topology. Multilevel converters can be beneficial for large systems in terms of cost and efficiency. Problems associated with shading and malfunction of PV units need to be addressed.

B. Non-insulated Voltage Source In this scheme [42], string of low voltage PV panels or one high-voltage unit can be coupled with the grid through DC to DC converter and voltage-source inverter. This topology is shown in Fig. 1.37. PWM-switching scheme can be used to generate AC output. Filter has been used to reject the switching components.

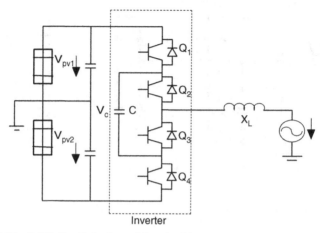

FIGURE 1.36   Half-bridge diode-clamped three-level inverter.

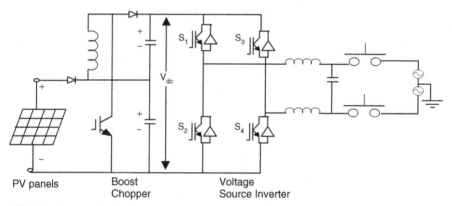

FIGURE 1.37   Non-insulated voltage source.

C. Non-insulated Current Source This type of configuration is shown in Fig. 1.38. Non-insulated current-source inverters [42] can be used to interface the PV panels with the grid. This topology involves low cost which can provide better efficiency. Appropriate controller can be used to reduce current harmonics.

D. Buck Converter with Half-bridge Transformer Link PV panels are connected to grid via buck converter and half bridge as shown in Fig. 1.39. In this, high-frequency PWM switching has been used at the low-voltage PV side to generate an attenuated rectified 100 Hz sine-wave current waveform [43]. Half-wave bridge is utilized to convert this output to 50 Hz signal suitable for grid interconnection. To step up the voltage, transformer has also been connected before the grid connection point.

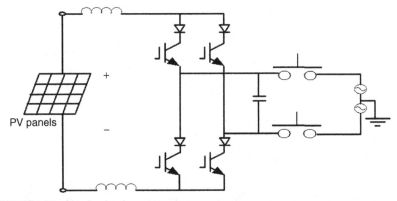

FIGURE 1.38   Non-insulated current source.

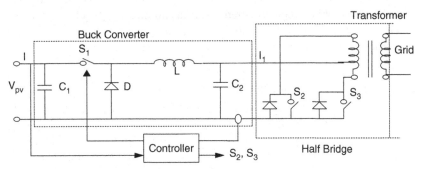

FIGURE 1.39   Buck converter with half-bridge transformer link.

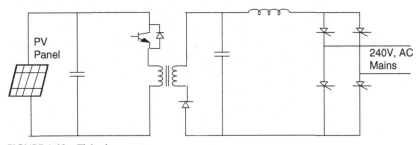

FIGURE 1.40   Flyback converter.

E. Flyback Converter This converter topology steps up the PV voltage to DC bus voltage. Pulse width modulation operated converter has been used for grid connection of PV system (Fig. 1.40). This scheme is less complex and has less number of switches. Flyback converters can be beneficial for remote areas due to less complex power conditioning components.

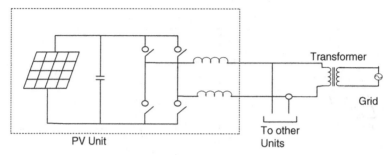

FIGURE 1.41 Converter using parallel PV units.

F. Interface Using Paralleled PV Panels Low voltage AC bus scheme [44] can be comparatively efficient and cheaper option. One of the schemes is shown in Fig. 1.41. A number of smaller PV units can be paralleled together and then connected to combine single low-frequency transformer. In this scheme, the PV panels are connected in parallel rather than series to avoid problems associated with shading or malfunction of one of the panels in series connection.

1.2.5.4 Power Control through PV Inverters

The system shown in Fig. 1.42 shows control of power flow on to the grid [45]. This control can be an analog or a microprocessor system. This control system generates the waveforms and regulates the waveform amplitude and phase to control the power flow between the inverter and the grid. The grid-interfaced PV inverters, voltage-controlled VSI (VCVSI), or current-controlled VSI (CCVSI) have the potential of bi-directional power flow. They cannot only feed the local load but also can export the excess active and reactive power to the utility grid. An appropriate controller is required in order to avoid any error in power export due to errors in synchronization, which can overload the inverter.

There are advantages and limitations associated with each control mechanism. For instance, VCVSIs provide voltage support to the load (here the VSI

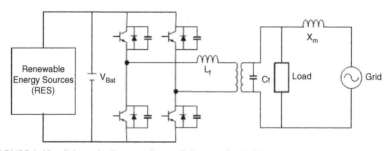

FIGURE 1.42 Schematic diagram of a parallel processing DGS.

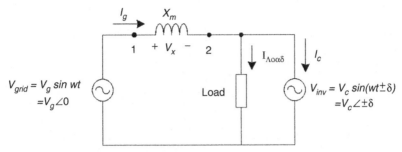

FIGURE 1.43   The equivalent circuit diagram of a VCVSI.

operates as a voltage source), while CCVSIs provide current support (here the VSI operates as a current source). The CCVSI is faster in response compared to the VCVSI, as its power flow is controlled by the switching instant, whereas in the VCVSI the power flow is controlled by adjusting the voltage across the decoupling inductor. Active and reactive power are controlled independently in the CCVSI, but are coupled in the VCVSI. Generally, the advantages of one type of VSI are considered as a limitation of the other type [46].

Figure 1.43 shows the simplified/equivalent schematic diagram of a VCVSI. For the following analysis it is assumed that the output low-pass filters (L_f and C_f) of VSIs will filter out high-order harmonics generated by PWMs. The decoupling inductor (X_m) is an essential part of any VCVSI as it makes the power flow control possible. In a VCVSI, the power flow of the distributed generation system (DGS) is controlled by adjusting the amplitude and phase (power angle (δ)) of the inverter output voltage with respect to the grid voltage. Hence, it is important to consider the proper sizing of the decoupling inductor and the maximum power angle to provide the required power flow when designing VCVSIs. The phasor diagram of a simple grid-inverter interface with a first-order filter are shown in Fig. 1.44.

Referring to Fig. 1.43, the fundamental grid current (I_g) can be expressed by Eq. (1.7):

$$I_g = \frac{V_g < 0 - V_c < \delta}{jX_m} = -\frac{V_c \sin \delta}{X_m} - j\frac{V_g - V_c \cos \delta}{X_m} \qquad (1.7)$$

where V_g and V_c are respectively the grid and the VCVSI's fundamental voltages, and X_m is the decoupling inductor impedance. Using per unit values ($S_{base} = V_{base}^2/Z_{base}$, $V_{base} = V_c$, and $Z_{base} = X_m$) where V_{base}, Z_{base}, and S_{base} are the base voltage, impedance and complex power values respectively. The grid apparent power can be expressed as Eq. (1.8).

$$S_{gpu} = -V_{gpu} \sin \delta + j(V_{gpu}^2 - V_{gpu} \cos \delta) \qquad (1.8)$$

Using per unit values, the complex power of the VCVSI and decoupling inductor are

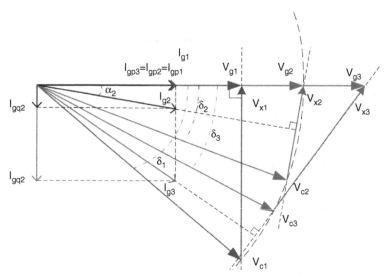

FIGURE 1.44 Phasor diagram of a VCVSI with resistive load and assuming the grid is responsible for supplying the active power [46].

$$S_{cpu} = -V_{gpu} \sin \delta + j(V_{gpu} \cos \delta - 1) \qquad (1.9)$$

$$S_{xpu} = j(V_{gpu}^2 - 2V_{gpu} \cos \delta + 1) \qquad (1.10)$$

where S_{gpu}, S_{cpu}, and S_{xpu} are per unit values of the grid, VCVSI, and decoupling inductor apparent power respectively, and V_{gpu} is the per unit value of the grid voltage.

Figure 1.45 shows the equivalent schematic diagram of a CCVSI. As a CCVSI controls the current flow using the VSI switching instants, it can be modeled as a current source and there is no need for a decoupling inductor (Fig. 1.45). As the current generated from the CCVSI can be controlled independently from the AC voltage, the active and reactive power controls are decoupled. Hence, unity power factor operation is possible for the whole range of the load. This is one of the main advantages of CCVSIs.

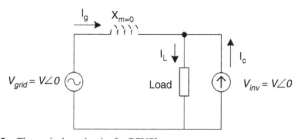

FIGURE 1.45 The equivalent circuit of a CCVSI.

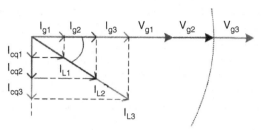

FIGURE 1.46 Phasor diagram of a CCVSI with inductive load and assuming grid is responsible for supplying the active power [46].

As the CCVSI is connected in parallel to the DGS, it follows the grid voltage. Figure 1.46 shows the phasor diagram of a CCVSI based DGS in the presence of an inductive load (considering the same assumption as VCVSI section). Figure 1.49 shows that when the grid voltage increases, the load's active power consumption, which supplied by the grid increases and the CCVSI compensates the increase in the load reactive power demand. In this case, the CCVSI maintains grid supply at unity power factor, keeping the current phase delay with respect to the grid voltage at a fixed value (θ). Therefore, the CCVSI cannot maintain the load voltage in the presence of a DGS without utilizing extra hardware and control mechanisms. This limitation on load voltage stabilization is one of the main drawbacks of CCVSI based DGS.

Assuming the load active current demand is supplied by the grid (reactive power support function), the required grid current can be rewritten as follows

$$I_g^* = \text{Re}\,[I_L] = \text{Re}\left[\frac{S_L}{V_g}\right] \tag{1.11}$$

where, S_L is the demanded load apparent power. For grid power conditioning, it is preferred that the load operate at unity power factor. Therefore, the CCVSI must provide the remainder of the required current Eq. (1.12)

$$I_c = I_L - I_g^* \tag{1.12}$$

For demand side management (DSM), it is desirable to supply the active power by the RES, where excess energy from the RES is injected into the DGS. The remaining load reactive power will be supplied by the CCVSI. Hence Eq. (1.12) can be rewritten as Eq. (1.13).

$$I_g^* = \text{Re}\,[I_L] - \text{Re}\,[I_c] = \text{Re}\left[\frac{S_L - P_{RES}}{V_g}\right] \tag{1.13}$$

When using a voltage controller for grid-connected PV inverter, it has been observed that a slight error in the phase of synchronizing waveform can grossly overload the inverter whereas a current controller is much less susceptible to voltage phase shifts [45]. Due to this reason, the current controllers are better suited for the control of power export from the PV inverters to the utility grid since they are less sensitive to errors in synchronizing sinusoidal voltage waveforms.

A prototype current-controlled type power conditioning system has been developed by the first author and tested on a weak rural feeder line at Kalbarri in Western Australia [47]. The choice may be between additional conventional generating capacity at a centralized location or adding smaller distributed generating capacities using RES like PV. The latter option can have a number of advantages like:

- The additional capacity is added wherever it is required without adding additional power distribution infrastructure. This is a critical consideration where the power lines and transformers are already at or close to their maximum ratings.
- The power conditioning system can be designed to provide much more than just a source of real power, for minimal extra cost. A converter providing real power needs only a slight increase in ratings to handle significant amounts of reactive or even harmonic power. The same converter that converts DC PV power to AC power can simultaneously provide the reactive power support to the week utility grid.

The block diagram of the power conditioning system used in the Kalbarri project has been shown in the Fig. 1.47. This CCVSI operates with a relatively narrow switching frequency band near 10 kHz. The control diagram indicates the basic operation of the power conditioning system. The two outer control loops operate to independently control the real and reactive power flow from the PV inverter. The real power is controlled by an outer MPPT algorithm with an inner DC link voltage control loop providing the real current magnitude request I_p^* and hence the real power export through PV converter is controlled through the DC link voltage regulation. The DC link voltage is maintained at a reference value by a PI control loop, which gives the real current reference magnitude as it's output. At regular intervals, the DC link voltage is scanned over the entire voltage range

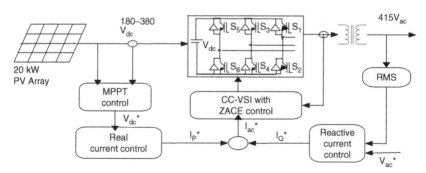

FIGURE 1.47 Block diagram of Kalbarri power conditioning system.

to check that the algorithm is operating on the absolute MPP and is not stuck around a local MPP. During the night, the converter can still be used to regulate reactive power of the grid-connected system although it cannot provide active power. During this time, the PI controller maintain a minimum DC link voltage to allow the power conditioning system to continue to operate, providing the necessary reactive power.

The AC line voltage regulation is provided by a separate reactive power control, which provides the reactive current magnitude reference I_Q^*. The control system has a simple transfer function, which varies the reactive power command in response to the AC voltage fluctuations. Common to the outer real and reactive power control loops is an inner higher bandwidth zero average current error (ZACE) current control loop. I_p^* is in phase with the line voltages, and I_Q^* is at 90° to the line voltage. These are added together to give one (per phase) sinusoidal converter current reference waveform (I_{ac}^*). The CCVSI control consists of analog and digital circuitry which acts as a ZACE transconductance amplifier in converting I_{ac}^* into AC power currents [48].

1.2.5.5   System Configurations

The utility compatible inverters are used for power conditioning and synchronization of PV output with the utility power. In general, four types of battery-less grid-connected PV system configurations have been identified:

- Central plant inverter.
- Multiple string DC/DC converter with single output inverter.
- Multiple string inverter.
- Module integrated inverter.

1.2.5.5.1   Central Plant Inverter

In the central plant inverter, usually a large inverter is used to convert DC power output of PV arrays to AC power. In this system, the PV modules are serially stringed to form a panel (or string) and several such panels are connected in parallel to a single DC bus. The block diagram of such a scheme is shown in Fig. 1.48.

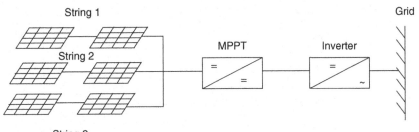

FIGURE 1.48   Central plant inverter.

1.2.5.5.2 Multiple String DC/DC Converter

In multiple string DC/DC converter, as shown in Fig. 1.49, each string will have a boost DC/DC converter with transformer isolation. There will be a common DC link, which feeds a transformer-less inverter.

1.2.5.5.3 Multiple String Inverters

Figure 1.50 shows the block diagram of multiple string inverter system. In this scheme, several modules are connected in series on the DC side to form a string. The output from each string is converted to AC through a smaller individual inverter. Many such inverters are connected in parallel on the AC side. This arrangement is not badly affected by the shading of the panels. It is also not seriously affected by inverter failure.

1.2.5.5.4 Module Integrated Inverter

In the module integrated inverter system (Fig. 1.51), each module (typically 50–300 W) will have a small inverter. No cabling is required. It is expected that high volume of small inverters will bring down the cost.

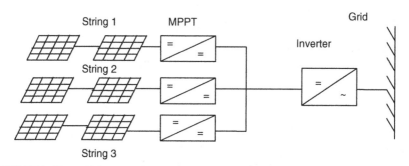

FIGURE 1.49 Multiple string DC/DC converter.

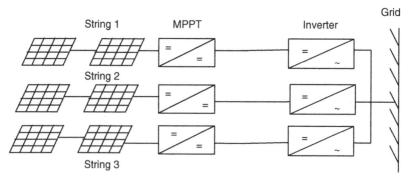

FIGURE 1.50 Multiple string inverter.

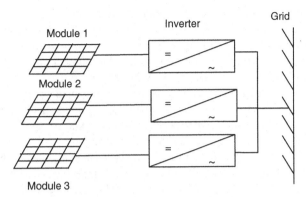

FIGURE 1.51   Module integrated inverter.

1.2.5.6   Grid-compatible Inverters Characteristics

The characteristics of the grid-compatible inverters are:

- Response time.
- Power factor.
- Frequency control.
- Harmonic output.
- Synchronization.
- Fault current contribution.
- DC current injection.
- Protection.

The response time of the inverters shall be extremely fast and governed by the bandwidth of the control system. Absence of rotating mass and use of semiconductor switches allow inverters to respond in millisecond time frame. The power factor of the inverters is traditionally poor due to displacement power factor and the harmonics. But with the latest development in the inverter technology, it is possible to maintain the power factor close to unity. The converters/inverters have the capability of creating large voltage fluctuation by drawing reactive power from the utility rather than supplying [49]. With proper control, inverters can provide voltage support by importing/exporting reactive power to push/pull towards a desired set point. This function would be of more use to the utilities as it can assist in the regulation of the grid system at the domestic consumer level.

Frequency of the inverter output waveshape is locked to the grid. Frequency bias is where the inverter frequency is deliberately made to run at 53 Hz. When the grid is present, this will be pulled down to the nominal 50 Hz. If the grid fails, it will drift upwards towards 53 Hz and trip on over frequency. This can help in preventing islanding.

Harmonics output from the inverters have been very poor traditionally. Old thyristor-based inverters are operated with slow switching speeds and could not be pulse width modulated. This resulted in inverters known as six-pulse or twelve-pulse inverters. The harmonics so produced from the inverters can be injected into the grid, resulting in losses, heating of appliances, tripping of protection equipments, and poor power quality. The number of pulses being the number of steps in a sine-wave cycle. With the present advent in the power electronics technology, the inverter controls can be made very good. Pulse width modulated inverters produce high quality sine waves. The harmonic levels are very low, and can be lower than the common domestic appliances. If the harmonics are present in the grid voltage waveform, harmonic currents can be induced in the inverter. These harmonic currents, particularly those generated by a voltage-controlled inverter, will in fact help in supporting the grid. These are good harmonic currents. This is the reason that the harmonic current output of inverters must be measured onto a clean grid source so that the only harmonics being produced by the inverters are measured.

Synchronization of inverter with the grid is performed automatically and typically uses zero crossing detection on the voltage waveform. An inverter has no rotating mass and hence has no inertia. Synchronization does not involve the acceleration of a rotating machine. Consequently the reference waveforms in the inverter can be jumped to any point required within a sampling period. If phase-locked loops are used, it could take up a few seconds. Phase-locked loops are used to increase the immunity to noise. This allows the synchronization to be based on several cycles of zero crossing information. The response time for this type of locking will be slower.

Photovoltaic panels produce a current that is proportional to the amount of light falling on them. The panels are normally rated to produce $1000\,W/m^2$ at $25°\,C$. Under these conditions, the short-circuit current possible from these panels is typically only 20% higher than the nominal current whereas it is extremely variable for wind. If the solar radiation is low then the maximum current possible under short-circuit is going to be less than the nominal full load current. Consequently PV systems cannot provide the short-circuit capacity to the grid. If a battery is present, the fault current contribution is limited by the inverter. With the battery storage, it is possible for the battery to provide the energy. However, inverters are typically limited between 100 and 200% of nominal rating under current limit conditions. The inverter needs to protect itself against the short circuits because the power electronic components will typically be destroyed before a protection device like circuit breaker trips.

In case of inverter malfunction, inverters have the capability to inject the DC components into the grid. Most utilities have guidelines for this purpose. A transformer shall be installed at the point of connection on the AC side to prevent DC from being entering into the utility network. The transformer can be omitted when a DC detection device is installed at the point of connection on the AC side in the inverter. The DC injection is essentially caused by the reference or power electronics device producing a positive half cycle that is different from the negative half cycle resulting in the DC component in the output. If the DC component can be measured, it can then be added into the feedback path to eliminate the DC quantity.

1.2.5.6.1 Protection Requirements

A minimum requirement to facilitate the prevention of islanding is that the inverter energy system protection operates and isolates the inverter energy system from the grid if:

- Over voltage.
- Under voltage.
- Over frequency.
- Under frequency exists.

These limits may be either factory set or site programmable. The protection voltage operating points may be set in a narrower band if required, e.g. 220–260 V. In addition to the passive protection detailed above, and to prevent the situation where islanding may occur because multiple inverters provide a frequency reference for one another, inverters must have an accepted active method of islanding prevention following grid failure, e.g. frequency drift, impedance measurement, etc. Inverter controls for islanding can be designed on the basis of detection of grid voltage, measurement of impedance, frequency variation, or increase in harmonics. This function must operate to force the inverter output outside the protection tolerances specified previously, thereby resulting in isolation of the inverter energy system from the grid. The maximum combined operation time of both passive and active protections should be 2 s after grid failure under all local load conditions. If frequency shift is used, it is recommended that the direction of shift be down. The inverter energy system must remain disconnected from the grid until the reconnection conditions are met. Some inverters produce high voltage spikes, especially at light load, which can be dangerous for the electronic equipment. IEEE P929 gives some idea about the permitted voltage limits.

If the inverter energy system does not have the above frequency features, the inverter must incorporate an alternate anti-islanding protection feature that is acceptable to the relevant electricity distributor. If the protection function above is to be incorporated in the inverter it must be type tested for compliance with these requirements and accepted by the relevant electricity distributor. Otherwise other forms of external protection relaying are required which have been type tested for compliance with these requirements and approved by the relevant electricity distributor. The inverter shall have adequate protection against short circuit, other faults, and overheating of inverter components.

1.3 POWER ELECTRONICS FOR WIND POWER SYSTEMS

In rural USA, the first wind mill was commissioned in 1890 to generate electricity. Today, large wind generators are competing with utilities in supplying clean power economically. The average wind turbine size has been 300–600 kW until recently. The new wind generators of 1–3 MW have been developed and are being installed worldwide, and prototype of even higher capacity is under development. Improved wind turbine designs and plant utilization have resulted in significant reduction in wind energy generation cost from 35 cents per kWh in 1980 to less than 5 cents per kWh in 1999, in locations where wind regime is favorable. At this generation cost, wind energy has become one of the least cost power sources. Main factors that have contributed to the wind power technology development are:

- High strength fiber composites for manufacturing large low-cost blades.
- Variable speed operation of wind generators to capture maximum energy.
- Advancement in power electronics and associated cost.
- Improved plant operation and efficiency.
- Economy of scale due to availability of large wind generation plants.
- Accumulated field experience improving the capacity factor.
- Computer prototyping by accurate system modeling and simulation.

The Table 1.2 is for wind sites with average annual wind speed of 7 m/s at 30 m hub height. Since 1980s, wind technology capital costs have reduced by 80% worldwide. Operation and maintenance costs have declined by 80% and the availability factor of grid-connected wind plants has increased to 95%. At present, the capital cost of wind generator plants has dropped to about $600 per kW and the electricity generation cost has reduced to 6 cents per kWh. It is expected to reduce the generation cost below 4 cents per kWh. Keeping this in view, the wind generation is going to be highly competitive with the conventional power plants. In Europe, USA, and Asia the wind power generation is increasing rapidly and this trend is going to continue due to economic viability of wind power generation.

The technical advancement in power electronics is playing an important part in the development of wind power technology. The contribution of power

TABLE 1.2 Wind Power Technology Developments

	1980	1999	Future
Cost per kWh	$0.35–0.40	$0.05–0.07	<$0.04
Capital cost per kW	$2000–3000	$500–700	<$400
Operating life	5–7 Years	20 Years	30 Years
Capacity factor (average)	15%	25–30%	>30%
Availability	50–65%	95%	>95%
Wind turbine unit size range	50–150 kW	300–1000 kW	500–2000 kW

electronics in control of fixed speed/variable speed wind turbines and interfacing to the grid is of extreme importance. Because of the fluctuating nature of wind speed, the power quality and reliability of the wind based power system needs to be evaluated in detail. Appropriate control schemes require power conditioning.

1.3.1   Basics of Wind Power

The ability of a wind turbine to extract power from wind is a function of three main factors:

- Wind power availability.
- Power curve of the machine.
- Ability of the machine to respond to wind perturbations.

The mechanical power produced by a wind turbine is given by

$$P_m = 0.5\rho C_p A U^3 \text{ W} \tag{1.14}$$

The power from the wind is a cubic function of wind speed. The curve for power coefficient C_p and λ is required to infer the value of C_p for λ based on wind speed at that time.

Where tip speed ratio, $\lambda = \frac{r\omega A}{U}$, ρ = Air density, Kg m^{-3}, C_p = power coefficient, A = wind turbine rotor swept area, m^2, U = wind speed in m/s.

The case of a variable speed wind turbine with a pitch mechanism that alters the effective rotor dynamic efficiency, can be easily considered if an appropriate expression for C_p as a function of the pitch angle is applied. The power curve of a typical wind turbine is given in Fig. 1.52 as a function of wind speed.

The C_p–λ curve for 150 kW windmaster machine is given in Fig. 1.53, which has been inferred from the power curve of the machine. The ratio of shaft power to the available power in the wind is the efficiency of conversion, known as the power coefficient C_p

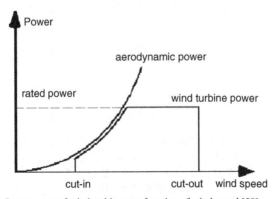

FIGURE 1.52   Power curve of wind turbine as a function of wind speed [50].

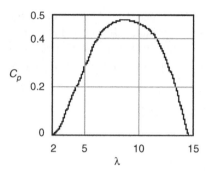

FIGURE 1.53 *Cp–λ* curve of wind machine [50].

$$C_p = \frac{P_m}{(1/2\rho A U^3)} \tag{1.15}$$

The power coefficient is a function of turbine blade tip speed to wind speed ratio (β). A tip speed ratio of 1 means the blade tips are moving at the same speed as the wind and when β is 2 the tips are moving at twice the speed of the wind and so on [51]. Solidity (σ) is defined as the ratio of the sum of the width of all the blades to the circumference of the rotor. Hence,

$$\sigma = Nd/(2\pi R) \tag{1.16}$$

where N = number of blades and d = width of the blades.

The power from a wind turbine doubles as the area swept by the blades doubles. But doubling of the wind speed increases the power output eight times. Figure 1.54 gives a family of power curves for a wind turbine. If the loading

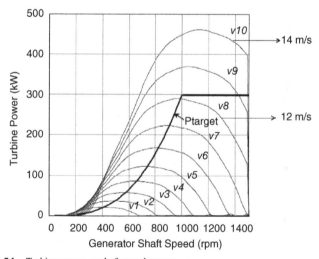

FIGURE 1.54 Turbine power vs shaft speed curves.

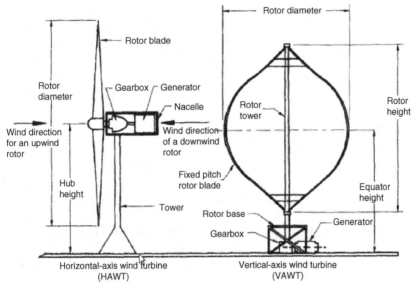

FIGURE 1.55   Typical diagram of HAWTs and VAWTs.

of the turbine is controlled such that the operating point is along the maximum power locus at different wind speeds, then the wind energy system will be more efficient.

1.3.1.1   Types of Wind Turbines

There are two types of wind turbines available Fig. 1.55:

- Horizontal axis wind turbines (HAWTs).
- Vertical axis wind turbines (VAWTs).

Vertical axis wind turbines (VAWTs) have an axis of rotation that is vertical, and so, unlike the horizontal wind turbines, they can capture winds from any direction without the need to reposition the rotor when the wind direction changes (without a special yaw mechanism). Vertical axis wind turbines were also used in some applications as they have the advantage that they do not depend on the direction of the wind. It is possible to extract power relatively easier. But there are some disadvantages such as no self starting system, smaller power coefficient than obtained in the horizontal axis wind turbines, strong discontinuation of rotations due to periodic changes in the lift force, and the regulation of power is not yet satisfactory.

The horizontal axis wind turbines are generally used. Horizontal axis wind turbines are, by far, the most common design. There are a large number of designs commercially available ranging from 50 W to 4.5 MW. The number of blades ranges from one to many in the familiar agriculture windmill. The best compromise for electricity generation, where high rotational speed allows use of a smaller and cheaper electric generator, is two or three blades. The mechanical and aerodynamic balance is better for three bladed rotor. In small wind turbines, three blades are common. Multiblade wind turbines are used for water pumping on farms.

Based on the pitch control mechanisms, the wind turbines can also be classified as:

- Fixed pitch wind turbines.
- Variable pitch wind turbines.

Different manufacturers offer fixed pitch and variable pitch blades. Variable pitch is desirable on large machines because the aerodynamic loads on the blades can be reduced and when used in fixed speed operation they can extract more energy. But necessary mechanisms require maintenance and for small machines, installed in remote areas, fixed pitch seems more desirable and economical. In some machines, power output regulation involves yawing blades so that they no longer point into the wind. One such system designed in Western Australia has a tail that progressively tilts the blades in a vertical plane so that they present a small surface to the wind at high speeds.

The active power of a wind turbine can be regulated by either designing the blades to go into an aerodynamic stall beyond the designated wind speed, or by feathering the blades out of the wind, which results in reducing excess power using a mechanical and electrical mechanism. Recently, an active stall has been used to improve the stability of wind farms. This stall mechanism can prevent power deviation from gusty winds to pass through the drive train [52].

Horizontal axis wind turbines can be further classified into fixed speed (FS) or variable speed (VS). The FS wind turbine generator (FSWT) is designed to operate at maximum efficiency while operating at a rated wind speed. In this case, the optimum tip-speed ratio is obtained for the rotor airfoil at a rated wind speed. For a VS wind turbine generator (VSWT), it is possible to obtain optimum wind speed at different wind speeds. Hence this enables the VS wind turbine to increase its energy capture. The general advantages of a VSWT are summarized as follows:

- VSWTs are more efficient than the FSWTs.
- At low wind speeds the wind turbines can still capture the maximum available power at the rotor, hence increasing the possibility of providing the rated power for wide speed range.

1.3.1.2 Types of Wind Generators

Schemes based on permanent magnet synchronous generators (PMSG) and induction generators are receiving close attention in wind power applications because of their qualities such as ruggedness, low cost, manufacturing simplicity, and low maintenance requirements. Despite many positive features over the conventional synchronous generators, the PMSG was not being used widely [23]. However, with the recent advent in power electronics, it is now possible to control the variable voltage, variable frequency output of PMSG. The permanent magnet machine is generally favored for developing new designs, because of higher efficiency and the possibility of a rather smaller diameter. These PMSG machines are now being used with variable-speed wind machines.

In large power system networks, synchronous generators are generally used with fixed-speed wind turbines. The synchronous generators can supply the active and reactive power both, and their reactive power flow can be controlled. The synchronous generators can operate at any power factor. For the induction generator, driven by a wind turbine, it is a well-known fact that it can deliver only active power, while consuming reactive power

Synchronous generators with high power rating are significantly more expensive than induction generators of similar size. Moreover, direct connected synchronous generators have the limitation of rotational speed being fixed by the grid frequency. Hence, fluctuation in the rotor speed due to wind gusts lead to higher torque in high power output fluctuations and the derived train. Therefore in grid-connected application, synchronous generators are interfaced via power converters to the grid. This also allows the synchronous generators to operate wind turbines in VS, which makes gear-less operation of the VSWT possible.

The squirrel-cage induction generators are widely used with the fixed-speed wind turbines. In some applications, wound rotor induction generators have also been used with adequate control scheme for regulating speed by external rotor resistance. This allows the shape of the torque-slip curve to be controlled to improve the dynamics of the drive train. In case of PMSG, the converter/inverter can be used to control the variable voltage, variable frequency signal of the wind generator at varying wind speed. The converter converts this varying signal to the DC signal and the output of converter is converted to AC signal of desired amplitude and frequency.

The induction generators are not locked to the frequency of the network. The cyclic torque fluctuations at the wind turbine can be absorbed by very small change in the slip speed. In case of the capacitor excited induction generators, they obtain the magnetizing current from capacitors connected across its output terminals [51, 53, 54].

To take advantage of VSWTs, it is necessary to decouple the rotor speed and the grid frequency. There are different approaches to operate the VSWT within a certain operational range (cut-in and cut-out wind speed). One of the approaches is dynamic slip control, where the slip is allowed to vary upto 10% [55]. In these cases, doubly-fed induction generators (DFIG) are used (Fig. 1.56). One limitation is that DFIG require reactive power to operate. As it is not desired that the grid supply this reactive power, these generators are usually equipped with capacitors. A gear box forms an essential component of

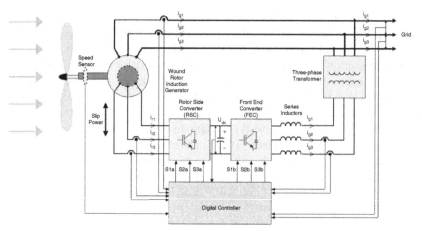

FIGURE 1.56   Variable speed doubly-fed induction generator (VSDFIG) system.

the wind turbine generator (WTG) using induction generators. This results in the following limitations:

- Frequent maintenance.
- Additional cost.
- Additional losses.

With the emergence of large wind power generation, increased attention is being directed towards wound rotor induction generators (WRIG) controlled from the rotor side for variable speed constant frequency (VSCF) applications. A wound rotor induction generator has a rotor containing a 3-phase winding. These windings are made accessible to the outside via slip rings. The main advantages of a wound rotor induction generator for VSCF applications are:

- Easier generator torque control using rotor current control.
- Smaller generator capacity as the generated power can be accessed from the stator as well as from the rotor. Usually the rotor power is proportional to the slip speed (shaft speed–synchronous speed). Consequently smaller rotor power converters are required. The frequency converter in the rotor (inverter) directly controls the current in the rotor winding, which enables the control of the whole generator output. The power electronic converters generally used are rated at 20–30% of the nominal generator power.
- Fewer harmonics exist because control is in the rotor while the stator is directly connected to the grid.

If the rotor is short-circuited (making it the equivalent of a cage rotor induction machine), the speed is primarily determined by the supply frequency and the nominal slip is within 5%. The mechanical power input ($P_{TURBINE}$) is converted into stator electrical power output (P_{STATOR}) and is fed to the AC supply. The rotor power loss, being proportional to the slip speed, is commonly referred to as the slip power (P_{ROTOR}). The possibility of accessing the rotor in a doubly-fed induction generator makes a number of configurations possible. These include slip power recovery using a cycloconverter, which converts the ac voltage of one frequency to another without an intermediate DC link [56–58], or back-to-back inverter configurations [59, 60].

Using voltage-source inverters (VSIs) in the rotor circuit, the rotor currents can be controlled at the desired phase, frequency, and magnitude. This enables reversible flow of active power in the rotor and the system can operate in sub-synchronous and super-synchronous speeds, both in motoring and generating modes. The DC link capacitor acts as a source of reactive power and it is possible to supply the magnetizing current, partially or fully, from the rotor side. Therefore, the stator side power factor can also be controlled. Using vector control techniques, the active and reactive powers can be controlled independently and hence fast dynamic performance can also be achieved.

The converter used at the grid interface is termed as the line-side converter or the front end converter (FEC). Unlike the rotor side converter, this operates at the grid frequency. Flow of active and reactive powers is controlled by adjusting the phase and amplitude of the inverter terminal voltage with respect to the grid voltage. Active power can flow either to the grid or to the rotor circuit depending on the mode of operation. By controlling the flow of active power, the DC bus voltage is regulated within a small band. Control of reactive power enables unity power factor operation at the grid interface. In fact, the FEC can be operated at a leading power factor, if it is so desired. It should be noted that, since the slip range is limited, the DC bus voltage is less in this case when compared to the stator side control. A transformer is therefore necessary to match the voltage levels between the grid and the DC side of the FEC. With a PWM converter in the rotor circuit, the rotor currents can be controlled at the desired phase, frequency, and magnitude. This enables reversible flow of active power in the rotor and the system can operate in sub-synchronous and super-synchronous speeds, both in motoring and generating modes (Fig. 1.57).

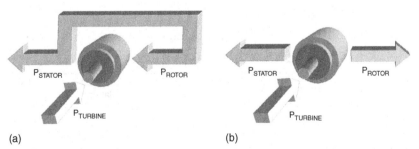

(a)   (b)

FIGURE 1.57 Doubly-fed induction generator power flow in generating mode: (a) sub-synchronous and (b) super-synchronous.

1.3.2 Types of Wind Power Systems

Wind power systems can be classified as:

- Stand-alone.
- Hybrid.
- Grid-connected.

1.3.3 Stand-alone Wind Power Systems

Stand-alone wind power systems are being used for the following purposes in remote area power systems:

- Battery charging.
- Household power supply.

1.3.3.1 Battery Charging with Stand-alone Wind Energy System

The basic elements of a stand-alone wind energy conversion system are:

- Wind generator.
- Tower.
- Charge control system.
- Battery storage.
- Distribution network.

In remote area power supply, an inverter and a diesel generator are more reliable and sophisticated systems. Most small isolated wind energy systems use batteries as a storage device to level out the mismatch between the availability of the wind and the load requirement. Batteries are a major cost component in an isolated power system.

1.3.3.2 Wind Turbine Charge Controller

The basic block diagram of a stand-alone wind generator and battery charging system is shown in Fig. 1.58.

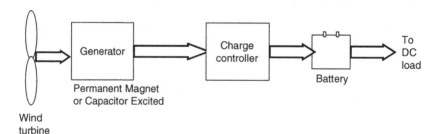

FIGURE 1.58 Block diagram for a stand-alone wind generator and battery charging system.

The function of charge controller is to feed the power from the wind generator to the battery bank in a controlled manner. In the commonly used permanent magnet generators, this is usually done by using the controlled rectifiers [61]. The controller should be designed to limit the maximum current into the battery, reduce charging current for high battery SOC, and maintain a trickle charge during full SOC periods.

1.3.4   Wind–diesel Hybrid Systems

The details of hybrid systems are already covered in Section 1.2.4. Diesel systems without batteries in remote area are characterized by poor efficiency, high maintenance, and fuel costs. The diesel generators must be operated above a certain minimum load level to reduce cylinder wear and tear due to incomplete combustion. It is a common practice to install dump loads to dissipate extra energy. More efficient systems can be devised by combining the diesel generator with a battery inverter subsystem and incorporating RES, such as wind/solar where appropriate. An integrated hybrid energy system incorporating a diesel generator, wind generator, battery or flywheel storage, and inverter will be cost effective at many sites with an average daily energy demand exceeding 25 kWh [62]. These hybrid energy systems can serve as a mini grid as a part of distributed generation rather than extending the grid to the remote rural areas. The heart of the hybrid system is a high quality sine-wave inverter, which can also be operated in reverse as battery charger. The system can cope with loads ranging from zero (inverter only operation) to approximately three times greater capacity (inverter and diesel operating in parallel).

Decentralized form of generation can be beneficial in remote area power supply. Due to high cost of PV systems, problems associated with storing electricity over longer periods (like maintenance difficulties and costs), wind turbines can be a viable alternative in hybrid systems. Systems with battery storage although provide better reliability. Wind power penetration can be high enough to make a significant impact on the operation of diesel generators.

High wind penetration also poses significant technical problems for the system designer in terms of control and transient stability [30]. In earlier stages, wind diesel systems were installed without assessing the system behavior due to lack of design tools/software. With the continual research in this area, there are now software available to assist in this process. Wind diesel technology has now matured due to research and development in this area. Now there is a need to utilize this knowledge into cost effective and reliable hybrid systems [63]. In Western Australia, dynamic modeling of wind diesel hybrid system has been developed in Curtin/MUERI, supported by the Australian Cooperative Research Centre for Renewable Energy (ACRE) program 5.21.

1.3.5 Grid-connected Wind Energy Systems

Small scale wind turbines, connected to the grid (weak or strong grid), have been discussed here. Wind diesel systems have been getting attention in many remote parts of the world lately. Remote area power supplies are characterized by low inertia, low damping, and poor reactive power support. Such weak power systems are more susceptible to sudden change in network operating conditions [64]. In this weak grid situation, the significant power fluctuations in the grid would lead to reduced quality of supply to users. This may manifest itself as voltage and frequency variations or spikes in the power supply. These weak grid systems need appropriate storage and control systems to smooth out these fluctuations without sacrificing the peak power tracking capability. These systems can have two storage elements. The first is the inertia of the rotating mechanical parts, which includes the blades, gearbox, and the rotor of the generators. Instead of wind speed fluctuation causing large and immediate change in the electrical output of the generator as in a fixed speed machine, the fluctuation will cause a change in shaft speed and not create a significant change in generator output. The second energy storage element is the small battery storage between the DC–DC converter and the inverter. The energy in a gust could be stored temporarily in the battery bank and released during a lull in the wind speed, thus reducing the size of fluctuations.

In larger scale wind turbines, the addition of inverter control further reduces fluctuation and increases the total output power. Thus the total output of the wind energy system can be stabilized or smoothed to track the average wind speed and can omit certain gusts. The system controller should track the peak power to maximize the output of the wind energy system. It should monitor the stator output and adjust the inverter to smooth the total output. The amount of smoothing would depend on SOC of the battery. The nominal total output would be adjusted to keep the battery bank SOC at a reasonable level. In this way, the total wind energy system will track the long-term variations in the wind speed without having fluctuations caused by the wind. The storage capacity of the battery bank need only be several minutes to smooth out the gusts in the wind, which can be easily handled by the weak grid. In the cases, where the weak grid is powered by diesel generators, the conventional wind turbine can cause the diesel engines to operate at low capacity. In case of strong wind application, the fluctuations in the output of the wind energy generator system can be readily absorbed by the grid. The main aim here is to extract the maximum energy from the wind. The basic block layout of such a system [65] is shown in the Fig. 1.59.

The function of the DC–DC converter will be to adjust the torque on the machine and hence ensure by measurement of wind speed and shaft speed that the turbine blades are operating so as to extract optimum power. The purpose of the inverter is to feed the energy gathered by the rotor and DC–DC converter, in the process of peak power tracking, to the grid system. The interaction between the two sections would be tightly controlled so as to minimize or

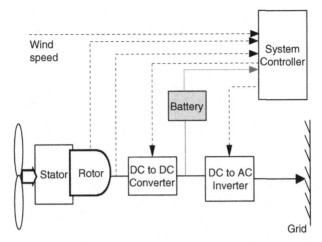

FIGURE 1.59   System block diagram of grid-connected wind energy system.

eliminate the need for a battery bank. The control must be fast enough so that the inverter output power set point matches the output of the DC–DC converter. For a wound rotor induction machine operating over a two to one speed range, the maximum power extracted from the rotor is equal to the power rating of the stator. Thus the rating of the generator from a traditional point of view is only half that of the wind turbine [65]. Since half the power comes from the stator and half from the rotor, the power electronics of the DC–DC converter and inverter need to handle only half the total wind turbine output and no battery would be required.

Power electronic technology also plays an important role in both system configurations and in control of offshore wind farms [66]. Wind farms connect in various configurations and control methods using different generator types and compensation arrangements. For instance, wind farms can be connected to the AC local network with centralized compensation or with a HVDC transmission system, and DC local network. Decentralized control with a DC transmission system has also been used [67].

1.3.5.1   Soft Starters for Induction Generators

When an induction generator is connected to a load, a large inrush current flows. This is something similar to the direct online starting problem of induction machines. It has been observed that the initial time constants of the induction machines are higher when it tries to stabilize initially at the normal operating conditions. There is a need to use some type of soft starting equipment to start the large induction generators. A simple scheme to achieve this is shown in the Fig. 1.60.

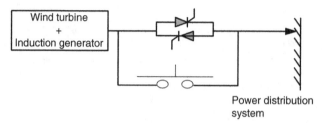

FIGURE 1.60   Soft starting for wind turbine coupled with induction generator.

Two thyristors are connected in each phase, back-to-back. Initially, when the induction generator is connected, the thyristors are used to control the voltage applied to the stator and to limit the large inrush current. As soon as the generator is fully connected, the bypass switch is used to bypass the soft starter unit.

1.3.6   Control of Wind Turbines

Theory indicates that operation of a wind turbine at fixed tip speed ratio (C_{pmax}) ensures enhanced energy capture [50]. The wind energy systems must be designed so that above the rated wind speed, the control system limit the turbine output. In normal operation, medium to large-scale wind turbines are connected to a large grid. Various wind turbine control policies have been studied around the world. Grid-connected wind turbines generators can be classified as:

- Fixed speed wind turbines.
- Variable speed wind turbines.

1.3.6.1   Fixed Speed Wind Turbines

In case of a fixed speed wind turbine, synchronous or squirrel-cage induction generators are employed and is characterized by the stiff power train dynamics. The rotational speed of the wind turbine generator in this case is fixed by the grid frequency. The generator is locked to the grid, thereby permitting only small deviations of the rotor shaft speed from the nominal value. The speed is very responsive to wind speed fluctuations. The normal method to smooth the surges caused by the wind is to change the turbine aerodynamic characteristics, either passively by stall regulation or actively by blade pitch regulation. The wind turbines often subjected to very low (below cut in speed) or high wind speed (above rated value). Sometimes they generate below rated power. No pitch regulation is applied when the wind turbine is operating below rated speed, but pitch control is required when the machine is operating above rated wind speed to minimize the stress. Figure 1.61 shows the effect of blade pitch angle on the torque speed curve at a given wind speed.

Blade pitch control is a very effective way of controlling wind turbine speed at high wind speeds, hence limiting the power and torque output of the wind

Curve index: pitch angle

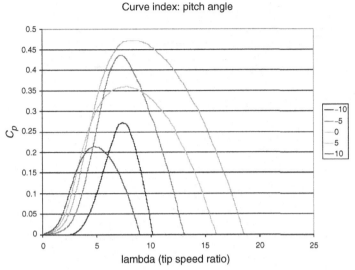

FIGURE 1.61 C_p/λ curves for different pitch settings.

machine. An alternative but cruder control technique is based on airfoil stall [50]. A synchronous link maintaining fixed turbine speed in combination with an appropriate airfoil can be designed so that, at higher than rated wind speeds the torque reduces due to airfoil stall. This method does not require external intervention or complicated hardware, but it captures less energy and has greater blade fatigue.

The aims of variable pitch control of medium- and large-scale wind turbines were to help in start-up and shutdown operation, to protect against overspeed and to limit the load on the wind turbine [68]. The turbine is normally operated between a lower and an upper limit of wind speed (typically 4.5–26 m/s). When the wind speed is too low or too high, the wind turbine is stopped to reduce wear and damage. The wind turbine must be capable of being started and run up to speed in a safe and controlled manner. The aerodynamic characteristics of some turbines are such that they are not self starting. The required starting torque may be provided by motoring or changing the pitch angle of the blade. In case of grid-connected wind turbine system, the rotational speed of the generator is locked to the frequency of the grid. When the generator is directly run by the rotor, the grid acts like an infinite load. When the grid fails, the load rapidly decreases to zero resulting in the turbine rotor to accelerate quickly. Over-speed protection must be provided by rapid braking of the turbine. A simple mechanism of one of blade pitch control techniques is shown in Fig. 1.62.

In this system, the permanent magnet synchronous generator (PMSG) has been used without any gearbox. Direct connection of generator to the wind turbine requires the generator to have a large number of poles. Both induction generators and wound filed synchronous generators of high pole number require

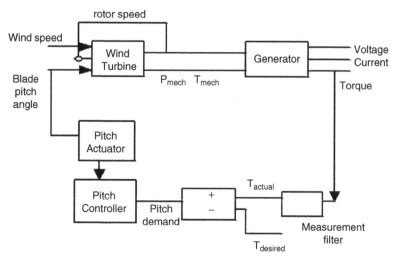

FIGURE 1.62   Pitch control block diagram of a PMSG.

a large diameter for efficient operation. Permanent magnet synchronous generators allow a small pole pitch to be used [69]. The power output, P_{mech}, of any turbine depends mainly upon the wind speed, which dictates the rotational speed of the wind turbine rotor. Depending upon the wind speed and rotational speed of turbine, tip speed ratio λ is determined. Based on computed λ, the power coefficient C_p is inferred. In the control strategy above, the torque output, T_{actual}, of the generator is monitored for a given wind speed and compared with the desired torque, T_{actual}, depending upon the load requirement. The generator output torque is passed through the measurement filter. The pitch controller then infers the modified pitch angle based on the torque error. This modified pitch angle demand and computed λ decides the new C_p resulting in the modified wind generator power and torque output. The controller will keep adjusting the blade pitch angle till the desired power and torque output are achieved.

Some of the wind turbine generator includes the gearbox for interfacing the turbine rotor and the generator. The general drive train model [68] for such a system is shown in Fig. 1.63. This system also contains the blade pitch angle control provision.

The drive train converts the input aerodynamic torque on the rotor into the torque on the low-speed shaft. This torque on the low-speed shaft is converted to high-speed shaft torque using the gearbox and fluid coupling. The speed of the wind turbine here is low and the gear box is required to increase the speed so as to drive the generator at rated rpm e.g. 1500 rpm. The fluid coupling works as a velocity-in-torque-out device and transfer the torque [68]. The actuator regulates the tip angle based on the control system applied. The control system here is based on a pitch regulation scheme where the blade pitch angle is adjusted to obtain the desired output power.

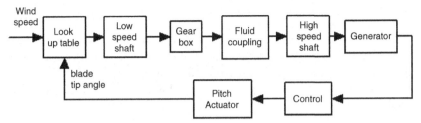

FIGURE 1.63   Block diagram of drive train model.

1.3.6.2 Variable Speed Wind Turbines

The variable speed constant frequency turbine drive trains are not directly coupled to the grid. The power-conditioning device is used to interface the wind generator to the grid. The output of the wind generator can be variable voltage and variable frequency, which is not suitable for grid integration and appropriate interfacing is required. The wind turbine rotor in this case is permitted to rotate at any wind speed by power generating unit.

A number of schemes have been proposed in the past which allow wind turbines to operate with variable rotor speed while feeding the power to a constant frequency grid. Some of the benefits that have been claimed for variable speed constant frequency wind turbine configuration is as follow [65]:

- The variable speed operation results in increased energy capture by maintaining the blade tip speed to wind speed ratio near the optimum value.
- By allowing the wind turbine generator to run at variable speed, the torque can be fixed, but the shaft power allowed to increase. This means that the rated power of the machine can be increased with no structural changes.
- A variable speed turbine is capable of absorbing energy in wind gusts as it speeds up and gives back this energy to the system as it slows down. This reduces turbulence induced stresses and allows capture of a large percentage of the turbulent energy in the wind.
- More efficient operation can be achieved by avoiding aerodynamic stall over most of operating range.
- Better grid quality due to support of grid voltage.

Progress in the power electronics conversion system has given a major boost to implementing the concept of variable speed operation. The research studies have shown that the most significant potential advancement for wind turbine technology was in the area of power electronic controlled variable speed operation. There is much research underway in the United States and Europe on developing variable speed wind turbine as cost effective as possible. In United States, the NASA MOD-0 and MOD-5B were operated as variable speed wind turbines [65]. Companies in United States and Enercon (Germany) made machines incorporate a variable speed feature. Enercon variable speed wind machine is already in operation in Denham, Western Australia.

The ability to operate at varying rotor speed, effectively adds compliance to the power train dynamics of the wind turbine. Although many approaches have been suggested for variable speed wind turbines, they can be grouped into two main classes: (a) discretely variable speed and (b) continuously variable speed [65, 70].

1.3.6.3 Discretely Variable Speed Systems

The discretely variable speed category includes electrical system where multiple generators are used, either with different number of poles or connected to the wind rotor via different ratio gearing. It also includes those generators, which can use different number of poles in the stator or can approximate the effect by appropriate switching. Some of the generators in this category are those with consequent poles, dual winding, or pole amplitude modulation. A brief summary of some of these concepts is presented below.

1.3.6.3.1 Pole Changing Type Induction Generators

These generators provide two speeds, a factor of two apart, such as four pole/eight pole (1500/750 rpm at a supply frequency of 50 Hz or 1800/900 rpm at 60 Hz). They do this by using one-half the poles at the higher speed. These machines are commercially available and cost about 50% more than the corresponding single speed machines. Their main disadvantage, in comparison with other discretely variable machines is that the two to one speed range is wider than the optimum range for a wind turbine [71].

1.3.6.3.2 Dual Stator Winding Two Speed Induction Generators

These machines have two separate stator windings, only one of which is active at a time. As such, a variety of speed ranges can be obtained depending on the number of poles in each winding. As in the consequent pole machines only two speeds may be obtained. These machines are significantly heavier than single speed machines and their efficiency is less, since one winding is always unused which leads to increased losses. These machines are commercially available. Their cost is approximately twice that of single speed machines [71].

1.3.6.3.3 Multiple Generators

This configuration is based on the use of a multiple generator design. In one case, there may simply be two separate generators (as used on some European wind turbines). Another possibility is to have two generators on the same shaft, only one of which is electrically connected at a time. The gearing is arranged such that the generators reach synchronous speed at different turbine rotor speeds.

1.3.6.3.4   Two Speed Pole Amplitude Modulated Induction Generator (PAM)

This configuration consists of an induction machine with a single stator, which may have two different operating speeds. It differs from conventional generators only in the winding design. Speed is controlled by switching the connections of the six stator leads. The winding is built in two sections which will be in parallel for one speed and in series for the other. The result is the superposition of one alternating frequency on another. This causes the field to have an effectively different number of poles in the two cases, resulting in two different operating speeds. The efficiency of the PAM is comparable to that of a single speed machine. The cost is approximately twice that of conventional induction generators.

The use of a discretely variable speed generator will result in some of the benefits of continuously variable speed operation, but not all of them. The main effect will be in increased energy productivity, because the wind turbine will be able to operate close to its optimum tip speed ratio over a great range of wind speeds than will a constant speed machine. On the other hand, it will perform as single speed machine with respect to rapid changes in wind speed (turbulence). Thus it could not be expected to extract the fluctuating energy as effective from the wind as would be continuously variable speed machine. More importantly, it could not use the inertia of the rotor to absorb torque spikes. Thus, this approach would not result in improved fatigue life of the machine and it could not be an integral part of an optimized design such as one using yaw/speed control or pitch/speed control.

1.3.6.4   *Continuously Variable Speed Systems*

The second main class of systems for variable speed operation are those that allow the speed to be varied continuously. For the continuously variable speed wind turbine, there may be more than one control, depending upon the desired control action [72–76]:

- Mechanical control.
- Combination of electrical/mechanical control.
- Electrical control.
- Electrical/power electronics control.

The mechanical methods include hydraulic and variable ratio transmissions. An example of an electrical/mechanical system is one in which the stator of the generator is allowed to rotate. All the electrical category includes high-slip induction generators and the tandem generator. The power electronic category contains a number of possible options. One option is to use a synchronous generator or a wound rotor induction generator, although a conventional induction generator may also be used. The power electronics is used to condition some or all the power to form a appropriate to the grid. The power electronics may also be used to rectify some or all the power from the generator, to control the rotational speed of the generator, or to supply reactive power. These systems are discussed below.

1.3.6.4.1 Mechanical Systems

A. Variable Speed Hydraulic Transmission One method of generating electrical power at a fixed frequency, while allowing the rotor to turn at variable speed, is the use of a variable speed hydraulic transmission. In this configuration, a hydraulic system is used in the transfer of the power from the top of the tower to ground level (assuming a horizontal axis wind turbine). A fixed displacement hydraulic pump is connected directly to the turbine (or possibly gearbox) shaft. The hydraulic fluid is fed to and from the nacelle via a rotary fluid coupling. At the base of the tower is a variable displacement hydraulic motor, which is governed to run at constant speed and drive a standard generator.

One advantage of this concept is that the electrical equipment can be placed at ground level making the rest of the machine simpler. For smaller machines, it may be possible to dispense with a gearbox altogether. On the other hand, there are a number of problems using hydraulic transmissions in wind turbines. For one thing, pumps and motors of the size needed in wind turbines of greater than about 200 kW are not readily available. Multiples of smaller units are possible but this would complicate the design. The life expectancy of many of the parts, especially seals, may well be less than five years. Leakage of hydraulic fluid can be a significant problem, necessitating frequent maintenance. Losses in the hydraulics could also make the overall system less efficient than conventional electric generation. Experience over the last many years has not shown great success with the wind machines using hydraulic transmission.

B. Variable Ratio Transmission A variable ratio transmission (VRT) is one in which the gear ratio may be varied continuously within a given range. One type of VRT suggested for wind turbines is using belts and pulleys, such as are used in some industrial drives [65, 77]. These have the advantage of being able to drive a conventional fixed speed generator, while being driven by a variable speed turbine rotor. On the other hand, they do not appear to be commercially available in larger sizes and those, which do exist, have relatively high losses.

1.3.6.4.2 Electrical/Mechanical Variable Speed Systems –Rotating Stator Induction Generator

This system uses a conventional squirrel-induction generator whose shaft is driven by a wind turbine through a gearbox [50, 77]. However, the stator is mounted to a support, which allows bi-directional rotation. This support is in turn driven by a DC machine. The armature of the DC machine is fed from a bi-directional inverter, which is connected to the fixed frequency AC grid. If the stator support allowed to turn in the same direction as the wind turbine, the turbine will turn faster. Some of the power from the wind turbine will be absorbed by the induction generator stator and fed to the grid through the inverter. Conversely, the wind turbine will turn more slowly when the stator support is driven in the opposite direction. The amount of current (and thus the torque) delivered to or from the DC machine is determined by a closed loop control circuit whose feedback signal is driven by a tachometer mounted on the shaft of the DC machine.

One of the problems with this system is that the stator slip rings and brushes must be sized to take the full power of the generator. They would be subjected to wear and would require maintenance. The DC machine also adds to cost, complexity, and maintenance.

1.3.6.4.3   Electrical Variable Speed Systems

A. High Slip Induction Generator This is the simplest variable speed system, which is accomplished by having a relatively large amount of resistance in the rotor of an induction generator. However, the losses increase with increased rotor resistance. Westwind Turbines in Australia investigated such a scheme on a 30 kW machine in 1989.

B. Tandem Induction Generator A tandem induction generator consists of an induction machine fitted with two magnetically independent stators, one fixed in position and the other able to be rotated, and a single squirrel-cage rotor whose bars extend to the length of both stators [65, 77]. Torque control is achieved by physical adjustment of the angular displacement between the two stators, which causes a phase shift between the induced rotor voltages.

1.3.6.4.4   Electrical/Power Electronics

The general configuration is shown in the Fig. 1.64. It consists of the following components:

- Wind generator.
- Rectifier.
- Inverter.

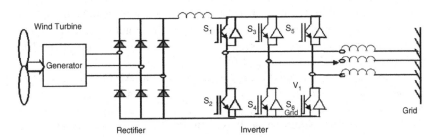

FIGURE 1.64   Grid-connected wind energy system through AC/DC/AC converter.

The generator may be DC, synchronous (wound rotor or permanent magnet type), squirrel-cage wound rotor, or brush-less doubly-fed induction generator. The rectifier is used to convert the variable voltage variable frequency input to a DC voltage. This DC voltage is converted into AC of constant voltage and frequency of desired amplitude. The inverter will also be used to control the active/reactive power flow from the inverter. In case of DC generator, the converter may not be required or when a cycloconverter is used to convert the AC directly from one frequency to another.

1.3.6.5 Types of Generator Options for Variable Speed Wind Turbines Using Power Electronics

Power electronics may be applied to four types of generators to facilitate variable speed operation:

- Synchronous generators.
- Permanent magnet synchronous generators.
- Squirrel-cage induction generators.
- Wound rotor induction generators.

1.3.6.5.1 Synchronous Generator

In this configuration, the synchronous generator is allowed to run at variable speed, producing power of variable voltage and frequency. Control may be facilitated by adjusting an externally supplied field current. The most common type of power conversion uses a bridge rectifier (controlled/uncontrolled), a DC link, and inverter as shown in Fig. 1.64. The disadvantage of this configuration include the relatively high cost and maintenance requirements of synchronous generators and the need for the power conversion system to take the full power generated (as opposed to the wound rotor system).

1.3.6.5.2 Permanent Magnet Synchronous Generators

The permanent magnet synchronous generator (PMSG) has several significant advantageous properties. The construction is simple and does not required external magnetization, which is important especially in stand-alone wind power applications and also in remote areas where the grid cannot easily supply the reactive power required to magnetize the induction generator. Similar to the previous externally supplied field current synchronous generator, the most common type of power conversion uses a bridge rectifier (controlled/uncontrolled), a DC link, and inverter as shown in Fig. 1.65 [78–80].

Figure 1.66 shows a wind energy system where a PMSG is connected to a three-phase rectifier followed by a boost converter. In this case, the boost converter controls the electromagnet torque and the supply side converter regulates the DC link voltage as well as controlling the input power factor. One drawback of this configuration is the use of diode rectifier that increases the

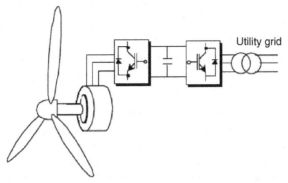

FIGURE 1.65   Grid-connected PMSG wind energy system through DC/AC converter.

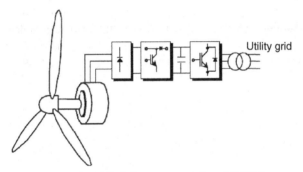

FIGURE 1.66   Grid-connected PMSG wind energy system through DC/AC converter with a boost chopper.

current amplitude and distortion of the PMSG. As a result, this configuration have been considered for small size wind energy conversion systems (smaller than 50 kW).

The advantage of the system in Fig. 1.65 with regardant to the system showed in Fig. 1.66 is, it allows the generator to operate near its optimal working point in order to minimize the losses in the generator and power electronic circuit. However, the performance is dependent on the good knowledge of the generator parameter that varies with temperature and frequency. The main drawbacks, in the use of PMSG, are the cost of permanent magnet that increase the price of machine, demagnetization of the permanent magnet material, and it is not possible to control the power factor of the machine

To extract maximum power at unity power factor from a PMSG and feed this power (also at unity power factor) to the grid, the use of back-to-back connected PWM voltage source converters are proposed [81]. Moreover, to reduce the overall cost, reduced switch PWM voltage source converters (four switch) instead of conventional (six switch) converters for variable speed drive systems can be used. It is shown that by using both rectifier and inverter current control or flux based control, it is possible to obtain unity power factor operation both at the WTG and the grid. Other mechanisms can also be included to maximize power extraction from the VSWT (i.e. MPPT techniques) or sensorless approaches to further reduce cost and increase reliability and performance of the systems.

1.3.6.5.3 Squirrel-cage Induction Generator

Possible architecture for systems using conventional induction generators which have a solid squirrel-cage rotor have many similarities to those with synchronous generators. The main difference is that the induction generator is not inherently self-exciting and it needs a source of reactive power. This could be done by a generator side self-commutated converter operating in the rectifier mode. A significant advantage of this configuration is the low cost and low maintenance requirements of induction generators. Another advantage of using the self-commutated double converter is that it can be on the ground, completely separate from the wind machine. If there is a problem in the converter, it could be switched out of the circuit for repair and the wind machine could continue to run at constant speed. The main disadvantage with this configuration is that, as with the synchronous generator, the power conversion system would have to take the full power generated and could be relatively costly compared to some other configurations. There would also be additional complexities associated with the supply of reactive power to the generator.

1.3.6.5.4 Wound Rotor Induction Generator

A wound rotor induction rotor has three-phase winding on the rotor, accessible to the outside via slip rings. The possibility of accessing the rotor can have the following configurations:

- Slip power recovery.
- Use of cycloconverter.
- Rotor resistance chopper control.

A. Slip Power Recovery (Static Kramer System) The slip power recovery configuration behaves similarly to a conventional induction generator with very large slip, but in addition energy is recovered from the rotor. The rotor power is first carried out through slip rings, then rectified and passed through a DC link to a line-commutated inverter and into the grid. The rest of the power comes directly from the stator as it normally does. A disadvantage with this system is that it can only allow super-synchronous variable speed operation. Its possible use in the wind power was reported by Smith and Nigim [82].

In this scheme shown in Fig. 1.67, the stator is directly connected to the grid. Power converter has been connected to the rotor of wound rotor induction generator to obtain the optimum power from variable speed wind turbine.

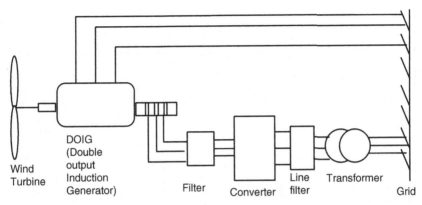

FIGURE 1.67 Schematic diagram of doubly-fed induction generator.

The main advantage of this scheme is that the power-conditioning unit has to handle only a fraction of the total power so as to obtain full control of the generator. This is very important when the wind turbine sizes are increasing for the grid-connected applications for higher penetration of wind energy and the smaller size of converter can be used in this scheme.

B. Cycloconverter (Static Scherbius System) A cycloconverter is a converter, which converts AC voltage of one frequency to another frequency without an intermediate DC link. When a cycloconverter is connected to the rotor circuit, sub- and super-synchronous operation variable speed operation is possible. In super-synchronous operation, this configuration is similar to the slip power recovery. In addition, energy may be fed into the rotor, thus allowing the machine to generate at sub-synchronous speeds. For that reason, the generator is said to be doubly fed [83]. This system has a limited ability to control reactive power at the terminals of the generator, although as a whole it is a net consumer of reactive power. On the other hand, if coupled with capacitor excitation, this capability could be useful from the utility point of view. Because of its ability to rapidly adjust phase angle and magnitude of the terminal voltage, the generator can be resynchronized after a major electrical disturbance without going through a complete stop/start sequence. With some wind turbines, this could be a useful feature.

C. Rotor Resistance Chopper Control A fairly simple scheme of extracting rotor power as in the form of heat has been proposed in [44].

1.3.6.6   Isolated Grid Supply System with Multiple Wind Turbines

The isolated grid supply system with a wind park is shown in Fig. 1.68. Two or more wind turbines can be connected to this system. A diesel generator can be connected in parallel. The converters, connected with wind generators will work

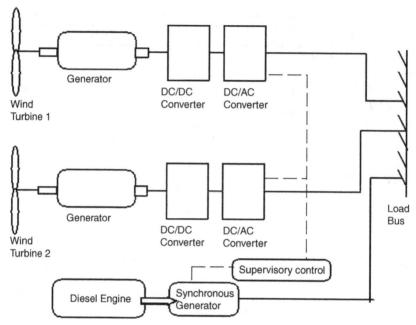

FIGURE 1.68   Schematic diagram of isolated grid system having a wind park.

in parallel and the supervisory control block will control the output of these wind generators in conjunction with the diesel generator. This type of decentralized generation can be a better option where high penetration of wind generation is sought. The individual converter will control the voltage and frequency of the system. The supervisory control system will play an important part in co-ordination between multiple power generation systems in a remote area power supply having weak grid.

1.3.6.7 Power Electronics Technology Development

To meet the needs of future power generation systems, power electronics technology will need to evolve on all levels, from devices to systems. The development needs are as follows:

- There is a need for modular power converters with plug-and-play controls. This is particularly important for high power utility systems, such as wind power. The power electronics equipment used today is based on industrial motor drives technology. Having dedicated, high power density, modular systems will provide flexibility and efficiency in dealing with different energy sources and large variation of generation systems architectures.
- There is a need for new packaging and cooling technologies, as well as integration with PV and fuel cell will have to be addressed. The thermal issues in integrated systems are complex, and new technologies such as direct

fluid cooling or microchannel cooling may find application in future systems. There is large potential for advancement in this area.

- There is a need for new switching devices with higher temperature capability, higher switching speed, and higher current density/voltage capability. The growth in alternative energy markets will provide a stronger pull for further development of these technologies.

REFERENCES

[1] G. Hille, W. Roth, H. Schmidt, Photovoltaic system, Fraunhofer Institute for Solar Energy Systems, Freiburg, Germany, 1995.

[2] D.P. Hodel, Photovoltaics-electricity from sunlight, U.S. Department of Energy Report, DOE/NBMCE.

[3] O.H. Wilk, Utility connected photovoltaic systems, Presented at International Energy Agency (IEA): Solar Heating and Cooling Program, Montreux, 1992.

[4] R.S. Wenham, M.A. Green, M.E. Watt, Applied photovoltaic, in: Centre for Photovoltaic Device and Systems, University of New Southwales, Sydney, 1998.

[5] W.B. Lawrance, H. Dehbonei, A versatile PV array simulation tools, Presented at ISES 2001 Solar World Congress, Adelaide, South Australia, 2001.

[6] M.A.S. Masoum, H. Dehbonei, E.F. Fuchs, Theoretical and experimental analyses of photovoltaic systems with voltage and current-based maximum power-point tracking, IEEE Trans. Energy Convers. 17 (2002) 514–522.

[7] B. Bekker, H.J. Beukes, Finding an optimal PV panel maximum power point tracking method, Presented at 7th AFRICON Conference, Africa, 2004.

[8] T. Noguchi, S. Togashi, R. Nakamoto, Short-current pulse based adaptive maximum-power-point tracking for photovoltaic power generation system, Presented at Proceedings of the 2000 IEEE International Symposium on Industrial Electronics, 2000.

[9] N. Mutoh, T. Matuo, K. Okada, M. Sakai, Prediction-data-based maximum-power-point-tracking method for photovoltaic power generation systems, Presented at 33rd Annual IEEE Power Electronics Specialists Conference, 2002.

[10] G.W. Hart, H.M. Branz, C.H. Cox, Experimental tests of open-loop maximum-power-point tracking techniques, Sol. Cells 13 (1984) 185–195.

[11] D.J. Patterson, Electrical system design for a solar powered vehicle, Presented at 21st Annual IEEE Power Electronics Specialists Conference, 1990.

[12] H.J. Noh, D.Y. Lee, D.S. Hyun, An improved MPPT converter with current compensation method for small scaled PV-applications, Presented at 28th Annual Conference of the Industrial Electronics Society, 2002.

[13] K. Kobayashi, H. Matsuo, Y. Sekine, A novel optimum operating point tracker of the solar cell power supply system, Presented at 35th Annual IEEE Power Electronics Specialists Conference, 2004.

[14] P. Midya, P.T. Krein, R.J. Turnhull, R. Reppa, J. Kimball, Dynamic maximum power point tracker for pholovoltaic applications, Presented at 27th Annual IEEE Power Electronics Specialists Conference, 1996.

[15] V. Arcidiacono, S. Corsi, L. Larnhri, Maximum power point tracker for photovoltaic power plants, Presented at Proceedings of the IEEE Photovoltaic Specialists Conference, 1982.

[16] Y.H. Lim, D.C. Hamill, Synthesis, simulation and experimental verification of a maximum power point tracker from nonlinear dynamics, Presented at 32nd Annual IEEE Power Electronics Specialists Conference, 2001.

[17] Y.H. Lim, D.C. Hamill, Simple maximum power point tracker for photovoltaic arrays, Electron. Lett. 36 (2000) 997–999.

[18] L. Zhang, Y. Bai, A. Al-Amoudi, GA-RBF neural network based maximum power point tracking for grid-connected photovoltaic systems, Presented at International Conference on Power Electronics Machines and Drives, 2002.

[19] C.C. Hua, J.R. Lin, Fully digital control of distributed photovoltaic power systems, Presented at IEEE International Symposium on Industrial Electronics, 2001.

[20] M.L. Chiang, C.C. Hua, J.R. Lin, Direct power control for distributed PV power system, Presented at Proceedings of the Power Conversion Conference, 2002.

[21] S. Jain, V. Agarwal, A new algorithm for rapid tracking of approximate maximum power point in photovoltaic systems, IEEE Power Electron. Lett. 2 (2004) 16–19.

[22] N. Femia, G. Petrone, G. Spagnuolo, M. Vitelli, Optimization of perturb and observe maximum power point tracking method, IEEE Trans. Power Electron. 20 (2005) 963–973.

[23] N.S. D'Souza, L.A.C. Lopes, X. Liu, An intelligent maximum power point tracker using peak current control, in: 36th Annual IEEE Power Electronics Specialists Conference, 2005, pp. 172–177.

[24] M.G. Simoes, N.N. Franceschetti, M. Friedhofer, A fuzzy logic based photovoltaic peak power tracking control, in: Proceedings of the 1998 IEEE International Symposium on Industrial Electronics, 1998, pp. 300–305.

[25] A.M.A. Mahmoud, H.M. Mashaly, S.A. Kandil, H.E. Khashab, M.N.F. Nashed, Fuzzy logic implementation for photovoltaic maximum power tracking, Presented at Proceedings of the 9th IEEE International Workshop on Robot and Human Interactive Communications, 2000.

[26] N. Patcharaprakiti, S. Premrudeepreechacharn, Maximum power point tracking using adaptive fuzzy logic control for grid-connected photovoltaic system, Presented at IEEE Power Engineering Society Winter Meeting, 2002.

[27] B.M. Wilamowski, X. Li, Fuzzy system based maximum power point tracking for PV system, Presented at 28th Annual Conference of the IEEE Industrial Electronics Society, 2002.

[28] M. Veerachary, T. Senjyu, K. Uezato, Neural-network-based maximum-power-point tracking of coupled-inductor interleaved-boost-converter-supplied PV system using fuzzy controller, IEEE Trans. Ind. Electron. 50 (2003) 749–758.

[29] K. Ro, S. Rahman, Two-loop controller for maximizing performance of a grid-connected photovoltaic-fuel cell hybrid power plant, IEEE Trans. Energy Convers. 13 (1998) 276–281.

[30] A. Hussein, K. Hirasawa, J. Hu, J. Murata, The dynamic performance of photovoltaic supplied dc motor fed from DC-DC converter and controlled by neural networks, Presented at Proceedings of the 2002 International Joint Conference on Neural Networks, 2002.

[31] X. Sun, W. Wu, X. Li, Q. Zhao, A research on photovoltaic energy controlling system with maximum power point tracking, Presented at Proceedings of the Power Conversion Conference, 2002.

[32] O. Wasynczuk, Dynamic behavior of a class of photovoltaic power systems, IEEE Trans. Power Appl. Syst. 102 (1983) 3031–3037.

[33] Y.C. Kuo, T.J. Lian, J.F. Chen, Novel maximum-power-point-tracking controller for photovoltaic energy conversion system, IEEE Trans. Ind. Electron. 48 (2001) 594–601.

[34] G.J. Yu, Y.S. Jung, J.Y. Choi, I. Choy, J.H. Song, G.S. Kim, A novel two-mode MPPT control algorithm based on comparative study of existing algorithms, Presented at Conference Record of the Twenty-Ninth IEEE Photovoltaic Specialists Conference, 2002.

[35] K. Kohayashi, I. Takano, Y. Sawada, A study on a two stage maximum power point tracking control of a photovoltaic system under partially shaded insolation conditions, Presented at IEEE Power and Energy Society General Meeting, 2003.

[36] D. Langridge, W. Lawrance, B. Wichert, Development of a photo-voltaic pumping system using a brushless DC motor and helical rotor pump, Sol. Energy 56 (1996) 151–160.

[37] H. Dehbonei, C.V. Nayar, A new modular hybrid power system, Presented at IEEE International Symposium on Industrial Electronics, Rio de Janeiro, Brazil, 2003.

[38] C.V. Nayar, S.J. Phillips, W.L. James, T.L. Pryor, D. Remmer, Novel wind/diesel/battery hybrid system, Sol. Energy 51 (1993) 65–78.

[39] W. Bower, Merging photovoltaic hardware development with hybrid applications in the U.S.A, Presented at Proceedings Solar '93 ANZSES, Fremantle, Western Australia, 1993.

[40] N. Mohan, M. Undeland, W.P. Robbins, Power Electronics, John Wiley and Sons, Inc., New York, 1995.

[41] M. Calais, V.G. Agelidis, M. Meinhardt, Multilevel converters for single phase grid-connected photovoltaic systems an overview, Sol. Energy 66 (1999) 525–535.

[42] K. Hirachi, K. Matsumoto, M. Yamamoto, M. Nakaoka, Improved control implementation of single phase current fed PWM inverter for photovoltaic power generation, Presented at Seventh International Conference on Power Electronics and Variable Speed Drives (PEVD'98), 1998.

[43] U. Boegli, R. Ulmi, Realisation of a new inverter circuit for direct photovoltaic energy feedback into the public grid, IEEE Trans. Ind. Appl. 22 (1986) 255–258.

[44] B. Lindgren, Topology for decentralised solar energy inverters with a low voltage A-bus, Presented at EPE99: European Power Electronics Conference, 1999.

[45] K. Masoud, G. Ledwich, Aspects of grid interfacing: current and voltage controllers, Presented at Proceedings of the AUPEC 99, 1999.

[46] H.K. Sung, S.R. Lee, H. Dehbonei, C.V. Nayar, A comparative study of the voltage controlled and current controlled voltage source inverter for the distributed generation system, Presented at Australian Universities Power Engineering Conference (AUPEC), Hobart, Australia, 2005.

[47] L.J. Borle, M.S. Dymond, C.V. Nayar, Development and testing of a 20 kW grid interactive photovoltaic power conditioning system in Western Australia, IEEE Trans. Ind. Appl. 33 (1999) 1–7.

[48] L.J. Borle, C.V. Nayar, Zero average current error controlled power flow for ac-dc power converters, IEEE Trans. Power Electron. 10 (1995) 725–732.

[49] H. Sharma, Grid integration of Photovoltaics, The University of Newcastle, Australia, 1998.

[50] L.L. Freris, Wind Energy Conversion Systems, Prentice Hall, New York, 1990.

[51] C.V. Nayar, J. Perahia, F. Thomas, Small Scale Wind Powered Electrical Generators, The Minerals and Energy Research Institute of Western Australia, Perth, Australia, 1992.

[52] R.D. Richardson, G.M. McNerney, Wind energy systems, Proc. IEEE 81 (1993) 378–389.

[53] J. Arillaga, N. Watson, Static power conversion from self excited induction generators, Proc. Inst. Elect. Eng. 125(8) (1978) 743–746.

[54] C.V. Nayar, J. Perahia, F. Thomas, S.J. Phillips, T.L. Pryor, W.L. James, Investigation of capacitor excited induction generators and permanent magnet alternators for small scale wind power generation, Renew. Energ. 125 (1991).

[55] T. Ackermann, L. Sörder, An overview of wind energy status 2002, Renew. Sust. Energ. Rev. 125(8) (2002) 67–128.

[56] S. Peresada, A. Tilli, A. Tonielli, Robust active-reactive power control of a doubly-fed induction generator, Presented at IECON '98, 1998.

[57] W.E. Long, N.L. Schmitz, Cycloconverter control of the doubly fed induction motor, IEEE Trans. Ind. Gen. Appl. 7 (1971) 162–167.

[58] A. Chattopadhyay, An adjustable-speed induction motor drive with a thyristor-commutator in the rotor, IEEE Trans. Ind. Appl. 14 (1978) 116–122.

[59] P. Pena, J.C. Clare, G.M. Asher, Doubly fed induction generator using back-to-back PWM converters and its application to variable speed wind-energy generation, IEEE Proc. Electr. Power Appl. 143 (1996) 231–241.

[60] H. Azaza, On the dynamic and steady state performances of a vector controlled DFM drive, Presented at IEEE International Conference on Systems, Man and Cybernetics, 2002.

[61] Bergey Windpower User Manual, 10 kW Battery Charging Wind Energy Generating System, Bergey Windpower Co., OK, USA, 1984.

[62] J.H.R. Enslin, F.W. Leuschner, Integrated hybrid energy systems for isolated and semi-isolated users, Presented at Proceedings of the Renewable Energy Potential in Southern African Conference, UCT, South Africa, 1986.

[63] D.G. Infield, Wind diesel systems technology and modelling—a review, Int. J. Renew. Energ. Eng. 1(1) (1999) 17–27.

[64] H. Sharma, S.M. Islam, C.V. Nayar, T. Pryor, Dynamic response of a remote area power system to fluctuating wind speed, Presented at Proceedings of the IEEE Power Engineering Society (PES 2000) Winter Meeting, 2000.

[65] W.L. James, C.V. Nayar, F. Thomas, M. Dymond, Variable Speed Asynchronous Wind Powered Generator with Dynamic Power Conditioning, Murdoch University Energy Research Institute (MUERI), Australia, 1993.

[66] F. Blaabjerg, Z. Chen, S.B. Kjaer, Power electronics as efficient interface in dispersed power generation systems, IEEE Trans. Power Electron. 19 (2004) 1184–1194.

[67] F. Blaahjerg, Z. Chen, P.H. Madsen, Wind power technology status, development and trends, Presented at Proceedings of the Workshop on Wind Power and Impacts on Power Systems, Oslo, Norway, 2002.

[68] J. Wilkie, W.E. Leithead, C. Anderson, Modelling of wind turbines by simple models, Wind Eng. 14 (1990) 247–273.

[69] A.L.G. Westlake, J.R. Bumby, E. Spooner, Damping the power angle oscillations of a permanent magnet synchronous generator with particular reference to wind turbine applications, IEEE Proc. Electr. Power Appl. 143 (1996) 269–280.

[70] J.F. Manwell, J.G. McGowan, B.H. Bailey, Electrical/mechanical options for variable speed turbines, Sol. Energy 46 (1991) 41–51.

[71] T.S. Andersen, H.S. Kirschbaum, Multi-speed electrical generator applications in wind turbines, Presented at Proceedings of the AIAA/SERI Wind Energy Conference, Boulder, 1980.

[72] E. Muljadi, C.P. Butterfield, Pitch-controlled variable-speed wind turbine generation, IEEE Trans. Ind. Appl. 37(1) (2001) 240–246.

[73] G. Riahy, P. Freere, Dynamic controller to operate a wind turbine in stall region, Presented at Proceeding of Solar'97—Australia and New Zealand Solar Energy Society, 1997.

[74] K. Tan, S. Islam, Optimum control strategies in energy conversion of PMSG wind turbine system without mechanical sensors, IEEE Trans. Energy Convers. 19(2) (2004) 392–399.

[75] K. Tan, S. Islam, Mechanical sensorless robust control of permanent magnet synchronous generator for maximum power operation, Presented at Australia University Power Engineering Conference, Australia, 2001.

[76] Q. Wang, L. Chang, An independent maximum power extraction strategy for wind energy conversion system, Presented at IEEE Canadian Conference on Electrical and Computer Engineering, Canada, Alberta, 1999.

[77] J. Perahia, C.V. Nayar, Power controller for a wind-turbine driven tandem induction generator, Electr. Mach. Power Syst. 19 (1991) 599–624.

[78] K. Tan, S. Islam, H. Tumbelaka, Line commutated inverter in maximum wind energy conversion, Int. J. Renew. Energ. Eng. 4(3) (2002) 506–511.

[79] E. Muljadi, S. Drouilhet, R. Holz, V. Gevorgian, Analysis of permanent magnet generator for wind power battery charging, Presented at Thirty-First IAS Annual Meeting, IAS '96, San Diego, CA, USA, 1996.

[80] B.S. Borowy, Z.M. Salameh, Dynamic response of a stand-alone wind energy conversion system with battery energy storage to a wind gust, IEEE Trans. Energy Convers. 12(1) (1997) 73–78.

[81] A.B. Raju, Application of Power Electronic Interfaces for Grid-Connected Variable Speed Wind Energy Conversion Systems, Department of Electrical Engineering, Indian Institute of Technology, Bombay, 2005.

[82] G.A. Smith, K.A. Nigim, Wind energy recovery by static Scherbius induction generator, IEE Proc. C Gen. Transm. Distrib. 128 (1981) 317–324.

[83] T.S. Anderson, P.S. Hughes, Investigation of Doubly Fed Induction Machine in Variable Speed Applications, Westinghouse Electric Corporation, Philadelphia, PA, 1983.

Chapter 2

Energy Sources

Omer C. Onar
Power Electronics and Electric Machinery Group, National Transportation Research Center, Oak Ridge National Laboratory, Oak Ridge, TN, USA

Alireza Khaligh
Electrical and Computer Engineering Department, University of Maryland at College Park, College Park, MD, USA

Chapter Outline

2.1 INTRODUCTION

In modern societies, development level and economic well-being of a society are directly measured by energy generation and consumption. Energy plays an important role on the economic health of a country that is reflected by the gross national product (GNP). The per capita GNP of a country is correlated to the per capita energy consumption. There is a steady demand to increase the energy generation capacity in all over the world since the global energy consumption is rising. The main reasons are the technological developments, industrial revolution, introduction of new loads and appliances, and increase in population. In a modern and industrialized community, energy is used in every single human activity. Some major examples are;

– Household applications: such as heating, cooking, lighting, water heating, and air conditioning.
– Transportation: passenger cars, busses, trains, trucks, ships, and aircrafts.
– Manufacturing heat and electricity as well as user-end or industrial products.
– Irrigation and fertilizing in agriculture.

The worldwide energy consumption has been growing steadily and rapidly right after the industrial revolution. Today's global energy consumption has reached to more than 532 EJ (Exajoule, 1 EJ = 10^{18} Joule = 277.778 TWh, Terawatt-hours). This amount of energy is consumed with an hourly rate of 16.87 TWh [1]. In other worlds, in the world, 16.87 TWh of energy is consumed hourly. Hourly global energy consumption in 1900 was ~ 0.7 TWh. From 1973 to

2010, the global annual energy consumption increased by more than two folds; from 255.687 EJ (6.107 Gtoe, Gigatons of oil equivalent) to more than 532 EJ (12.717 Gtoe).

United States is the second largest energy consumer in the world, after China surpassed United States as top energy consumer. United States ranks seventh in energy consumption per capita. The amount of energy consumed in United States was 91.8 EJoule or 25,500 TWh in 2014, with equivalent hourly rate of 2.91 TW [2,3]. If the hourly consumed power is 2.91 TW, 1 year of consumption corresponds to 25,500 TWh, by 2.91 TW × 24 h × 365 days. The energy consumption in the United States reduced by 4.7% as compared to the consumption levels of 2004. This is mostly due to the advanced manufacturing technologies, more efficient building materials and technologies, and the 2009 economic crises.

Between 1980 and 2006, the worldwide energy consumption annual growth rate was 2%. The total energy consumption in 2012 for the United States is given in Table 2.1 according to the US Energy Information Administration [4]. According to the table,

This total consumption is shared among several sectors. Around 10.97% of this 95.02 Btu is consumed by residential sector including lighting, heating, air-conditioning, and household appliances. 8.72% is consumed by the commercial sector including heating, cooling, lighting, and office equipment. Industrial sector, covering manufacturing, construction, and agriculture, consumes about 24.87% of the total energy. By 28.12%, transportation is the most energy consuming sector, including light-duty vehicles, commercial light trucks and

TABLE 2.1 Total Energy Consumption

Source	Total Consumed Energy (quadrillion Btu)	Percentage
Oil	35.87	37.75
Gas	26.20	27.57
Coal	17.34	18.25
Nuclear	8.05	8.47
Hydropower	2.67	2.82
Biomass	2.53	2.66
Renewables[a]	1.97	2.07
Other[b]	0.39	0.41
Total	95.02 quadrillion Btu = 27,847.61 TWh = 100.25 EJ	

[a] *Wind, PV, solar thermal, grid-connected electricity from landfill gas and biowaste.*
[b] *Nonbiogenic waste, liquid hydrogen, methanol, some domestic inputs to the refineries.*

larger freight trucks, and air and rail transportation. The electricity delivery-related losses count for 27.32% of the total energy consumed, attributed to all residential, commercial, industrial, and transportation sectors [4]. This delivery-related loss amount includes the transmission and distribution losses.

The overall energy consumption percentages by sector are shown in Fig. 2.1.

According to Fig. 2.1, electricity delivery-related losses are considerably high. Almost every 1/3 of the generated energy is lost in electricity generation and delivery. This is due to the fact that efficiency of a typical power plant is around 38% [5]. Figure 2.1 clearly reveals that more than a quarter of the produced energy is lost in low efficient conventional energy conversion systems and transmission lines. This issue encourages the increased focus on high efficiency renewable energy systems, which may also eliminate or reduce the transmission and/or distribution losses if they are built as localized energy generation units. As compared to the efficiency of conventional power plants, new generation of power plants using gas turbines or microturbines may reach a substantially higher efficiency of 55%; however, they still rely on another fossil fuel which is natural gas [6].

After the invention of steam engines, coal began to be the main source of energy for the eighteenth and nineteenth centuries. Since the automobiles were invented and electricity usage became more common, coal left its popularity to the oil during the twentieth century. From 1920s to 1973, oil was the main resource fueling the industry and transportation and its price steadily dropped during these years. Oil kept its expansiveness till 1970s. In the oil crises of 1973 and 1979, price per barrel boosted from 5 to 45 dollars and oil was not the best and most common source of energy production anymore [7]. From these dates, coal and nuclear became the major sources for electric power generation. In those years, energy conversation and increasing the energy efficiency gained importance. However, the use of fossil fuels has continued over the past 30

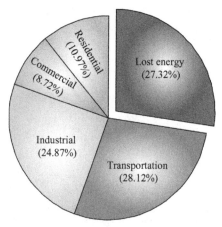

FIGURE 2.1 Percentage share of sectors on energy consumption.

years and their contribution to overall energy production has increased. During past decade, coal has become the fastest growing fossil fuel, since it has large remaining reserves [7]. On the other hand, renewable energy sources have gained interest due to the depletion of fossil fuels, increased oil prices, national energy security, and climate change concerns related to the carbon emissions. Therefore, government support on development, deployment, incentives, and commercialization of renewable energies are ever increasing. For instance, in March 2007, it is agreed by the European Union members that at least 20% of their nations' energy should be produced from renewable sources by 2020 while United States also has a national renewable energy target of 20%. In Canada, there are nine provincial renewable energy targets although there is no national target that has been set yet. This is also a part of environmental concerns such as global warming [8] and building a sustainable energy economy by reducing the nations' dependence on imported fossil fuels. However, although there is a large availability of renewable sources, their contribution to globally consumed energy is relatively poor.

In developed countries, such as Germany and Japan, gross national product is 6 kW per person and 11.4 kW per person in the United States. Bangladesh has a relatively lower consumption with 0.2 kW per person while it is around 0.7 kW per person for a developing country such as India. 25% of overall world's energy is consumed by the United States while its share of global energy generation is 22% [1]. Energy consumption in United States is shared by four broad sectors according to the US Department of Energy and the Energy Information Administration. Accordingly, the largest user is the transportation sector, currently consuming 28.12% of the total energy. Next largest amount of energy is consumed for industry followed by residential and commercial users. Energy consumption shares of United States for four major energy consumption sectors (industrial, transportation, residential, and commercial) are demonstrated in Table 2.2 [4,9,10].

Most of the electric power generation of the United States is provided by conventional thermal power plants. Most of these power plants are operated by coal. However, from 1990 to 2000, the number of natural gas or other types of gas-operated power plants were increased significantly. 270 GW of new gas-operated thermoelectric power plants were built in United States from 1992 to 2005. Only 14 GW of capacity belonged to new nuclear and coal-fired power plants, with 2.315 GW of this amount being nuclear while the remaining is the coal-fired power plants [11]. The significant shift to the gas-operated power plants is due to the deregulation, political, and economic factors; however, nuclear and coal are considerably capital intensive. On the other hand, there is a great potential for renewable energies in the United States. For instance, US wind power capacity is close to 20 GW, which is sufficient to supply power to 4.5 million typical households [12]. Although there is a great availability of the solar power, solar power percentage of total capacity is about 0.11% retrieved by plants that are currently in operation; including the new Nevada Solar One

TABLE 2.2 US Nation-Wide Energy Consumption Sectors

Sector	Major Use Shares
Transportation: Transportation accounts for the 28.12% of the total energy consumption. This energy is shared among light-duty vehicles, trucks including commercial light trucks and freight trucks, passenger and freight rail, shipping, and air. "Other" converse military, boats, busses, lubricants, and pipeline fuel	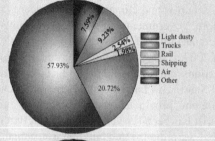
Industrial: This sector mainly consumes 24.87% of total energy for the manufacturing, producing, and processing goods such as chemical refining, metal production, paper, and cement production and several other industrial processes	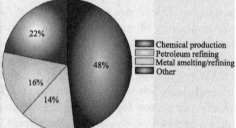
Residential: This sector consumes the 10.97% of total energy for household power requirements	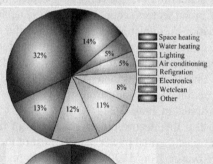
Commercial: This sector includes the business, government, and other service providing institutions, facilities, and their equipments corresponding to 17% of total energy consumption	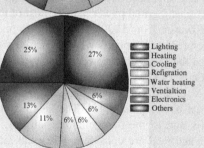

TABLE 2.3 Energy Generation in United States in 2012

Source	Total Capacity(MW)	Annual Production (Billions kWh)	Number of Plants in Operation	% Of Annual Production
Coal	336.3	1514.04	557	36.97
Natural gas	488.2	1237.79	1758	30.23
Nuclear	107.9	769.33	66	18.79
Hydro	78.2	276.24	1426	6.75
Renewables	80.5	218.33	1956	5.33
Petroleum	53.8	23.19	1129	0.57
Other	2	13.79	64	0.34
Storage	20.9	− 4.95	41	− 0.12
Import-export		47.26		1.15
Total	1168	4095	6997	100

plant with 64 MW capacity and the largest solar thermal power station in the Mojave Desert with a total generation capacity of 354 MW, which is the world's largest solar plant [13].

The existing power generation infrastructure in the United States has a generation capacity of 1.168 GW in total. In 2012, 4047.76 billion kWh of energy was generated in the United States. Table 2.3 summarizes the U.S. electric power generation for 2012 in terms of total power capacity (MW), annual production (billion kWh), and number of power plants in operation [14]. Please note that number of plants refer to the number of facilities and not the number of units; i.e., number of wind farms instead of number of turbines or number of nuclear power plants instead of number of reactors.

According to Table 2.3, it is seen that the most contributing power source to US energy production is the coal-fired thermoelectric power plants. Second greatest source in annual energy production is the natural gas-fired power plants due to the fact that natural gas is cleaner than coal, natural gas is more environmental friendly, gas-fired power plants are more efficient, and gas-fired units have faster dynamics (easier up and downregulation).

Other than the nuclear power plants and fossil fueled power plants, hydro-electric power plants have a considerable contribution to the annual production although most of them are small hydro units. Renewable energy sources of wind and solar still have insignificant number of units in operation and their annual contribution is considerably insignificant to the nation's total energy production although the available potential capacity of these sources are really high.

The energy production rates of energy sources are given in Fig. 2.2 [14].

TABLE 2.4 Electrical Energy Generation from Renewables in United States in 2012

Source	Total Capacity(MW)	Annual Production (Billions kWh)	Number of Plants in Operation	% Of Annual Production
Hydro	78.2	276.24	4023	6.75
Wind	59.6	140.82	947	3.44
Wood	8.5	37.8	351	0.92
Biomass	5.5	19.82	1766	0.48
Geothermal	3.7	15.56	197	0.38
Solar	3.2	4.33	553	0.11
Total	158.7	494.57	7837	12.08

The electrical energy usage of the United States increased by 5% from 2002 to 2012 [15]. However, it decreased by 2.2% since 2007 when it peaked. This is mostly due to the conservation efforts and more efficient buildings, industrial processes, and appliances. At least for the next decade, coal, nuclear, and natural gas will remain in top three fuels for electric energy generation of the United States. From 2008 to 2012, within 4 years, the contribution of coal in annual energy generation reduced from about 50% to 37% and natural gas contribution increased. Nuclear power contributions will likely stay about the same due to the decommissioning of the older power plants nearing end of their lifetimes and bringing new plants online. Hydro, wind, and solar will most likely keep increasing in order to meet the national renewable energy targets as well as the targets set by states. The environmental concerns on global warming and sustainability

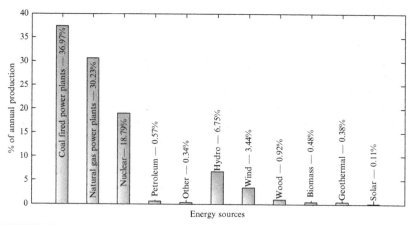

FIGURE 2.2 Percentage of the energy sources to the annual energy generation.

along with the political considerations on the supplies' security will shift the overall energy consumption away from the imported fuels. Nowadays, many researchers and politicians call for immediate actions for long-term sustainable energy solutions. Based on a growing consensus, peak oil may be reached in the near future and this will result in severe oil price increases [16]. If a long-term solutions cannot be developed prior to the peak oil scenario, the world economy may face a grinding halt.

This chapter focuses on naturally available energy sources and deals with the efficient utilization of these sources. Moreover, renewable and sustainable energy generation techniques are discussed in this chapter. In addition, the operating principles, efficient utilization, and grid-connection issues as well as power electronic interfaces for these renewable energy sources are demonstrated. While inventing new methodologies to maximize the efficient usage of traditional sources such as coal, oil, and natural gas, it is of great importance to develop new technologies to produce emerging sources of energy from renewables.

Consequently, by efficient use of conventional energy sources and utilizing alternate sources of energy, the reserves of the conventional energy resources can last for longer, global warming can be slowed down, and environmental pollution can be reduced [17].

2.2 AVAILABLE ENERGY SOURCES

Primary sources of energy are fossil fuels such as oil, natural gas, coal, and other sources such as nuclear, solar, wind, hydroelectricity, or potential sources available in oceans. The energy that has not been converted or transformed into another form is known as primary energy source. More convenient form of energy such as electrical energy is obtained by transforming primary energies in energy conversion processes. These converted forms are called as secondary energy sources.

2.2.1 Coal

Coal is the most abundant fossil fuel of the world with around 948 billion short tons of reserves. It is expected to sustain for the next   150 years at the current production rate [4]. Coal is the fastest growing fossil fuel to meet the energy demand of the global community. However, coal is the dirtiest energy source with numerous pollutants and high emissions [11]. The United States has the largest estimated and proven recoverable reserves of coal of the world. More than 81% of the coal in United States was used in power plants to generate electric power.

2.2.2 Oil

It is estimated that there is 57 ZJ of oil reserves on Earth. This amount includes the available but not necessarily recoverable reserves. Other estimates vary from 8 ZJ including currently proven and recoverable reserves to a maximum of

110 ZJ including nonrecoverable reserves [18]. World's current oil consumption is 85 million barrels per day (mbd) and it is estimated that the peak consumption will be 93 mbd in 2020. Oil and its chemical derivatives are mainly used for transportation and electric power generation. The total reserves are estimated to be 1,481,526 million barrels. If the current demand would remain static, then the remaining oil supplies would last for about 120 years.

2.2.3 Natural Gas

It is estimated that there are 850 trillion cubic meters of estimated remaining recoverable reserves of natural gas. According to the U.S. Energy Information Administration, there are 5,977,000 millions of cubic meter natural gas reserves in the United States [19]. Natural gas has become one of the major sources of electric power generation through the steam turbines and gas turbines due to their higher efficiency. Natural gas is cleaner than any other fossil fuels and produces fewer pollutants per generated unit energy. Burning natural gas produces about 30% less carbon dioxide than burning petroleum and about 45% less than burning coal for an equivalent amount of heat [20]. Some of the natural gas power plants are operated in combined cycle mode to obtain higher efficiencies. In this operation, gas turbines are combined with the steam turbines in order to get the benefit of waste heat using steam turbines.

2.2.4 Hydropower

Hydroelectric power plants supplies about 16.4% of the world's electric annually [21] and is expected to grow more than 3% every year for the next 25 years. The hydroelectric power may not be a long-term effective solution, since, most of the potential sites are already in use or they are not feasible to be exploited due to environmental and economical concerns. In addition, the life span of hydroelectric power plants is limited, due to the soil erosion and accumulation. Because of these concerns, the construction of large hydroelectric power plants has stagnated. The new trend in all over the world has been building smaller hydro power units called as "micro-hydro" since they can be as a part of distributed generation, opening up many locations for power generation and they have less or negligible environmental effects [22,23]. On the other hand, hydroelectric power plants have no emissions since no fuel is burnt. Hence, hydropower is a clean energy source in compare to fossil fuel-based energy sources. In 2010, the worldwide hydroelectric energy generation reached to 3427 TWh.

2.2.5 Nuclear Power

Nuclear power plants provided about 5.7% of the world's energy and contributed to the world's electric power generation by 13% in 2012 [14,24,25]. Total power capacity of the established nuclear power plants were about 372 GW by November 2007 [24]. The remaining uranium resources are estimated to be 2500 ZJ by the International Atomic Energy Agency [26]. Since there is plenty of available sources and developed technology, the contribution of nuclear power to the future's energy demand is not limited. However, there are political and environmental constraints, which restrict the growth of nuclear power plants. The cost of generating nuclear power is approximately equal to that of the coal power. Moreover, nuclear power has zero pollutant emissions such as CO, CO_2, NO, and SO_2.

2.2.6   Solar

Earth receives around 174,000 TW of solar energy resource per year. As an energy source, less than 0.02% of available solar resources are capable of entirely replacing all nuclear power and fossil fuels [27,28]. Although it is still expensive in compare to conventional energy conversion techniques, the fastest growing energy source in 2007 were grid-connected photovoltaic systems. The total installed capacity reached to 8.7 GW by increasing all photovoltaic installations by 83% in 2007 [29]. In the United States, from 2013 to 2014, the total power of all installations increased from 930 to 2106 MW. High cost of manufacturing solar cells, reliance on weather conditions, storage, and grid-connection problems are the major barriers of further development of solar generation. On the other hand, efficiency of solar photovoltaic cells continuously increases with the developments in material science and technologies. Nowadays, research level solar photovoltaics have reached to about 40% of efficiencies.

2.2.7   Wind

Wind is one of the greatest available potential energy sources. The available wind power is estimated to be 300 TW [30] to 870 TW [31]. Only 5% of the available energy is capable of supplying the current worldwide energy demands. However, due to the fewer obstacles, most of this wind energy is available on the open oceans on which construction of wind turbines and energy transmission is relatively difficult and expensive. From 2006 to 2007, the installed wind turbines' capacity was increased by 27% to total of 94 GW according to the Global Wind Energy Council [32]. However, the actual generated power is less than the nominal capacity since the nominal capacity represents the peak output and actual output is around 40% of the nominal capacity due to efficiency issues and lower wind speeds [33]. In 2010, wind energy production exceeded 2.5% of the world electric energy generation and it grows about 25% per year.

2.2.8   Ocean

Energy of ocean can be categorized in three major methods; *ocean wave power*, *ocean tidal power*, and *ocean thermal power*. All of these three methods can be installed as on-shore or off-shore applications. It is estimated that theoretical potential is equivalent to 4-18 million tons of oil equivalent.

Wave energy harvesting is a concept that the kinetic energy of waves of the deep water or waves hitting the shores is captured and converted to electrical energy. The kinetic energy of waves is converted to electrical energy using several different methods. It is estimated that the deep water wave power resources vary from 1 to 10 TW [34], while the total power of the waves hitting the shores may add an additional power of 3 TW [30]. Capturing this entire amount of power is not practical and feasible. It is estimated that 2 TW of this power can be usefully captured [35,36].

Ocean tides occur due to the tidal forces of the moon and the sun, in combination with the Earth's rotation. Tidal power has a great potential for future energy generation since it is cleaner in compare to fossil fuels and it is more predictable in compare to other renewable energies such as wind and solar. The kinetic energy of the moving water can be captured by tidal stream or tidal current turbines. Alternatively, the barrages can be used to capture the potential energy created due to the height difference between the low and high tides. Various methods can be employed for the realization of these concepts. The total estimated tidal power potential is 3.7 TW [37]. However, only around 0.8 TW of this amount is available due to the dissipation of tidal fluctuations. The amount of energy generated from ocean tides was 0.3 GW at the end of 2005 [38], which is much less than the available potential.

The other way of generating power from the oceans is the ocean thermal energy conversion (OTEC). In this method, the temperature difference between the warm shallow water and the cold deep water is used to drive a heat engine, which in turn drives an electric generator [39]. The efficiency of OTEC power plants is relatively low [39,40] due to the power requirements of the auxiliary OTEC devices such as water intake and discharge pumps. Moreover, this technique is expensive since the efficiency is low and greater capacities of installations are required to produce reasonable amounts of energy [41].

2.2.9 Hydrogen

Hydrogen is an energy carrier [42,43], in other words it is an intermediate medium for energy storage and carriage. Hydrogen is the most abundant element of the Earth (approximately corresponding 75% of the elemental mass of the universe) [44] and it is the simplest and lightest element of all chemical elements with an atomic number of 1. Hydrogen exists in nature in combination with other elements such as carbon and nitrogen in fossil fuels, biological materials, or with oxygen in water [45]. Hydrogen can be combusted in air or it can react with oxygen using fuel cells to produce energy. The resultant combustion energy or electrical energy does not cause any CO or CO_2 emissions. However, splitting hydrogen from the combination of other elements requires additional energy. The main source of global hydrogen production is natural gas (48%). Other sources of hydrogen production are oil (30%), coal (18%), and water electrolysis (4%) [46]. Currently, most of the hydrogen is produced from gas derivatives such as natural gas, ethane, methane, ethanol, or methanol. Hydrogen production from fossil fuels, known as reformation, contains several pollutant emissions. Although electrolysis is clean, this method has various challenges and still has very poor efficiencies and high production costs. Biological or fermentative reactions can be another method of hydrogen production; however, this method has some obstacles such as the amount of products are not significant [46,47]. Using hydrogen in hydrogen combustion engines is several percents more efficient than the conventional internal combustion engines. On the other hand, using hydrogen in the fuel cells is twice or three times more efficient than that of the internal combustion engines. However, there are several challenges for the commercialization of fuel cells such as the size, weight, cost, and durability. Other major technical difficulties related to hydrogen are the production, delivery, and storage issues.

2.2.10   Geothermal

Geothermal energy is the utilization of heat stored in the inner layers of the Earth or collecting the absorbed heat derived from underground. The geothermal energy production has reached to 37.3 GW at the end of 2005 [48]. 9.3 GW of this amount is used for electric power generation while the rest of it is used for residential or commercial heating purposes. Enhanced geothermal systems (EGS) is a technique that extends the potential for the use of geothermal energy. In this technique, the heat is extracted by building subsurface fractures to which water can be added through injection wells. Through this technique, the electrical generation capacity can reach to about 138 GW [49]. The overall EGS capacity of the world is calculated to be more than 13 YJ, where 200 ZJ of this amount is extractable. By the technological improvements and investments, this amount is projected to increase over 2 YJ [50]. However, in contrary to this enormous potential, geothermal supplies less than 1% of the world's energy demand as of 2008 [21]. The electricity generation potential is estimated to be from 35 to 2000 GW while the current installed capacity is 10.715 GW with the highest capacity is in the United States by more than 3 GW. Geothermal energy has high availability (average daily availabilities more than 90%) and in fact has no pollutant emissions since it does not require any fuel or combustion. Furthermore, geothermal power stations do not rely on weather conditions. In addition, it is considered to be a sustainable source of energy since the extracted heat is relatively small in compare to the heat reservoir's size. In other words, geothermal heat energy is replenished from deeper layers of the Earth, therefore it is not exhaustible.

2.2.11   Biomass

Biomass is a fuel that is also called biofuel, and the bioenergy is the energy enclosed in the biomass. Today, biomass has a small contribution to the overall energy supply, although it was the major fuel till the nineteenth century. In 2005, electric power from biomass was about 44 GW while more than 230 GW biomass power is used for heating [38]. As a sustainable energy source, biomass is a developing industry in many countries such as Brazil, United States, Germany, and many others. As an alternative to the fossil fuels, biomass production is significantly increasing worldwide. The biodiesel production increased by 85% to 1.03 billion gallons in 2005 and biodiesel became the world's fastest growing renewable source of energy. Bioethanol production was also increased by 8% and reached 8.72 billion gallons during 2005 [38]. Even though it is commonly believed that biomasses may be carbon-neutral, their current farming methods cause substantial carbon emissions [51,52]. As of 2012, there are 351 wood and biomass fired electric power plants in the United States with a total power generation capacity of 8.5 GW and annual production of 37.8 billion kWh.

2.3 ELECTRIC ENERGY GENERATION TECHNOLOGIES

Electric energy generation is a process that the energy sources or energy potential is converted to the electrical energy. Energy generation can be done in various techniques. Due to the upcoming emerging challenges in the global energy supply systems, energy from the conventional sources need to be highly efficient. In addition, there should be an increase in utilization of energy generation from alternative and renewable energy sources. In following sections, the energy sources and their conversion to electric energy are described.

2.3.1 Thermoelectric Energy

Thermoelectric power plants are mainly coal-fired power stations. In a thermoelectric power plant, coal or other fuels are burnt in order to heat up the water in the boiler. In this system, the high-pressurized steam rotates a steam turbine, which is coupled to an electric generator. After the steam passes through the turbine, it is cooled and condensed back to water in the condenser. This is known as Rankine Cycle [53]. More than 80% thermal power plants in all over the world operate based on this cycle. In the Rankine Cycle, there are four processes in which the working fluid's state is changed as shown in Fig. 2.3 [54]. These processes can be described as follows [55,56]:

Process 1-2: When the fluid is condensed and converted to liquid form, the liquid is pumped from low to high pressure. The pumping process requires a small amount of energy.

Process 2-3: The high pressure liquid that pumped into the boiler is heated at constant pressure until it becomes saturated dry vapor. The boiler is energized by a heat source such as a coal furnace.

Process 3-4: During this process, saturated vapor passes through the steam turbine. The heat energy is converted to mechanical energy. While the steam

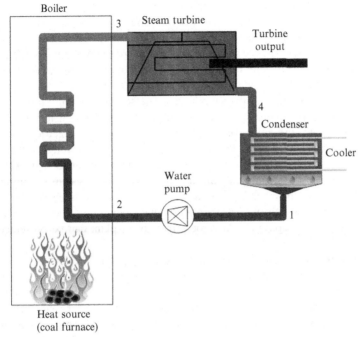

FIGURE 2.3 Rankine Cycle block diagram.

passes through the turbine, it may partly get condensed since this process decreases the pressure and temperature of the vapor.

Process 4-1: In this process, the vapor is condensed at a constant pressure and temperature, in a condenser. As a result, wet vapor is converted to saturated liquid. The cooler helps keeping the temperature constant as the vapor changes its phase from steam to liquid.

These four processes of the Rankine Cycle are shown in Fig. 2.3.

Since coal is the most abundant energy source of the world, coal-fired power plants have been widely used in electric power generation in all over the world [57]. Coal is a cheap energy source and coal-fired power plants have mature technology. Therefore, the generation cost is less and thermoelectric power plants can be constructed anywhere close to fuel and water supply. Although the consumption sites might be relatively far away from the coal mines or water supplies, fuel and water can be transported to the generation plants. Since the coal has been the backbone of the electric power industry since late 1800s, approximately 49% of the electric power generated in the world is supplied by coal-fired thermoelectric power plants [58]. Today, energy generation from coal corresponds to about 27% of the world's total energy generation.

In a simple form, the operation of a coal-fired power plant can be similar to Rankine Cycle. In this form, the plant consists of a boiler, a steam turbine driving an electric generator, a condenser, and a feed-water pump. Coal is first pulverized and burnt in the steam generation furnaces. The water in the boiler tubes is heated and steam is generated in this way at high pressures. The steam generation process is composed of three sub processes which are economizing, boiling, and superheating. In the economizer, the water is heated to a point that it is close to the boiling point. Then, the steam is raised in the boiler. Finally, the steam is further heated and dried at the superheater. The steam at its final form is then conveyed to the steam turbine. The mechanical force pushing the turbine blades yields the steam turbine to rotate which in turn drives the electric generator producing electricity. The cooler steam with lower pressure is released from the turbine. This steam is conveyed to the condenser to be liquefied. This water is pumped back to the steam generator and the closed loop system is completed [55]. Considering the other auxiliary devices and peripheral components such as cooling tower, coal conveyor, ash&waste management units, and many others, the schematic of a coal-fired thermoelectric power plant can be presented as Fig. 2.4.

The components of the thermoelectric power plant are described in Table 2.5.

The operation of the coal-fired power plant begins with the coal conveyor. From an exterior stack, coal is conveyed through a coal hopper to the pulverizing fuel mill where it is grounded and converted to a fine powder. The pulverized coal is mixed preheated air. The air is taken by an air intake pipe and pumped to be mixed with pulverized coal. This preheated air is supplied by the forced draught fan. In the boiler, where the air-fuel mixture is ignited at high pressure, generated heat increases the temperature of the water. The water is then changes its phase to steam where it flows vertically up the boiler tubes. This steam is passed to the boiler drum where its remaining water content is separated. This dry steam is then passed through a manifold from the drum into the superheater. In the superheater, further pressure and temperature increases, steam reaches about 200 bar and 570 °C. The turbine process of the power plant comprises three stages; high pressure turbine, intermediate pressure turbine, and low pressure

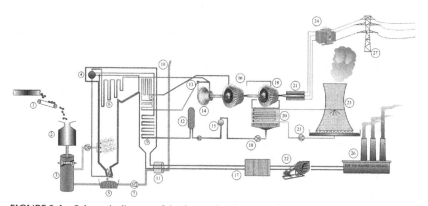

FIGURE 2.4 Schematic diagram of the thermoelectric power plant.

TABLE 2.5 Thermoelectric Power Plant Components

1. Coal conveyor	10. Air intake pipe	19. Low pressure turbine
2. Coal hopper	11. Air preheater	20. Condenser
3. Pulverization mill	12. Feed heater	21. Electric generator
4. Boiler drum	13. Steam governor	22. Induced draught fan
5. Ash hopper	14. High pressure turbine	23. Cooling water pump
6. Superheater	15. Deaerator	24. Power transformer
7. Forced draught fan	16. Intermediate pressure turbine	25. Cooling tower
8. Reheater	17. Precipitator	26. Chimney stack
9. Economizer	18. Boiler feed pump from condenser	27. Transmission network

turbine. First, the steam passes to the high pressure turbine through the pipes. Both the manual turbine control and the automatic set-point following can be provided by a steam governor valve. The temperature and the pressure of the steam decrease when it is exhausted from the high pressure turbine. This steam is returned to the boiler reheater for further use. The reheated steam passes to the intermediate pressure turbine. The steam released from the intermediate pressure turbine is passed directly to the low pressure turbine. Now the steam is cooler and just above its boiling point. This steam is then condensed in the condenser by contacting thermally with the cold water tubes of the condenser. As a result, the steam is converted back into water and the condensation causes a vacuum effect inside the condenser chest. The condensed water is prewarmed by the feed-heater using the heat of the steam released from the high pressure turbine and then in the economizer. Then, this prewarmed water is deaerated and passed by a feed-water pump, which completes the closed cycle. The cooling tower cools down the water from the condenser creating an intense and visible plume. Finally, the water is pumped back to the cooling water cycle. The induced draft fan draws the exhaust gas of the boiler. Here, an electrostatic precipitator is used. Finally, this exhaust gas is vented through the chimneys of the power plant.

In the thermoelectric power plants; load following capability, efficiency, fuel and water management, and emissions are important issues. In addition, the active and reactive outputs of the power plant's generators and frequency and voltage regulations have impact on the power plant operation.

2.3.2   Hydroelectric Energy

Hydroelectric energy is generated by the kinetic and potential energy of flowing or falling water under the effect of gravitational force. Hydroelectric is the most mature and widest utilized form of renewable energies. Hydroelectric energy has approximately 17% contribution to the overall world energy generation [59]. No fuel is burnt at hydroelectric power plants; therefore they do not have greenhouse gas emissions. The operating cost is relatively low since the water running the plant is supplied free by the nature. It is a renewable source of energy since the rainfall naturally replenishes and enriches the water reservoirs.

Hydroelectric energy is generally obtained from the potential energy of dammed or reservoired water. When the water falls from a certain height of the reservoir output, it looses its potential energy and gains kinetic energy. The water flow drives a water turbine that is coupled to an electric generator which in turn generates electricity. This generated energy is a function of the water volume and the difference between the source and outflow of the water [60]. This height difference between the water output and turbine is called as "head." The potential energy of the water is proportional to the head. In order to generate greater amounts of energy, the head can be increased by running the water for hydraulic turbine through a large and long pipe called as penstock [61]. The cross-sectional view of a hydraulic dam and the hydroelectric power plant components are represented in Fig. 2.5.

In Table 2.6, these components are explained.

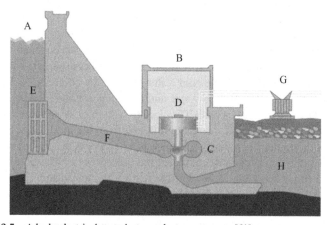

FIGURE 2.5   A hydroelectric dam and power plant components [61].

TABLE 2.6 Hydroelectric Power Plant Components

A. Reservoir	E. Water intake
B. Intake	F. Penstock
C. Water turbine	G. Transformer and transmission lines
D. Electric generator	H. River (or lower reservoir)

Electric power generation in a hydroelectric power plant can be approximately calculated as [62],

$$P = (hrg)\, \eta_t \eta_g \qquad (2.1)$$

where P is the generated power (kW), h is the height (m), r is the water flow rate (m^3/s), and g is the gravitational acceleration (m/s^2). In Eq. 2.1, the term hrg represents the potential energy of the water. η_g and η_g represent the efficiency of the water turbine and the generator, respectively. These efficiency rates are required for the water potential energy conversion into the electrical energy.

The other methods of electric generation by hydroelectricity are the pumped storage hydroelectric power plants and run-of-the-river plants. In pumped storage method, the water is pumped into higher elevations by using the excess generation capacity during the periods when electrical demand and cost are relatively lower. The water is released back into lower elevations through a turbine when the electric power demand and cost are relatively higher. In this method, water acts as an energy carrier in order to compensate the generation-consumption difference in a commercial device by improving the daily load factor [60–62]. In run-of-the-river plants, water reservoirs are not used and the kinetic energy of the flowing water through a river is captured using waterwheels.

2.3.3   Solar Energy Conversion and Photovoltaic Systems

Solar energy is one of the fastest growing renewable energy sources, which is plentiful and has the greatest availability among other energy sources. The amount of solar energy supplied from solar to the Earth in 1 h is capable of satisfying the total energy requirements of the Earth for 1 year [63]. Furthermore, solar energy does not produce pollutants or harmful byproducts, it is free of emissions. Solar energy is applicable to many fields such as vehicular, residential, space, and naval applications.

2.3.3.1   Photovoltaic Effect and Semiconductor Structure of PVs

Photovoltaic (PV) effect is known as a physical process in which that a PV cell converts the sunlight into electricity. When a PV cell is subject to the sunlight, the absorbed amount of light generates electric energy while remaining sunlight can be reflected or passed through. The electrons in the atoms of the PV cell are energized by the energy of the absorbed light. With this energy, these electrons move from their normal positions in the semiconductor PV material and they create an electrical flow, i.e., electric current through an external electric circuit connected to the PV cell terminals. The built-in electric field which is a specific electric feature of the PV cells provides the voltage potential difference that drives the current through an external load [64]. Two layers of different semiconductor materials are placed in contact with each other in order to induce the built-in electric field within a PV cell. The first layer which is *n-type* has

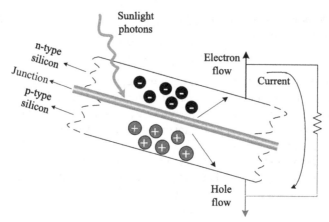

FIGURE 2.6 p-n junction structure and current flow in a PV cell.

abundance of electrons and it is negatively charged. The other layer which is *p-type* has abundance of holes and it is positively charged. Since the *n-type* silicon has excess electrons and *p-type* silicon has excess holes, contacting these layers together creates a p/n junction at their interface, thereby creating an electric field. In this contact, excess electrons move from the n-type side to the p-type side. As a result, a positive charge is built-up along the n-type side of the interface and negative charge along the p-type side. Thus an electric field is created at the surface where the layers meet, called the p/n junction. This electric field is due to the flow of electrons and holes. This electric field causes the electrons to move from the semiconductor toward the negative surface to carry current. At the same time, the holes move in the opposite direction, toward the positive surface, where they wait for incoming electrons [64]. The basic structure of a p-n junction in a PV cell is illustrated in Fig. 2.6.

2.3.3.2 PV Cell/Module/Array Structures

A photovoltaic (PV) or solar cell is the basic building block of a PV (or solar electric) system. An individual PV cell is usually quite small, typically producing about 1 or 2 W of power [65]. PV cells can be connected together to form a larger unit called modules in order to increase the power output of PV cells. Modules can be connected together and form larger units that are called arrays to generate more electric power. The output voltage of a PV system can be boosted by connecting the cells or modules in series. On the other hand, the output current can reach higher values by connecting them in parallel.

2.3.3.3 Active and Passive Solar Energy Systems

Based on the solar tracking capability, solar energy systems can be categorized in two types, passive and active systems [66,67]. In passive solar energy systems there are not any moving mechanisms for the panels. In this technique, the energy is absorbed and retained and spaces are designed that naturally circulate air to transfer energy and referencing the position of a building to the sun to enhance energy capture. On the other hand, in active solar energy systems, typically there are electrical and mechanical components such as tracking mechanisms, sensors, motors, pumps, heat exchangers, and fans to capture sunlight and process it into usable forms such as heating, lighting, and electricity. The panel positions are controlled in order to maximize exposure to the sun.

2.3.3.4   Components of a Complete Solar Electrical Energy System

In Fig. 2.7, the block diagram of a solar energy system is demonstrated. In this system, the sunlight is captured by the PV array. The photodiode or photo-sensor signals determine the sun tracking motor positions. This sun tracking control helps following the daily and seasonal solar position changes to face the sun directly and capture the most available sunlight. A DC/DC converter is employed at the PV panels' output in order to operate at the maximum power point (MPP) based on the current-voltage (*I-V*) characteristics of the PV array [68]. This DC/DC converter is controlled to operate at the desired current and voltage output of the PV array. A DC/AC inverter is usually connected to the output of this MPPT DC/DC converter in order to feed the AC loads for grid interconnection. A battery pack can be connected to the DC bus of the system to provide extra power that might not be available from the PV module during night and cloudy periods. The battery pack can also store energy when the PV module generates more power than the demanded. A grid connection is also useful to draw/inject power from/to the utility network to take the advantage of excess power or to recharge the batteries using grid power during the peak-off periods of the utility network.

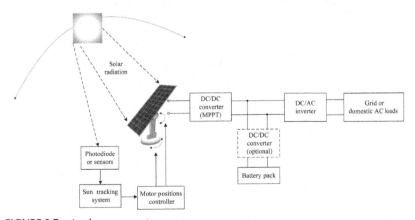

FIGURE 2.7   A solar energy system.

2.3.3.5 I-V Characteristics of Photovoltaic (PV) Systems, PV Models, and Equivalent PV Circuit

PV systems have a special current-voltage characteristic. As more current is drawn from the PV system, the system output voltage decreases. These characteristic curves differ at different solar insulation and temperature conditions hence the curves can be obtained by varying the load resistance (varying the output current) and measuring the output voltage for many different current values. *I-V* curve passes through two points for zero voltage and zero current.

- The short-circuit current (I_{SC}): I_{SC} is the current produced when the positive and negative terminals of the cell are short-circuited, and the voltage between the terminals is zero, which corresponds to zero load resistance.
- The open-circuit voltage (V_{OC}): V_{OC} is the voltage across the positive and negative terminals under open-circuit conditions, when the current is zero, which corresponds to infinite load resistance.

In order to extract maximum power from a PV system, for a constant ambient condition there is only one current-voltage pair. On the *I-V* curve, the maximum-power point (P_m) occurs when the product of current and voltage is the maximum. Although the current is maximum, no power is produced at the short-circuit current due to zero voltage. In addition, no power is produced at the open-circuit voltage due to zero current. The MPP is somewhere between these two points. Maximum power is generated at about the "knee" of the curve. This point represents the maximum efficiency of the solar device in converting sunlight into electricity [69]. A typical *I-V* curve characteristic of a PV system is given in Fig. 2.8.

PV systems exhibit nonlinear *I-V* characteristics [70]. There are various models available to mathematically model the *I-V* characteristics of the PV systems. An equivalent circuit expressing the PV model characteristics is shown in Fig. 2.9. This model is known as single-diode model and is one of the most common equivalent circuits representing PV system behaviors.

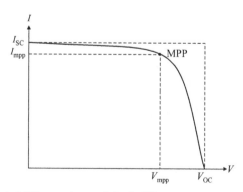

FIGURE 2.8 A typical *I-V* curve characteristic of a PV system.

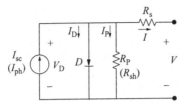

FIGURE 2.9 Single-diode model of solar cell equivalent circuit.

In this model, open-circuit voltage and short-circuit current are the key parameters. Illumination or solar radiation affects the short-circuit current, while the open-circuit voltage is affected by the material and temperature. In this model, I_{SC} is the short-circuit current, I_s is the diode reverse saturation current, m is the diode ideality factor (generally various between 1 and 5), and V_T is the temperature voltage expressed as $V_T = kT/q$, which is 25.7 mV at 25 °C. The equations defining this model are,

$$I_D = I_S \left[e^{\frac{V}{mV_T}} - 1 \right],$$
(2.2)

$$I = I_{SC} - I_D,$$
(2.3)

and

$$V = mV_T \cdot \ln \left[\frac{I_{SC} - I}{I_S} + 1 \right]$$
(2.4)

The I-V characteristic of the solar cell can be alternatively defined by [71],

$$I = I_{ph} - I_D$$

$$= I_{ph} - I_0 \left[\exp \left(\frac{q(V + R_s I)}{A k_B T} \right) - 1 \right] - \frac{V + R_s I}{R_{sh}}$$
(2.5)

where V is the PV output voltage (V), I is the PV output current (A), I_{ph} is the photocurrent (A), I_D is the diode current (A), I_0 is the saturation current (A), A is the ideality factor, q is the electronic charge (C), k_B is the Boltzmann's constant (J K^{-1}), T is the junction temperature (K), R_s is the series resistance (Ω), and R_{sh} is the shunt resistance (Ω).

2.3.3.6 Sun Tracking Systems

Incident solar radiation is the most important parameter for the power generated by solar energy systems. Sun changes its position during the day from morning to night. Moreover, the sun orbit differs from one season to another. By properly following the sun, through utilizing sun tracking systems, the incident solar irradiance can be effectively increased [72]. A sun tracker is an electromechanic component used for orienting a solar photovoltaic panel, concentrating solar reflector, or lens toward the sun. Solar panels require a high degree of accuracy to ensure that the concentrated sunlight is directed precisely to the photovoltaic device. Solar tracking systems can substantially improve the amount of power produced by a system by enhancing morning and afternoon performance. For instance, the orientation of PV panels can increase the solar-electric energy conversion efficiency between 20% and 50% [73–77]. A fixed system oriented to a fixed sun facing direction will have a relatively low annual production because they do not move to track the sun which yields significant increase of incident irradiation. An efficient sun tracking system should be capable of movement from north to south and from east to west as shown in Fig. 2.10.

2.3.3.7 Maximum Power Point Tracking Techniques

The conditions of radiation and temperature affect the current-voltage (I-V) characteristics of solar cells. The voltage and current should be controlled to track the maximum power of the PV systems in order to operate the PV systems

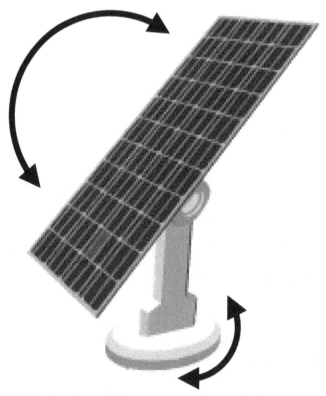

FIGURE 2.10   Rotations of a sun tracking system.

at the point of (V_{max}, I_{max}). Maximum power point tracking (MPPT) techniques are used to extract maximum available power from the solar cells by controlling the voltage and current. Systems composed of various PV modules located at different positions should have individual power conditioning systems to ensure the MPPT for each module [78]. In this section, most common and applicable MPPT techniques are described.

2.3.3.7.1 Incremental Conductance-Based MPPT Technique

The incremental conductance (INC) technique is the most commonly used MPPT for PV systems [71] (79–81). The technique is based on the fact that the sum of the instantaneous conductance I/V and the INC $\Delta I/\Delta V$ is zero at the MPP, negative on the right side of the MPP, and positive on the left side of the MPP.

If the change of current and change of voltage is zero at the same time, no increment or decrement is required for the reference current. If there is no change for the current, while the voltage change is positive, reference current should be increased. Similarly, if there is no change for the current while the voltage change is negative, reference current should be decreased. Contrarily, the change of the current might not be zero. If the current change is not zero, while $\Delta V/\Delta I = - V/I$, the PV is operating at MPP. If the current change is not zero and $\Delta V/\Delta I \neq - V/I$, then $\Delta V/\Delta I > - V/I$. If $\Delta V/\Delta I \neq - V/I$ and $\Delta V/\Delta I > - V/I$, the reference current should be decreased. However, if $\Delta V/\Delta I \neq - V/I$ and $\Delta V/\Delta I < - V/I$, the reference current should be increased in order to track the MPP.

Practically, due to the noise and errors, satisfying the condition of $\Delta I/\Delta V = - I/V$ may be very difficult [82]. Therefore, this condition can be satisfied with good approximation by,

$$|\Delta I/\Delta V + I/V| < \varepsilon \qquad (2.6)$$

where ε is a positive small value. Based on this algorithm, the operating point is either located in BC interval or it is oscillating among the AB and CD intervals as shown in Fig. 2.11.

Selecting the step size (ΔV_{ref}), shown in Fig. 2.11, is a tradeoff of accurate steady tracking and dynamic response. If larger step sizes are used for quicker dynamic responses, the tracking accuracy decreases and the tracking point oscillates around the MPP. On the other hand, when small step sizes are selected, the tracking accuracy will increase. In the mean time, the time duration required to reach the MPP will increase [83].

The normalized IV, PV (power-voltage), and absolute derivative of PV characteristics of a PV array are shown in Fig. 2.12.

From these characteristics, it is seen that the $|dP/dV|$ decreases as the MPP is approached and it gets greater when the operating point gets away from the MPP. This relation can be given by,

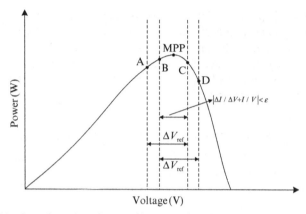

FIGURE 2.11 Operating point trajectory of incremental conductance-based MPPT.

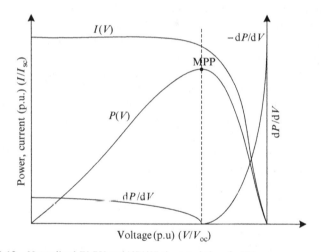

FIGURE 2.12 Normalized *IV*, PV, and |d*P*/d*V*| characteristics of a PV array.

$$\begin{cases} dP/dV < 0, & \text{right of MPP} \\ dP/dV = 0, & \text{at MPP} \\ dP/dV > 0, & \text{left of MPP} \end{cases} \tag{2.7}$$

In order to obtain the operating MPP, dP/dV should be calculated. The dP/dV can be obtained by only measuring the incremental and instantaneous conductance of the PV array, i.e., $\Delta I/\Delta V$ and I/V [79].

$$\frac{dP}{dV} = \frac{d\,(IV)}{dV} = I + V\frac{dI}{dV}. \tag{2.8}$$

2.3.3.7.2 Other MPPT Techniques

In the Perturb and Observe technique, the current drawn from the PV array is perturbed in a given direction and if the power drawn from the PV array increases, the operating point gets closer to the MPP and, thus, the operating current should be further perturbed in the same direction [84]. If the current is perturbed and this results in a decrease in the power drawn from the PV array, this means that the point of operation is moving away from the MPP and, therefore, the perturbation of the operating current should be reversed.

The *P-V* and *I-V* characteristics of a roof-mounted PV array are monotonously increasing or decreasing under a stable insulation conditions. The *I-V* characteristic is a function of voltage, insulation level, and temperature. From these characteristics, MPPT controllers can be developed based on the linearized *I-V* characteristics [85–87].

Fractional open-circuit voltage-based method [88–95], fractional short-circuit-based method [95,96], fuzzy logic controller-based method [97–106], neural network-based method [107–112], ripple correlation-based method [113], current sweep-based method [114], and DC-link capacitor droop control-based method [115,116] are the other applicable methods for MPPT.

2.3.3.8 *Power Electronic Interfaces for PV Systems*

Power electronic interfaces are either used to convert the DC energy to AC energy to supply AC loads or connection to the grid or to control the terminal conditions of the PV module to track the MPP for maximizing the extracted energy. They also provide wide operating range, capability of operation over different daily and seasonal conditions, and reaching the highest possible efficiency [117]. There are various ways to categorize power electronic interfaces for solar systems. In this book, power electronic interfaces are categorized as power electronic interfaces for grid-connected PV systems and stand-alone PV systems.

2.3.3.8.1 Power Electronic Interfaces for Grid-Connected PV Systems

The power electronic interfaces for grid-connected PV systems can be classified into two main criteria: classification based on inverter utilization and classification based on converter stage and module configurations.

2.3.3.8.1.1 *Topologies Based on Inverter Utilization:* The centralized inverter system is illustrated in Fig. 2.13.

In this topology, PV modules are connected in series and parallel to achieve the required current and voltage levels. Only one inverter is used in this topology at the common DC bus. In this topology, the inverter's power losses are higher than string inverter or multi-inverter topologies due to the mismatch between the modules and necessity of string diodes that are connected in series. In this

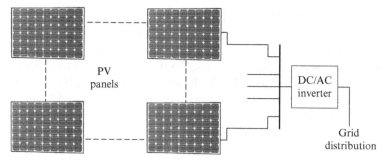

FIGURE 2.13 Conventional PV system technology using centralized inverter system topology.

topology, voltage boost may not be required since the voltage of series connected string voltages is high enough [118].

In string inverters topology, the single string of modules connected to the separate inverters for each string [119]. In this topology, voltage boosting may not be required if enough number of components are connected in series in each string.

In the multi-string invert topology, several strings are interfaced with their own integrated DC/DC converter to a common DC/AC inverter [120,121] as shown in Fig. 2.14.

Therefore, this is a flexible design with high efficiency. In this topology each PV module has its integrated power electronic interface with utility. The power loss of the system is relatively lower due to the reduced mismatch among the modules, but the constant losses in the inverter may be the same as for the string inverter. In addition, this configuration supports optimal operation of each module, which leads to an overall optimal performance [118]. This is due to the fact that each PV panel has its individual DC/DC converter and maximum power levels can be achieved separately for each panel.

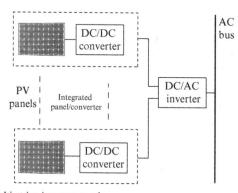

FIGURE 2.14 Multi-string inverters topology.

2.3.3.8.1.2 Topologies Based on Module and Stage Configurations:
The power electronic conditioning circuits for solar energy systems can be transformer-less, or they can utilize high-frequency transformers embedded in a DC/DC converter, which avoids bulky low-frequency transformers. The number of stages in the presented topologies refers to the number of cascaded converters/inverters in the system.

Isolated DC/DC converters consist of a transformer between the DC/AC and AC/DC conversion stages [122]. This transformer provides isolation between the PV source and load. A typical topology is depicted in Fig. 2.15.

In the topology shown in Fig. 2.15, the outputs of the PV panel and DC/DC converter are DC voltages. The two stage DC/DC converter consists of a DC/AC inverter, a high-frequency transformer, and a rectifier. In this topology, a capacitor can be used between the bottom leg of the high-frequency inverter and the transformer, forming an LC resonant circuit with the equivalent inductance of the transformer. This resonance circuit reduces the switching losses of the inverter. Alternatively, only two switches are enough if a push-pull converter is used; however, this topology requires a middle terminal outputted transformer [118].

The topologies shown in Fig. 2.16(a) and (b) are two stage single-module topologies, in which a DC/DC converter is connected to a DC/AC converter for grid connection. The DC/DC converter deals with the MPP tracking and the DC/AC inverter is employed to convert the DC output to AC voltage for grid connection. These are nonisolated converters since they are transformer-less.

Instead of using a full-bridge inverter for the DC/AC conversion stage, a half-bridge inverter can also be used. In this way, number of switching elements can be reduced and controller can be simplified, however, for the DC bus, two

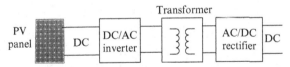

FIGURE 2.15   Isolated DC/DC converter topology.

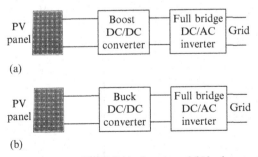

FIGURE 2.16   (a) Boost converter with full-bridge inverter and (b) buck converter with full-bridge inverter.

series connected capacitor is required to obtain the midpoint. This midpoint of two series connected capacitors will be used as the negative terminal of the AC network of the half-bridge configuration.

The single-stage inverter for multiple modules is depicted in Fig. 2.17, which is the simplest grid-connection topology [123]. The inverter is a standard voltage source PWM inverter, connected to the utility through an LCL filter. The input voltage, generated by the PV modules, should be higher than the peak voltage of the utility. The efficiency is about 97%. On the other hand, all the modules are connected to the same MPPT device. This may cause severe power losses during partial shadowing. In addition, a large capacitor is required for power decoupling between PV modules and the utility [124].

A topology for multi-module multi-string interfaces is shown in Fig. 2.18 [121,125]. The inverter in Fig. 2.18 consists of up to three boost converters, one for each PV string, and a common half-bridge PWM inverter. The circuit can also be constructed with an isolated current- or voltage-fed push-pull or full-bridge converter [126], and a full-bridge inverter toward the utility. The voltage across each string can be controlled individually [121,126].

As an alternative to the topology shown in Fig. 2.18, other types of DC/DC converter can be employed to the first stage, such as isolated DC/DC converters.

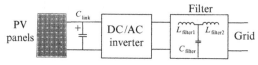

FIGURE 2.17   Single-stage inverter for multiple modules.

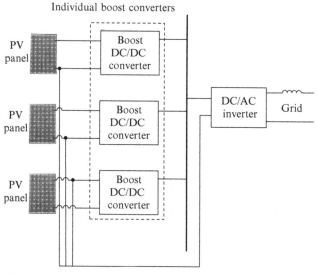

FIGURE 2.18   Topology of the power electronics of the multi-string inverter.

2.3.3.8.2 Power Electronic Interfaces for Stand-Alone PV Systems

The stand-alone PV systems composed of a storage device and its controller for sustainable satisfaction of the load power demands [127]. The storage device with the controller should provide the power difference when the available power from the PV panel is smaller than the required power at the load bus [128]. When the available power from the PV panel is more than the required power, the PV panel should supply the load power and the excess power should be used to charge the storage device. A simple PV panel/battery connection topology is shown in Fig. 2.19.

In this simple topology, the DC/DC converter between the battery and the PV panel is used to capture the all available power from the PV panel. In this system, battery pack acts as an energy buffer, charged from the PV panel and discharged through the DC/AC inverter to the load side. The charging controller determines the charging current of the battery, depending on the MPP of the PV panels at a certain time. When there is no solar insulation available, the DC/DC converter disables and the stored energy within the battery supplies the load demands. The battery size should be selected so that it can supply all the power demands during a possible no-insulation period. In addition, it could be fully charged during the insulated periods to store the energy for future use. Since the combined model produces AC electrical energy, it should be converted to AC electrical energy for domestic electrical loads. The combined system requires a DC/AC inverter, which also used to match the different dynamics of the combined energy system and various loads. The proper response of the PV/battery system to the overall load dynamics can be achieved by generating appropriate switching signals to the inverter while modulating for both active and reactive powers. The load bus voltage can be controlled by the modulation index control of the inverter; while the load control can be achieved by the phase angle control of the inverter.

2.3.4 Wind Turbines and Wind Energy Conversion Systems

Wind turbines are devices that are capable of capturing the kinetic energy of winds. This kinetic energy is converted to the mechanical energy to rotate the turbine which is coupled to an electric generator. In this way, kinetic energy of the wind can be converted into a usable form of energy, i.e., electrical energy.

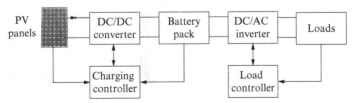

FIGURE 2.19 PV/battery connection for stand-alone applications.

Wind turbines can be installed stand-alone to power remote or isolated locations, or they can be grid connected, to supply power to the utility grid. Wind power is renewable, widely distributed, plentiful, and it is a clean way of energy conversion. Additionally, it contributes in reducing the greenhouse gas emissions; since it can be used as an alternative to fossil fuel-based power generation [129]. Although wind energy has a great potential to significantly contribute the world's power generation, only 3% of worldwide power requirement is supplied by wind turbines [130].

Several key parameters, such as air density, area of the blades, wind speed, and rotor area, need to be considered in order to efficiently capture wind energy. Wind force is converted into a torque that rotates the blades of wind turbine. The wind force is stronger in higher air densities. In other words, kinetic energy of the wind depends on air-density and heavier winds carry more kinetic energy. At normal atmospheric pressure and at 15 °C the weight of the air is 1.225 kg/m^3, but if the humidity increases, the density decreases slightly. The other fact, which determines the air density is whether the air is warm or cold. Warmer winds are less dense than cold ones, so at high altitudes the air is less dense [131]. Besides, the area of the blades (air swept area), the diameter of the blade, plays important role in captured wind energy. Under the same conditions more wind can be captured with longer blades and bigger rotor area of wind turbine [130, 131]. The other parameter is the wind speed. It is expected that wind kinetic energy arises as wind speed increases [131].

Kinetic energy of the wind can be expressed as

$$E_k = \frac{1}{2}m \cdot v^2 = \frac{1}{2}\rho \cdot R^2 \cdot \pi \cdot d \cdot v^2 \qquad (2.9)$$

where E_k represents kinetic energy of the wind, m stands for the mass of the wind, v is wind speed, ρ is air density, A is rotor area, R is blade length, while d stands for thickness of the "air disc" shown in Fig. 2.20.

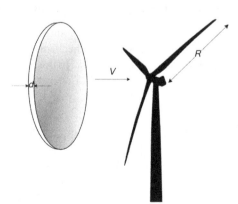

FIGURE 2.20 Kinetic energy of wind.

Hence, the overall wind power (P) is [131]:

$$P = \frac{E_k}{t} = \frac{1}{2}\rho \cdot R^2 \cdot \pi \cdot v^3 \tag{2.10}$$

From Eq. 2.10 it can be seen that the power content of the wind varies with the cube (the third power) of the average wind speed as shown in Fig. 2.21.

2.3.4.1 Wind Turbine Power

2.3.4.1.1 Betz's Law

The theoretical maximum power that can be extracted from the wind is demonstrated by the Betz's law [132,133]. The wind turbines extract the kinetic energy of the wind. Higher wind speeds results in higher extracted energy. It should be noted that the wind speed after turbine (after passes through turbine) is much lower than before it comes to turbine (before energy is extracted) since the wind looses its speed by transferring its kinetic energy to the wind turbine. That means wind speed before wind approaches (in front of) the turbine, and its speed after (behind) turbine are different. Figure 2.22 shows both speeds. The wind after the turbine has less amount of energy due to decreased speed of wind.

The decreased wind speed, after turbine, provides information on amount of possible extracted energy from the wind. The extracted power from the wind can be calculated using Eq. 2.7.

$$P_{\text{extracted}} = \frac{E_k}{t} = \frac{1}{2}\rho \cdot R^2 \cdot \pi \cdot \frac{v_a + v_b}{2} \cdot \left(v_b{}^2 - v_a{}^2\right) \tag{2.11}$$

where $P_{\text{extracted}}$ shows maximum extracted power from the wind, v_A and v_b are wind speeds after and before passing through the turbine. ρ is the air density and R demonstrates the radius of the blades.

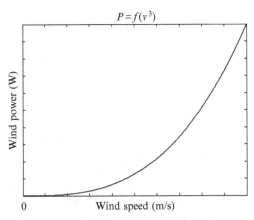

FIGURE 2.21 Specific wind power due to wind speed variation.

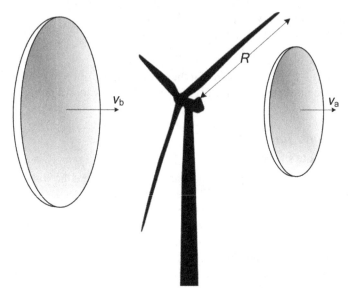

FIGURE 2.22   Wind speed before and after turbine.

The relation of total amount of power P_{total} to the extracted power $P_{extracted}$ can be calculated as

$$\frac{P_{extracted}}{P_{total}} = \frac{\frac{v_a + v_b}{2} \cdot (v_b^2 - v_a^2)}{v_b^3} = \frac{1}{2}\left(1 - \frac{v_a^2}{v_b^2}\right) \cdot \left(1 + \frac{v_a}{v_b}\right) \qquad (2.12)$$

For the maximum power extraction, the ratio of the wind speed after and before the turbine can be calculated using $\dfrac{d\,(P_{extract}/P_{total})}{d\,(v_a/v_b)} = 0$.

Solving Eq. (2.12) for v_a/v_b, yields,

$$\frac{v_a}{v_b} = \frac{1}{3} \qquad (2.13)$$

As a result, Eq. (2.12) reaches its maximum value for $\dfrac{v_a}{v_b} = \dfrac{1}{3}$.

$$\left. \frac{P_{extracted}}{P_{total}} \right|_{\frac{v_a}{v_b} = \frac{1}{3}} \approx 59.3\% \qquad (2.14)$$

Equation (2.14) shows that the maximum extracted power from the wind is 59.3% of the total available power. In other words, it is not possible to extract all 100% of wind energy since the wind speed after turbine cannot be 0.

Betz's law indicates that the maximum theoretical extracted wind power is 59%. However, in practice, the real efficiency of wind turbine is slightly different (lower) due to the other nonideal properties.

2.3.4.2   Different Electrical Machines in Wind Turbines

There are many types of electrical machines that are used in wind turbines. There is no clear criterion for choosing particular machine to work as wind generator. Based on the installed power, site of turbine, load type, and simplicity of control the wind generator can be chosen. Squirrel cage induction or Brushless DC (BLDC) generators are usually used for small wind turbines in household applications. Doubly fed induction generators (DFIGs) are usually used for megawatt size turbines. Synchronous machines and permanent magnet synchronous machines (PMSM) can also be used for wind turbine applications.

2.3.4.2.1   Brushless DC Machines

BLDC machines are very popular in many applications due to the recent advances in their development. In addition, the development of fast semiconductor switches, cost-effective DSP processors, and other microcontrollers have influenced the development of the motor/generator drives. BLDC machines are widely used because of their simple control, efficiency, compactness, lightweight, ease of cooling, less noise, and low maintenance [134,135]. Usually BLDC machines are used in small wind turbines (up to 15 kW).

The simplified equivalent circuit of the BLDC generator connected to a diode rectifier is shown in Fig. 2.23. This is the simplest way of using BLDC machine for wind applications, because there is no switch to control the phase current. The full bridge rectifies the induced voltages of variable frequency (because of variable wind speeds). Basically, the waveform of the induced EMF (electromotive force) is converted to DC voltage regardless of the input waveform. Usually these types of wind turbines are connected to batteries; therefore rectified electrical power is used to charge the battery.

Three-phase active synchronous rectifiers can be used with BLDC generators. In this case, the controlled rectifier is used for BLDC-phase current control. Usually hysteresis regulators are used to control current. In synchronous rectifiers, active switching devices such as IGBTs or MOSFETs are used. By employing a PWM control strategy for the synchronous rectifier, MPPT of the wind turbine can be achieved. An inverter can be placed at the DC bus for grid interconnection or powering the AC loads.

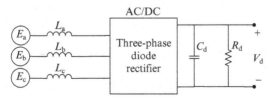

FIGURE 2.23   Diode rectifier connected to BLDC generator.

2.3.4.2.2 Permanent Magnet Synchronous Machines

For both fixed and variable speed applications PMSM can be used. The permanent magnet synchronous generator (PMSG) is very efficient and suitable for wind turbine applications. PMSGs allow direct-drive energy conversion for wind applications. Direct-drive energy conversion helps eliminating the gearbox between the turbine and generator; thus, these systems are less expensive and less maintenance is required [136,137]. However, lower speed determined by the turbine shaft is the operating speed for the generator.

A wind power system (WPS) where a PMSG is connected to a full-bridge rectifier followed by a boost converter is shown in Fig. 2.24. In this case, the boost converter controls the electromagnetic torque. The supply side converter regulates the DC-link voltage and controls the input power factor. One drawback of this configuration is the use of diode rectifier that increases the current amplitude and losses. The grid-side converter can be used to control active and reactive power being supplied to the grid. Automatic voltage regulator (AVR) obtains the information of speed of turbine, DC-link voltage, current, and grid-side voltage and current. It calculates PWM pattern (control scheme) for converter. This configuration has been considered for small size (less than 50 kW) WPS [138].

Instead of using a diode rectifier cascaded by a DC/DC converter, both rectifier and inverter can be controllable. A PMSG where the PWM rectifier is placed between the generator and the DC link, and PWM inverter is connected to the utility is shown in Fig. 2.25. In this case, the back-to-back converter can be used as the interface between the grid and the stator windings of the PMSG [139]. The turbine can be operated at its maximum efficiency and the variable speed operation of PMSG can be controlled by using a power converter which

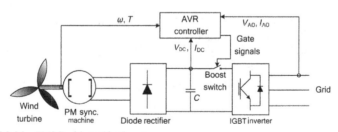

FIGURE 2.24 PMSG with rectifier/inverter.

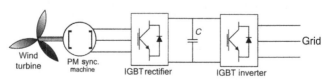

FIGURE 2.25 PMSG with back-to-back inverter.

is utilized to regulate the maximum power flow. The stator terminal voltage can be controlled in several ways [140]. In this system, utilizing the field oriented control (FOC) allows the generator to operate near its optimal working point in order to minimize the losses in the generator and power electronic circuit. However, the performance depends on the knowledge of the generator parameter that varies with temperature and frequency. The main drawbacks are the cost of permanent magnets that increases the price of the machine; and demagnetization of the permanent magnets. In addition, it is not possible to control the power factor of the machine [134].

2.3.4.2.3 Squirrel Cage Induction Machines

The three-phase induction machines are commonly used in industrial motor applications. However, they can also be effectively used as generators in electrical power systems. The main issue with induction machines as electric power generators is the need for an external reactive power source that will excite the induction machine, which is certainly not required for synchronous machines in similar applications. If induction machine is connected to the grid, required reactive power can be provided by the power system. Induction machine may be used in cogeneration with other synchronous generators or the excitation might be supplied from capacitor banks (only for stand-alone self-excited generators application) [141–146]. The reactive power required for excitation can be supplied using static VAr compensators [147,148] or static compensators (STATCOMs) [149].

Due to its low cost, brushless rotor construction does not need a separate source for excitation. No maintenance and self protection against severe over loads, short circuits, and self-excited induction generators are used in wind turbine applications [142–146]. The only drawback of these types of generators can be their inherent generated voltage and frequency regulation under varied loads [150].

Common structure of a squirrel cage induction generator with back-back converters is shown in Fig. 2.26. In this structure, stator winding is connected to utility through a four-quadrant power converter. Two PWM VSI are connected back-to-back through a DC link. The stator-side converter regulates the electromagnetic torque and supplies reactive power, while the grid-side converter controls the real and reactive power delivered from the system to the utility and regulates the DC link. This topology has several practical advantages, and one

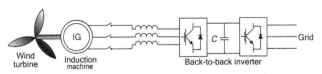

FIGURE 2.26 Induction machine controlled by back-to-back inverter.

of them is possibility of fast transient response for speed variations. In addition, the inverter can operate as a VAR/harmonic compensator [151].

On the other hand, main drawback is the complex control system. Usually FOC is used to control this topology, where its performance relies on the generator parameters, which vary with temperature and frequency. Hence, in order to supply the magnetizing power requirements, i.e., to magnetize the machine, the stator-side converter must be oversized 30-50% with respect to rated power.

2.3.4.2.4 Doubly Fed Induction Generator

Figure 2.27 presents a topology consists of a DFIG with AC/DC and DC/AC converters; i.e., a four-quadrant AC/AC converter using IGBTs connected to the rotor windings. In the DFIG topology, the induction generator is not a squirrel cage machine and the rotor windings are not short circuited. Instead they are used as the secondary terminals of the generator which provides the capability of controlling the machine power, torque, speed, and reactive power. To control the active and reactive power flow of the DFIG topology, rotor and grid-side converters should be controlled separately [152–155].

Wounded rotor induction machines can be supplied from both rotor and stator sides. The speed and the torque of the wounded rotor induction machine can be controlled by regulating voltages from both rotor and stator sides of machine. The DFIG can be considered as a synchronous/asynchronous hybrid machine. In the DFIG, similar to the synchronous generator, the real power depends on the rotor voltage magnitude and angle. In addition, the induction machine slip is also a function of the real power [156]. DFIG topology offers several advantages in compare to systems using direct-in-line converters [157,158]. These benefits are;

– The main power is transferred through the stator windings of the generator which is directly connected to the grid. Around 65-75% of the total power is transmitted through stator windings. The remaining power is transmitted

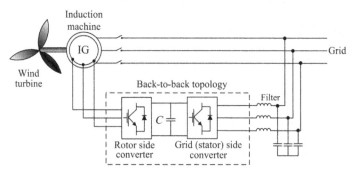

FIGURE 2.27 Doubly fed induction machine topology.

using the rotor windings, i.e., through the converters, which is about 25% of the total power. Since the inverter rating is 25% of total system power the inverter cost and size can considerably be reduced.

- While the generator losses are the same in both topologies (direct-in-line and DFIG), the inverter losses can be reduced from around 3% to 0.75%, because the inverter is supposed to only transfer 25% of the total power. Therefore, approximately 2-3% efficiency improvement can be obtained.
- DFIG topology offers a decoupled control of generator active and reactive powers [159,160].
- Cost and size of the inverter and EMI filters can be reduced since the inverter size is reduced. In addition, the inverter harmonics are lowered because the inverter is not connected to the main stator windings.

In the rotor circuit, two voltage-fed PWM converters are connected back-to-back while the stator windings are directly connected to the AC grid side as shown in Fig. 2.27. The direction and magnitude of power between the rotor windings and stator windings can be controlled by adjusting the switching of the PWM signals of the inverters [161–163]. This is very similar to connecting a controllable voltage source to the rotor circuit [164]. This can also be considered as a conventional induction generator without a zero rotor voltage.

To take the benefits of variable speed operation, the optimum operating point of the torque-speed curve should be tracked precisely [165]. By controlling the torque of the machine, speed can be adjusted. Thus, using the instantaneous rotor speed value and by controlling the rotor current i_{ry} in stator flux-oriented reference frame, the desired active power can be obtained. Operation at the desired active power results in the desired speed and torque [153]. On the other hand, the grid-side converter is controlled to keep the DC-link voltage fixed, independent of the direction of rotor power flow. By using supply voltage vector-oriented control, the decoupled control of active and reactive power flow between rotor and grid can be obtained.

Using DFIG, the over-sizing problem can be solved. Still speed range of turbine is wide enough, thus a power converter, which is rated for much lower powers, can be placed in rotor side only and stator is connected to grid directly. Since power flowing through rotor is usually around 25-30% of power going through stator, the power electronic interface is designed for only 25-30% of total power. This is the most important advantage of DFIG.

2.3.4.2.5 Synchronous Generators

Synchronous generators are commonly used for variable speed wind turbine applications, due to their low rotational synchronous speeds that produce the voltage at grid frequency. Synchronous generators can be an appropriate selection for variable speed operation of wind turbines [166,167]. They do not need a pitch control mechanism. The pitch control mechanism increases the cost of the turbine and causes stress on turbine and generator [168]. Synchronous generators in variable speed operation will generate variable voltage and variable frequency power. Using an AVR for the excitation of the field voltage, the output voltage of the synchronous generator can be controlled. However, induction generators require controlled capacitors for voltage control. In addition, their operating speed should be over synchronous speed in order to operate in generating mode [169].

Multi-pole synchronous generators can be used more efficiently since the gear can be eliminated and direct drive of the turbine and generator can be achieved [170,171]. However, synchronous generators without multipoles require gearboxes in order to produce the required frequency for grid connection. On the other hand, a DC voltage source or an AC/DC converter is required for synchronous generators in wind applications in order to produce the required excitation voltage for the field windings. The synchronous generator connection with wind turbine is shown in Fig. 2.28.

2.3.4.3 Energy Storage Applications for Wind Turbines

The batteries and other DC energy storage devices can be connected to the DC links of any topologies. The main purpose of batteries is to assists the generator to meet the load demand. When the load current is smaller than generator current, the extra current is used to charge the battery energy storage. On the other hand, when the load current is larger than generator current, the current is supplied from the battery to the load. With this strategy, the voltage and frequency of the generator can be controlled for various load conditions. Energy storage decreases system inertia, improves the behavior of the system in the case of disturbances, compensates transients, and therefore, improves the efficiency [172]. However, it brings an initial cost to the system and requires periodical maintenance depending on the storage devices. Therefore, the voltage and frequency control can be modified by using batteries as the controllable load of the VSI as presented in Fig. 2.29. In this way, the load can be regulated by controlling the power flow to the batteries. A bidirectional inverter/converter can be used for power flow from/to the batteries. As another alternative, the battery voltage can be converted to AC voltage with another individual inverter to provide power to AC loads. Although an induction generator is shown in Fig. 2.29, these energy storage systems are applicable to any other topologies.

Storage systems can be connected in various forms to the wind turbine systems [173–177]. Generally, a bidirectional DC/DC converter is required for the integration of the storage system to the DFIG system [178]. In this topology, one of the converters regulates the storage power; whereas the other is responsible for DC bus voltage control. The bidirectional energy storage topology for DFIGs in wind applications is shown in Fig. 2.30.

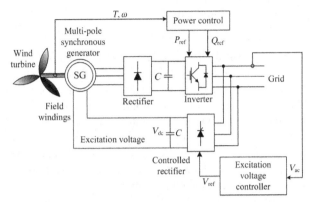

FIGURE 2.28 Multi-pole synchronous generator for wind turbine applications.

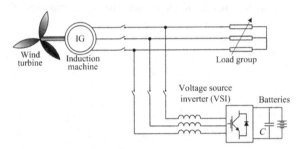

FIGURE 2.29 Voltage and frequency control using energy storage.

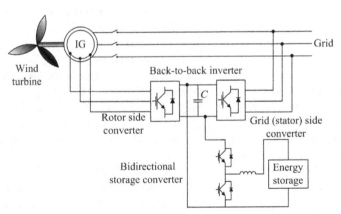

FIGURE 2.30 Energy storage with bidirectional converter in DFIG systems.

2.3.5   Ocean Energy Harvesting

2.3.5.1   Ocean Wave Energy

Ocean waves are a plentiful, clean, and renewable source of energy. The total power of waves breaking around the world's coastlines is estimated at 2-3 million MW. The west coasts of the United States and Europe and the coasts of Japan and New Zealand are good sites for harnessing wave energy [179]. Wave energy conversion is one of the feasible future energy technologies; however, it is not mature enough. Therefore, construction cost of wave power plants is considerably high. These energy systems are not developed and maturated commercially due to the problems of dealing with sea conditions, complexity, and difficulty of interconnection and transmission of electricity.

A wave power absorber, a turbine, a generator, and power electronic interfaces are the main components of a typical ocean wave energy harvesting technique. The kinetic energy of the ocean waves are captured by absorbers. The absorbed kinetic energy of the waves is either conveyed to turbines or the absorber directly drives the generator. The shaft of the electric generator is driven by the turbine. Turbines are generally used within the systems including rotational generators. Linear motion generators are used in systems without turbines, which can be directly driven by the power absorber or movement of the device. Due to the varying amplitude and period of the ocean waves, both linear and rotational generators generate variable frequency—variable amplitude AC voltage. This AC voltage can be rectified to DC voltage to take the benefit of DC energy transmission through the salty ocean water. DC transmission in salty water does not require an additional cable for the negative polarity. Thus, it will be more cost effective than transmitting the power in AC form, which requires three-phase cables. Transmission cable length varies depending on the location of the application, which is either near-shore or off-shore. However, the main idea and the principles are same for both types of applications. After the DC power is transmitted from ocean to the land, a DC/DC converter or a tap-changing transformer can be used for voltage regulation. Depending on the utilized voltage regulation system, a DC/AC inverter is used before or after the voltage regulator. The voltage synchronization is provided by the inverter and the output terminals of the inverter can be connected to the grid.

In Fig. 2.31, a system level diagram of the ocean wave energy harvesting technology is shown. At an in-water substation, wave energy conversion devices (including the absorber, turbine, and generator) are interconnected. The substation consists of the connection equipments and controllers for individual devices. The outputs of the generators are connected to a common DC bus using DC/AC converters for transforming power before transmission to the shore. A transmission line connects the cluster to shore. An on-shore inverter converts the DC voltage to a 50 or 60 Hz AC voltage for grid connection. An optional shore transformer with tap changer or a DC/DC converter compensates the voltage

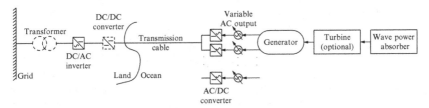

FIGURE 2.31 System level diagram of ocean wave energy harvesting.

variations. A group of absorber, turbine, and generators could be used in a farm structure thus the captured energy can be increased.

Alternatively, the land converters might be moved off-shore to come up with space limitations that may occur in land stations. This brings some complexity to the system and may require more maintenance, which is harder to deal, in compare to land-side converters. Moreover, the transformer can be installed off-shore. This would increase the power transmission capability, since the higher voltage transmission will result in less transmission losses. However, in this case the advantage of DC transmission will not exist. As a different option, boost DC/DC converters can be used after the AC/DC converter of the generator. This allows a high voltage DC transmission link. In this case, both transmission losses will be kept at minimum and only single-line DC transmission through the ocean water will be required.

2.3.5.1.1 Energy of Ocean Waves

The total potential and kinetic energy of an ocean wave can be expressed as,

$$E = \frac{1}{2}\rho g A^2 \tag{2.15}$$

where g is the acceleration of gravity (9.8 m/s^2), ρ is the density of water (1000 kg/m^3), and A is the wave amplitude (m).

The power of a wave in a period is equal to the energy E multiplied by the speed of wave propagation, v_g, for deep water

$$v_g = \frac{L}{2T} \tag{2.16}$$

where T is the wave period (s) and L is the wave length (m) [180].

$$P_w = \frac{1}{2}\rho g A^2 \frac{L}{2T} \tag{2.17}$$

The dispersion relationship describes the connection between the wave period T and the wave length L as,

$$L = \frac{gT^2}{2\pi}. \tag{2.18}$$

If Eq. 2.18 is substituted in Eq. 2.17, the power or energy flux of an ocean wave can be calculated as

$$P_w = \frac{\rho g^2 T A^2}{8\pi}. \tag{2.19}$$

Instead of using the wave amplitude, wave power can also be rewritten as a function of wave height, H (m). Considering that the wave amplitude is the half of the wave height, the wave power becomes:

$$P_w = \frac{\rho g^2 T H^2}{32\pi}. \tag{2.20}$$

2.3.5.1.2 Ocean Wave Energy Harvesting Technologies

In general, ocean wave energy harvesting technologies can be classified in two types with respect to their distance from the shore; Off-Shore Ocean Wave Energy Harvesting Technologies and On-Shore Ocean Wave Energy Harvesting Technologies. These are discussed in details in following subsections.

2.3.5.1.2.1 Off-Shore Ocean Wave Energy Harvesting Technologies:

Off-shore applications are located away from the shore and they generally use a floating body as wave power absorber and another body that is fixed to the ocean bottom. Generally, linear generators with buoys are used in off-shore applications. Linear generators are directly driven by the movement of a floating body on the ocean. Salter cam and buoys with air-driven turbines are the only applications involving rotational generators in off-shore applications.

2.3.5.1.2.1.1 Air-Driven Turbine-Based Off-Shore Technologies.

In air-driven turbine systems for off-shore applications, the primary conversion is from wave to the pressurized air. Secondary conversion stage is the conversion to mechanical energy by rotating shaft of the turbine. The last stage is converting mechanical rotation into electric power by electric generators.

The operating principle of an off-shore application which consists of a floating buoy with an air chamber and an air-driven generator is shown in Fig. 2.32. In this system, the water level inside channel of the buoy increases when the waves hit the body. This increase in water level, applies a pressure to the air in the air chamber. When the air is pressurized, it applies a force to the ventilator turbine and rotates it. This turbine drives the electric generator and electricity is generated at the output terminals of this generator. When the waves are pulled back to the ocean, the air in the air chamber is also pulled back since the water level in the buoy channel decreases. Due to the syringe effect, this time turbine shaft rotates in the contrary direction but produces electricity. There should be very good mechanical insulation through the air chamber and the ventilating generator to achieve higher efficiencies. However, this brings design complexity and additional cost to the system.

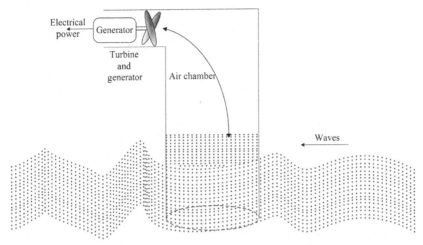

FIGURE 2.32   Spinning the air-driven turbines using wave power [181].

In another method, water level increases and air is taken out from the upper outlets while spinning the turbines as shown in Fig. 2.33(a). Contrarily, when the waves are pulled back to the sea, water level decreases. This results in sucking the air back from the upper inlets while spinning the turbines as shown in Fig. 2.33(b).

2.3.5.1.2.1.2   Direct-Drive Permanent Magnet Linear Generator-Based Buoy Applications. The height differences of the wave top and bottom levels yields an up and down motion for the piston which is the transaxle of the linear generator. When the wave is floating on the ocean surface, the buoy follows the motion of the wave. Buoy can move vertically on a pillar, which is connected to a hull. On the surface of the hull, permanent magnets are mounted, while outside of the hull contains the coil windings. The pillar and stator are connected together on a concentrate foundation standing on the seabed of the ocean. The hull and mounted magnets, called rotor or piston, are the moving parts of the generator. Since the motion is linear, this generator is called a linear generator.

In Fig. 2.34, the linear generator in the floating buoy and fixed pillar are shown.

When the wave rises, the buoy will drive the generator piston through a stiff rope. When the wave subsides, the generator will be driven by the spring that stores the mechanical energy in the first case. Thus, electric generation is provided during both up and down motion. Due to the existence of variable frequency in the current and voltage from the stator, an AC/DC rectifier followed by a DC/AC converter is required to make the grid connection possible. Instead of placing moving parts to the ocean bottom, the permanent magnets and the stator windings can be placed at the sea level [182]. Fig. 2.35 shows the *x-y* plane of the cross-section view of one pole of the longitudinal flux surface mounted linear permanent magnet generator.

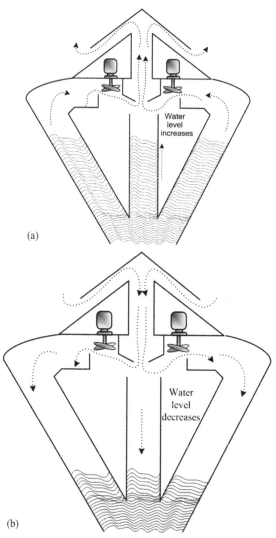

FIGURE 2.33 Air pressure ring buoy. (a) Water level increases and air is taken out from the upper outlets and (b) water level decreases and air is pulled back from upper inlets.

2.3.5.1.2.1.3 *Salter Cam Method.* Salter cam method implementation is shown in Fig. 2.36. Salter cam rolls around a fixed inner cylinder by activation of an incoming wave. Through the differential rotation between the cylinder and the cam, power can be captured. The motion of the cam is converted from wave to a hydraulic fluid. Then hydraulic motor is used to convert the pressurized hydraulic fluid to rotational mechanical energy. Finally, rotational mechanical energy is converted to electricity by utilizing electric generators. Flywheels or

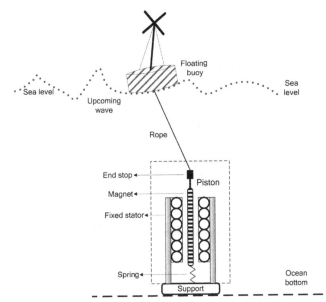

FIGURE 2.34 Linear generator-based buoy type wave energy harvesting method.

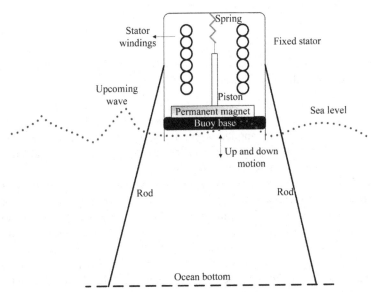

FIGURE 2.35 Schematic of a longitudinal flux permanent magnet generator used for wave energy conversion.

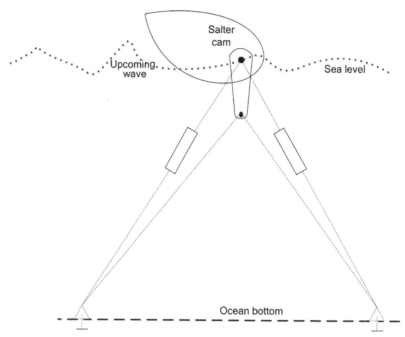

FIGURE 2.36   A schematic illustrating the fixation of Salter cam wave energy conversion device.

pressurized liquids can be used as an intermediate step in order to reduce the intermittencies of the wave power.

2.3.5.1.2.2   Near-Shore Ocean Wave Energy Harvesting Technologies:
Near-shore topologies are applied within the surfing zone of the ocean or right on the shore. Near-shore applications have some advantages and disadvantages in compare to the off-shore applications.

2.3.5.1.2.2.1   Channel/Reservoir/Turbine-Based Near-Shore Wave Energy Harvesting Method.
Wave currents can be tapered into a narrow channel to increase their power and size in order to harness the wave energy. As shown in Fig. 2.37, waves can be channeled into a catch basin and used directly to rotate the turbines. Since this method requires building a reservoir to collect the water coming with the waves to drive the turbine, it is more expensive in compare to the other buoy-shaped off-shore applications. However, it requires less maintenance in compare to the off-shore applications, since all components of the wave energy conversion system are located on land. Additionally, since a reservoir collects the ocean water, the intermittencies can be eliminated. This will create a convenient platform for voltage and frequency regulation. Building these types of plants in the locations, where they have regular and sustaining wave regimes, is more advantageous.

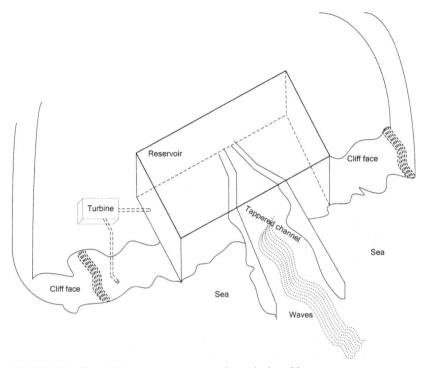

FIGURE 2.37 Channeled ocean wave to a reservoir to spin the turbines.

2.3.5.1.2.2.2 Air-Driven Turbine-Based Near-Shore Wave Energy Harvesting Method Using oscillating water columns that generates electricity from the wave-driven rise and fall of water in a cylindrical shaft or pipe is another way to harness the wave energy. The air is driven into and out of the top of the shaft due to the rising and falling water, powering an air-driven turbine which is shown in Fig. 2.38.

The general structure of the near shore air-driven turbine is shown in Fig. 2.38(a). Waves push the air through the ventilator which drives the electrical machine as shown in Fig. 2.38(b). The wave retreats from the wave chamber inside the channel which decreases the pressure as shown in Fig. 2.38(c). This method is advantageous because of the capability of using not only wave power but also the power from the tidal motions. However, mechanical isolation should be provided within the wave and air chambers in order to obtain better efficiencies. This will also bring some more cost and design complexity to the system.

2.3.5.2 Ocean Tidal Energy

The generation of electrical power from ocean tides is similar to the traditional hydroelectric generation. A dam, known as a barrage, across an inlet is required for the simplest tidal power plants. In a tidal power plant, usually a tidal pond created by a dam and a powerhouse, which contains a turbo generator, and a sluice gate to allow the bidirectional tidal flow. During the flood tide, the rising

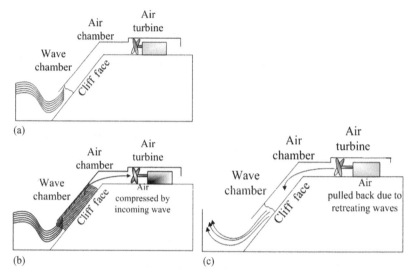

FIGURE 2.38 Air-driven turbines using the wave power. (a) Upcoming wave starts filling the chamber, (b) air is compressed by rising water, and (c) air is pulled back by retreating waves [181].

tidal waters fill the tidal basin after opening the gate of the dam. When the dam is filled to capacity, the gates are closed. The tidal basin is released through a turbo generator after the ocean waters have receded. Power can be generated during ebb tide, flood tide, or both. When the water is pulled back ebb tide occurs, and when the water level increases near the shore flood tide happens [183]. Tidal power can be economical at sites where mean tidal range exceeds 16 ft. [183,184].

Tidal current is not affected by climate change, lack of rain, or snowmelt. Therefore, tidal energy harvesting is practical since the tidal current is regular and predictable. Moreover, environmental and physical impacts and pollution issues are negligible. Tidal power can additionally be used for water electrolysis for hydrogen production applications and desalination. However, tidal power generation is an immature technology, which needs further investigations and developments.

Tidal turbines can be used for tidal energy harvesting, similar to the wind turbines. Tidal turbines can be located where there is a strong tidal flow. These turbines have to be much stronger than wind turbines since the water is about 800 times as dense as air. They will be heavier and more expensive; however, they will be able to capture more energy at much higher densities [181]. In Fig. 2.39, a typical tidal turbine is shown.

Usually, tidal fences having multiple turbines are mounted in the entrance of channels which are affected by ocean tides. Tidal water is forced to pass through a fence structure, which is called caisson in this fence application. Unlike barrage stations, basins are not required for fence applications and they can be used in

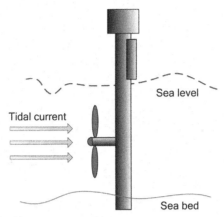

FIGURE 2.39 Tidal turbine.

a channel between the mainland and a nearby off-shore island, or between two islands. Tidal fences can be mounted at the entrance of channels that ocean water gets inside the land via a bay (Fig. 2.40(a)), or between the main land and an island (Fig. 2.40(b)), or simply between two islands (as shown in Fig. 2.40(c)). Since they do not require flooding the basin, tidal fences have much less impact on the environment. In addition, they are significantly cheaper to install; however, the caisson may disrupt the movement of large marine animals and shipping [184].

2.3.5.3 Power Electronic Interfaces for Ocean Energy Harvesting Applications

Both in ocean wave and ocean tidal energy harvesting applications, the generators may produce alternating currents and voltages that have varying magnitude and frequency. Therefore, output power of the ocean energy converters need further conditioning prior to the grid connection. In addition, the frequency of the output voltage should be regulated to be the same as grid frequency. Output power conditioning, amplitude, phase, and frequency of the conversion system can be regulated by utilizing power electronic converters.

Block diagram of a typical power conditioning system for a grid-connected ocean energy conversion system is shown in Fig. 2.41. Ocean wave or ocean tide potential and kinetic energies rotate the water turbine or a power absorber directly drives a linear generator with up and down motions. The varying wave and tides result in variable frequency and amplitude of the generator output. The AC power produced by the generator is converted into DC power via three-phase bridge rectifiers followed by a DC/AC inverter. The output of the DC/AC inverter generally contains harmonics, which should be filtered. Finally, output power can be connected to grid and transmitted to consumers after its voltage is increased and isolated by a power transformer.

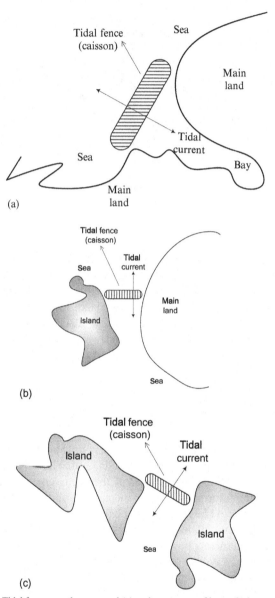

FIGURE 2.40 Tidal fences can be mounted (a) at the entrance of bays, (b) between the main land and an island, and (c) between two islands.

During the intermittencies, power cannot be generated. Therefore, an energy storage system should be connected to the generator output or output of a conversion stage in the power conditioning system. Stored energy can be

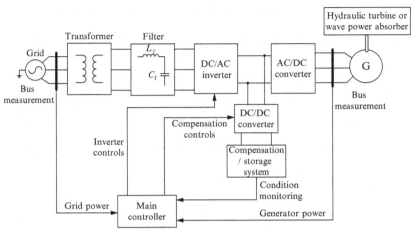

FIGURE 2.41   Grid connection and controls of tidal current power conditioning system.

supplied to the grid during the intermittency periods. Hence, it is ensured that continual power is supplied to the grid.

2.3.6   Geothermal Energy Systems

Geothermal energy is the thermal energy that is stored in the inner layers of the Earth composed of rocks and fluids. The temperature of the inner layers of the Earth gets hotter as the depth increases. In deeper layers it is even extremely hotter due to the hot molten rock called magma [185,186].

Geothermal energy can be utilized by several methods. It can be used as direct heat for electric power generation. In the direct heat utilization, applications can be categorized as hydrothermal, agricultural, or industrial [187]. Hydrothermal resources have low to moderate temperatures between 20 and 150 °C. These resources can be utilized to provide direct heating for residential, industrial, and commercial sectors [188]. These applications include but not limited to water and space heating, greenhouse and agricultural heating, cleaning, textile processes, and food dehydration. Agricultural production is one of the utilization methods of direct use of geothermal energy. It is used to warm the greenhouses in order to provide cultivation. Industrial utilization examples can be food and cloth processing, manufacturing paper, pasteurizing milk, drying fish, vegetables and fruits, and even for refrigeration and air conditioning.

Other than the direct use of geothermal energy, it can be used as the heat and steam source for electric power plants. Instead of burning fossil fuels and generating heat for water boiling, geothermal power plants use the readily obtained heat or steam. Natural hot water and/or steam from the inner layers of the Earth are used to drive the turbines and generators to produce electricity. Absolutely, no fuel firing is required for heating or steam generation for geothermal power plants. Therefore, geothermal power plants do not have emissions and they are environmentally clean. Moreover, since the extracted heat is replaced by the thermal energy of the Earth's inner layers, geothermal energy is sustainable and renewable.

Schematic of a geothermal power plant is presented in Fig. 2.42.

The components of the geothermal power plant are described in Table 2.7.

The operating principle of a geothermal power plant is very similar to that of a coal-fired power plant. However, in geothermal power plants hot steam

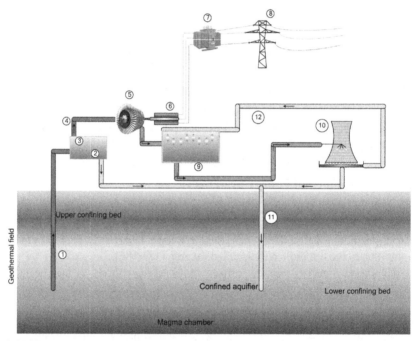

FIGURE 2.42 A geothermal power plant.

TABLE 2.7 Geothermal Power Plant Components

1. Production well	7. Power transformer
2. Water-steam mix	8. High voltage transmission lines
3. Separator	9. Condenser
4. Steam	10. Cooling tower
5. Steam turbine	11. Injection well
6. Generator	12. Water flow cycle

and/or water is obtained from the deeper Earth layers instead of burning any fuel. Production well is used to draw hot water and/or steam from deeper layers. This mixture is separated by the separator in order to get the dry steam. Through a steam governor, a steam turbine is rotated by this high pressure and high temperature steam. Since the steam turbine is coupled to an electric generator, the mechanical steam power is converted into electric power. The output voltage of the generator is increased by a power transformer. The high voltage output of the power transformer can then be connected to the high voltage transmission network. The steam looses its temperature and pressure after it goes through the steam turbine. Thus, the output flow of the steam turbine is condensed in the condensers. Condensed water is cooled down through the cooling tower and cool water helps condense the low pressure/temperature steam in the condenser. The cooled water is then injected back to the inner Earth layers to get hot again. If the geothermal field is rich of hot water reservoirs, this cooled but still relatively warm water can be used for other heating purposes.

Geothermal energy is abundant, secure, reliable, and a renewable source of energy. It has high availability and capacity factor in compare to other renewables. It is not a source of pollution for the environment, i.e., their CO_2 emissions are less than 0.2% of the cleanest fossil fuel-fired power plant, SO_2 emissions are less than 1%, and particulate emissions are less than 0.1%. It has an inherent energy storage capability and requires very small land area for establishment [186].

Geothermal power plants can be classified into three main generation technologies; dry stream power plant, flash stream power plant, and binary cycle power plant.

Dry steam power plants are the most common geothermal power plants since they are simple and cost effective. These power plants are applicable to the geothermal fields where the geothermal steam is not mixed with water. In this method, production wells are drilled down to the aquifer to get superheated and pressurized steam. This steam is brought to the surface at high speeds. When the expanding steam passes through the turbine, the generator generates electricity [189,190]. The low pressure steam output from the turbine is ventilated to the atmosphere in simple power plants. However, exhaust steam from the turbine is condensed in more complex power plants. The condensate can be reinjected to the reservoir by the injection wells and/or it can be used as makeup cooling water.

Flash steam power plants use a flash steam technology where the hydrothermal source is in liquid form. This fluid is sprayed into a flash tank which is at a much lower pressure than the fluid. Therefore, the fluid immediately vaporizes rapidly into steam [189,191]. This generated steam is used to rotate the steam turbines that are coupled to electric generators. The production well is kept under high pressure in order to prevent the geothermal fluid flashing inside the well [186]. Instead of using a single-flashing system, dual-flashing systems are also used. The brine from the high pressure steam is piped into a low-pressure separator/flash tank where the pressure is additionally reduced to generate lower pressure steam in the dual-flash systems. In order to generate additional electric power, this lower pressure steam is piped into a lower pressure stage of the turbine. The steam exhaust from the high and low pressure turbines is condensed. Just like the dry steam plants, the condensate is then used as makeup cooling water or reinjected to the reservoir.

Although the dual-flash power plants have higher capital cost, they have higher thermoelectric efficiency. Resource characteristics, power plant output, thermodynamic and economic factors, and equipment availability are the factors affecting the decision to build and operate a single-flash or dual-flash geothermal power plants [192]. Generally, dual-flash system is preferred if the fluid temperature is between 175 and 260 °C, while single-flash systems are efficient enough for the fluid temperatures higher than 260 °C.

Binary cycle power plants are preferred when the geothermal resource is insufficiently hot to produce steam. Sometimes the resource may have other chemical components causing impurities and flashing may not become possible [186,193]. In these cases binary cycle power plants are preferred. In binary cycle geothermal power plants, isobutene, isopentane, or pentane is used as the secondary fluid which has a lower boiling point than water. Since a separate working fluid is used, the cycle called "binary." The geothermal fluid (water) is passed through a heat exchanger in order to heat up the secondary fluid. Secondary fluid vaporizes and expands through the turbines that are coupled to electric generators. After passing through the turbines, the working fluid is condensed and recycled for the next cycle. Moreover, the fluid remaining in the tank of flash steam plants can be reutilized in binary cycle plants. In a closed-cycle system, all of the geothermal fluid is injected back to the ground. Usually, binary cycle plants are more efficient than the flash plants in low to moderate temperatures of geothermal fluids. Furthermore, corrosion problems are avoided since a pure working fluid is used.

2.3.7 Nuclear Power Plants

In nuclear power plants, energy is extracted from atomic nuclei by the controlled nuclear reactions. There are several available methods such as nuclear fission, nuclear fusion, and radioactive decay. The most common method is the nuclear fission. Similar to the conventional fossil fuel-fired power plants, nuclear reactors generate heat in order to produce steam. However, unlike many conventional thermal power plants, nuclear power plants convert the energy released from the atoms' nucleus generally via nuclear fission, instead of burning fossil fuels. This energy is used for steam production which is utilized to operate the turbines that are coupled to electric generators. In this way, the mechanical work of the high pressure steam is converted into electricity [194,195].

The fission of an atom occurs when a relatively large fissile atomic nucleus such as uranium-235 or plutonium-239 absorbs a neutron. The atom is then splitted by the fission into two or more smaller nuclei with kinetic energy, gamma radiation, and free neutrons [194]. Other fissile atoms may absorb a portion of these neutrons and create more fission, which release more neutrons, and so on [195]. By using neutron moderators and neutron poisons, this nuclear chain reaction can be controlled in order to adjust the potion of neutrons that will cause more fission. Manual or automatic control systems are used for this purpose or to shut down the reactor if unsafe conditions are detected [196].

Heat generation by the reactor core from fission involves several stages. The kinetic energy of the fission products is converted into thermal energy when a collision happens between the nuclei and nearby atoms. The reactor absorbs some of the gamma radiation produced during fission in the form of heat. Neutron absorption activates some materials and the radioactive decay of fission products produces heat. Even after the reactor is shut down, this decay heat source may remain for some time. A nuclear reaction can generate heat power that is 1,000,000 times that of the equal mass of coal.

After the fission process, the heat released from the reactor is removed by a cooling system. This heat is conveyed to another part of the power plant, in which the thermal energy is utilized to generate electricity. The hot coolant in general is used as the heat source for a boiler. The boiler generates the pressurized steam which mechanically drives the steam turbines. The steam turbines rotate the electrical generators [197]. A simple operating schematic of a nuclear power plant is depicted in Fig. 2.43.

By utilizing different coolants and fuels and integrating different control methodologies, many different reactor designs can be accomplished. In order

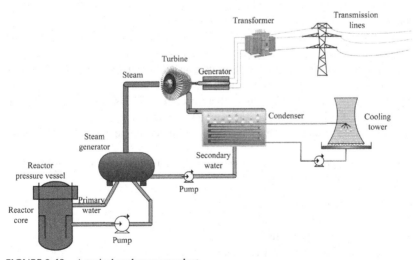

FIGURE 2.43 A typical nuclear power plant.

to meet a specific need, some of these designs can be employed for various applications. Space and naval applications are some of these specific applications. In these applications, generally highly enriched uranium is used as the fuel which increases the reactors power density and efficiency [198]. Currently, researchers are investigating new nuclear power generation techniques, known as the Generation IV reactors. These new designs will have the possibility to offer cleaner and more secure fission reactors with less risk of the proliferation of nuclear weapons. New designs such as ESBWR offer passively safe plants and other designs are believed to be almost foolproof are being pursued or are available to be built [199]. In near future, it is expected that the fusion reactors will be viable, which will reduce or eliminate many safety risks associated with nuclear fission [200].

2.3.8   Fuel Cell Power Plants

Since the beginning of twenty-first century, fuel cell technology has been rapidly developed and has shown an invasive improvement for the applications ranging from portable electronic devices to vehicular power systems and MW size power plants [201–203]. Fuel cells are promising future energy conversion devices due to their high efficiency, excellent performance, low or zero emissions, and wide application area.

Fuel cell power plants are electrochemical devices that produce electrical energy directly from a chemical reaction. Fuel cells use fuel on the anode side and oxidant on the cathode side. The chemical reaction occurs on the electrolyte. The reactants, i.e., fuel and oxidant flow into the cell while the reaction product (water) flows out of the cell. Many fuel and oxidant types can be used for fuel cells. Generally, hydrogen as the fuel and oxygen as the oxidant, from the air, can be used. On the other hand, alcohols and hydrocarbons can be other fuel types for different fuel cells, while other oxidants may be chlorine and chlorine dioxide [204–206].

Just like a battery, a fuel cell is composed of an electrolyte and a pair of electrodes. However, unlike the batteries the reactants are continuously replenished during the operation, therefore, the cell is not required to be recharged. Ideally, fuel cells operate and continue to produce energy as long as the reactants are appropriately supplied to the anode and cathode sides.

There are many kinds of fuel cells categorized by their electrolyte type. Most common fuel cell types are:

- Proton exchange membrane fuel cells (PEMFCs),
- Phosphoric acid fuel cells (PAFCs),
- Direct methanol fuel cells (DMFCs),
- Solid oxide fuel cells (SOFCs), and
- Molten carbonate fuel cells (MCFCs).

PEMFCs are generally used for residential, vehicular, and portable applications. Solid electrolyte structure reduces the corrosion, they can operate at low temperature and they have quick start-up and faster response times. PAFCs are typically used for transportation, heating, and electric utility applications. They may reach high efficiency points in electric cogeneration applications [207,208]. Currently, DMFCs are considered as a replacement alternative for batteries for small portable devices' power requirements. DMFC can be considered advantageous since methanol can be used directly without any reformer or fuel processor. However, they have relatively low efficiencies and slow response times since the reaction rate for the methanol is slow on presently available catalysts. On the other hand, DMFCs can be competitive with batteries since the simplicity, high storage density, and liquid methanol portability may compensate the relatively low efficiency [207]. MCFCs and SOFCs are generally used as large power plants for electric utility applications. Both of these two technologies have higher efficiencies, fuel flexibility, and inexpensive catalysts [208]. However, they operate at really high temperatures generally between 600 and 1000 °C. This high temperature issue avoids these two fuel cell technologies to be best candidates for portable or vehicular applications.

Especially PEMFCs are considered to be one of the most promising fuel cell technologies among these next-generation fuel cell power plants. This is due to their high efficiency and compact structure [209,210]. The operating principle of PEMFCs is focused in this section.

The operating principle along with the basic components of a PEMFC is presented in Fig. 2.44.

After the fuel is supplied to the anode side, the fuel is oxidized resulting in releasing electrons. The anode reaction for a fuel cell can be expressed as:

$$2H_2 \rightarrow 4H^+ + 4e^- \tag{2.21}$$

These released electrons are transported to the cathode side through an external circuit. The hydrogen protons are traveled through the proton exchange

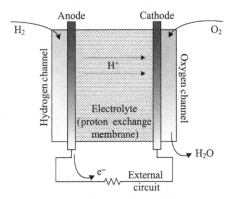

FIGURE 2.44 Components of a PEM fuel cell and its operating principle.

membrane to the cathode side. The oxidant (i.e., oxygen) is reduced at the cathode side, using the electrons coming from the external circuit. Therefore, the cathode reaction is,

$$O_2 + 4e^- \rightarrow 2O^{2-}. \qquad (2.22)$$

The hydrogen protons travel through the membrane, balance the flow of electrons through the external circuit. Therefore, the overall reaction equation becomes,

$$2H_2 + O_2 \rightarrow 2H_2O + \text{Electric power}. \qquad (2.23)$$

A typical single-fuel cell has a theoretical output voltage of 1.2 V. They generate ideally 0.6 Å/cm^2. In order to reach higher voltage outputs from a fuel cell system, cells are connected in series in the form of a string. For higher current outputs the cells or the cell strings should be connected in parallel. Unfortunately, the output voltage of a fuel cell or a fuel cell system decreases as the current drawn from the fuel cell is increased. This voltage drop at the fuel cell output is due to the ohmic, activation, and concentration losses [211].

A typical current-voltage characteristic curve of a fuel cell is shown in Fig. 2.45, which is also known as the polarization curve.

The output voltage of a fuel cell is less then its theoretical value even in the open-circuit conditions. This is due to the fact that the open-circuit voltage is calculated based on the ideal burning enthalpy of the hydrogen. The activation loss is generally effective at the low current densities. Activation loss is due to the electrode kinetics in which the electrochemical reaction of hydrogen and oxygen is slow. Activation loss causes a nonlinear voltage drop as the current starts to be drawn from the fuel cell. Ohmic losses are due to the electron flow through the electrolyte and electrodes and the equivalent resistance

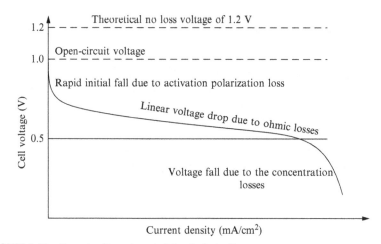

FIGURE 2.45 Current-voltage characteristic of a fuel cell.

of the external circuit. Ohmic losses are directly proportional to the current density and they increase linearly as the current increases. Concentration losses are due to the inability of maintaining the initial fuel concentration on the electrodes. Fuel and oxidant should be supplied sufficiently and continuously in order to meet sustained load demands. If the current is more than a certain value, fuel cell fails to meet the new power demand and the output voltage dramatically decreases. Therefore, this loss is quite severe at the high current densities.

Due to the polarization and current-voltage characteristics of the fuel cells, power conditioning devices such as DC/DC and/or DC/AC converters are required to maintain a fixed and stable DC voltage for the load bus. The power conditioning is also useful for converting the fuel cell output to an appropriate magnitude and type. Power conditioning unit (PCU) not only controls the fuel cell output voltage but also it delivers a high power factor in grid-connected applications. PCU can reduce or eliminate the harmonics and help operating effectively under all conditions. A fuel cell power plant operating together with a PCU is presented in Fig. 2.46.

In the stationary or vehicular applications, fuel cell power plant may not be sufficient to satisfy all of the load demands [210]. Especially during transient load changes or peak demand periods, fuel cell needs to be operated with an auxiliary power device such as battery packs or ultra-capacitors. By operating fuel cell cascaded with batteries and/or ultra-capacitors, steady-state, peak power demands, and transient load changes can be controlled more efficiently. In the topology of Fig. 2.47, the fuel cell power plant is operated with auxiliary power devices.

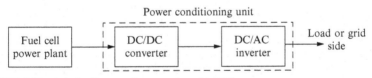

FIGURE 2.46 A fuel cell power plant operation with PCU.

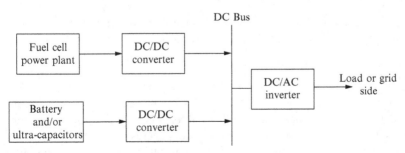

FIGURE 2.47 Fuel cell power plant operation with auxiliary power devices.

2.4 OTHER UNCONVENTIONAL ENERGY SOURCES AND GENERATION TECHNOLOGIES

Thermal depolymerization is used to convert various waste products, usually, plastic and biomass, into light crude oil by using hydrous pyrolysis. Long-chain polymers of oxygen, hydrogen, and carbon decompose into short-chain petroleum hydrocarbons once pressure and heat applied [212].

Oil sands can be in form of loose sand or partially consolidated sandstone containing naturally occurring mixtures of sand, clay, and water, and bitumen. The natural or crude bitumen is a highly viscous, sticky mixture, and very thick that it will not flow unless heated of diluted with other hydrocarbons such as condensed natural gas or light crude oil. With advanced techniques, the profitable extraction and processing can be enabled [213].

Syngas or synthetic gas is a fuel gas mixture that is composed of carbon monoxide, carbon dioxide, and hydrogen. This mixture is used as intermediates in producing synthetic natural gas, methanol, or ammonia [214]. With the Fischer-Tropsch process or with the Mobil methanol to gasoline process, syngas is also used as intermediate in producing synthetic petroleum for use as fuel or lubricant [215]. Syngas is combustible and can be used in internal combustion engines as a fuel. However, the energy density is relatively lower.

Synthetic fuel, also known as synfuel, is a liquid fuel that is obtained from syngas [216]. By using Fischer-Tropsch, synfuel can be produced through methanol or direct coal liquefaction.

Downdraft, also known as "energy tower" is a device that generates electric power by spraying water on hot air at the top of the tower, making the cooled air fall through the tower and rotate a turbine at the tower's bottom [217].

Magnetohydrodynamics is a technique that can harvest the energy of electrical currents that are induced as a result of moving conductive fluid. These induced currents also create electromagnetic forces on the fluid and also changes the magnetic field itself [218].

Piezoelectricity is the electric charge that is induced in certain materials in response to applied mechanical stress such as straining, squeezing, applying pressure, or flexing [219].

Other alternative energy generation techniques include sulfur-iodine cycle, pyrolysis, osmotic power, vibration energy harvesting, and electromagnetic energy harvesting. Some of the other alternative energy carriers include liquid nitrogen, ethanol, methanol, peat, and hydrogen.

SUMMARY

In order to meet the future energy requirements, the energy should be generated and utilized wisely. Increasing the demand for energy, decreasing conventional fossil fuel energy sources, and environmental concerns are driving forces toward renewable energy sources. However, the conventional sources will be utilized until modern, clean, and renewable technologies replace them. Therefore, a comprehensive strategy that supports a diversity of resources over the next century should be developed. Sustainable and long-term energy solutions in numerous forms are required to restructure the future's increasing energy demand.

REFERENCES

[1] Key World Energy Statistics 2012, Report of Energy Information Administration, International Energy Annual 2012, Available online: http://www.iea.org/publications/freepublications/publication/kwes.pdf.

[2] Energy Consumption: Consumption per Capita, Technical Report by the World Resources Institute, Available online: http://www.wri.org/.

[3] World per Capita Total Primary Energy Consumption, 1980-2008, Report of Energy Information Administration, International Energy Annual 2014.

[4] International Energy Outlook 2008, Technical Report by the Energy Information Administration, Official Energy Statistics from the US Government, Available online: http://www.eia.doe.gov/.

[5] Energy Efficiency Measures and Technological Improvements, e8 Continuum of Action, Annual Activity Report 2007 to 2008 (e8 is an "Energy Organization" comprising 9 leading electricity companies from the G8 countries), 2010.

[6] Coal Facts 2006 Edition with 2005 Data, Technical Annual Report of World Coal Institute, 2006.

[7] D. Yergin, The Prize: The Epic Quest for Oil, Money, and Power, Free Press, New York, 1993

[8] Official European Parliament Resolution on the Road Map for Renewable Energy in Europe, September 2007, Available online: http://www.europarl.europa.eu.

[9] 2008 Buildings Energy Data Book, Prepared for the Buildings Technologies Program Energy Efficiency and Renewable Energy of US Department of Energy, D&R International Ltd., Under Contract to National Energy Technology Laboratory, November 2008.

[10] Manufacturing Trend Data 1998 and 2002, Technical Report by the Energy Information Administration, Official Energy Statistics from the US Government, Available online: http://www.eia.doe.gov/emeu/efficiency/mecs_trend_9802/mecs9802_table2b.html.

[11] The WNA Market Report, The Global Nuclear Fuel Market: Supply and Demand 2007-2030, Technical Report by the World Nuclear Association (WNA), 2007.

[12] C.R. de Auza, Installed U.S. Wind Power Capacity Surged 45% in 2007, American Wind Energy Association Market Report, January 2007.

[13] SEGS III, IV, V, VI, VII, VIII & IX, NEXTera Energy Sources, Available online: http://www.nexteraenergyresources.com/content/where/portfolio/contents/segs_viii.shtml.

[14] National Energy Survey of 2012, Technical Report by the Energy Information Administration, Official Energy Statistics from the US Government, Available online: http://www.eia.doe.gov/.

[15] 2008-2017 Regional & National Peak Demand and Energy Forecast Bandwidths, Technical Report by the Load Forecasting Working Group of the Reliability Assessment Subcommittee, North American Electric Reliability Corporation (NERC), August 2008.

[16] G. Russel and D. Ann, Oil officials see limit looming in production, Wall Str. J., November 19, (2007)

[17] S. S. Devgan, Impact of environmental factors on the economic evaluation of renewable energy alternative generation, in: Proceedings of the 33rd Southeastern Symposium on System Theory, March 2001, pp. 123–126

[18] World Consumption of Primary Energy by Energy Type and Selected Country Groups, 1980-2004, Energy Information Administration, US Department of Energy, July 2006.

[19] Natural Gas Reserves Summary as of December 31, 2007, Technical Report by the Energy Information Administration, Official Energy Statistics from the US Government.

[20] Natural Gas and the Environment, Available online: http://www.naturalgas.org.

[21] Key World Energy Statistics 2012, Annual Technical Report of the International Energy Agency.

[22] M. Brower, Cool Energy: Renewable Solutions to Environmental Problems, revised ed., The MIT Press, Cambridge, MA, 1992

[23] Environmental Impacts of Renewable Energy Technologies, Briefing paper of the Union of Concerned Scientists, August 2005.

[24] World Nuclear Power Reactors, Technical Report of the Uranium Information Center, 2007.

[25] Nuclear Power in the World Today, Briefing Paper 7, Uranium Information Center, 2003.

[26] Global Uranium Resources to Meet Projected Demand, Latest Edition of 'Red Book' Predicts Consistent Supply up to 2025, International Atomic Energy Agency, June 2006.

[27] International Energy Annual 2006, World Energy Reserves, Technical Report by the Energy Information Administration, Official Energy Statistics from the US Government, Report released June-December 2008.

[28] International Energy Annual 2006, World Energy Consumption in Standard U.S. Physical Units, Technical Report by the Energy Information Administration, Official Energy Statistics from the US Government, Report released June-December 2008.

[29] 2007 World PV Industry Reports, Annual World Solar PV Industry Report by Marketbuzz™, 2008.

[30] J. W. Tester, E. M. Drake, M. J. Driscoll, M. W. Golay, W. A. Peters, Sustainable Energy: Choosing Among Options, The MIT Press, Cambridge, MA, 2005

[31] W. A. Hermann, Quantifying global exergy resources, J. Energy 31 (12) (2006), pp. 1685–1702

[32] Global Wind Energy Markets, Technical Report by the Global Wind Energy Council, February 2008.

[33] A. Inoue, M. H. Ali, R. Takahashi, T. Murata, J. Tamura, A calculation method of the total efficiency of wind generator, in: Proceedings of the Power Electronics and Drives Systems (PEDS), vol. 2, November 2005, pp. 1595–1600

[34] J. Brooke, Engineering Committee on Oceanic Resources Working Group on Wave Energy Conversion, Wave Energy Conversion, Elsevier, London, 2003

[35] T. W. Thorpe, An overview of wave energy technologies, status, performance, and costs, in: Proceedings of the Wave Power: Moving Towards Commercial Viability, London, November 1999

[36] J. Cruz, M. Gunnar, S. Barstow, D. Mollison, J. Cruz, Green Energy and Technology, Ocean Wave Energy, Springer Science and Business Media, Berlin, 2008

[37] W. Munk and C. Wunsch, Abyssal recipes II: energetics of tidal and wind mixing, Deap Sea Res. 45 (1998), pp. 1977–2010

[38] Renewables, Global Status Report 2006, Technical Report by the REN21, Renewable Energy Policy Network for the 21st Century, 2006.

[39] D. E. Lennard, Ocean thermal energy conversion—past progress and future prospects, in: IEEE Proceedings on Physical Science, Measurement and Instrumentation, Management and Education Reviews, vol. 134, May 1987, pp. 381–391

[40] L. R. Berger and J. A. Berger, Countermeasures to microbiofouling in simulated ocean thermal energy conversion with surface and deep ocean waters in Hawaii, Appl. Environ. Microbiol. 51 (6) (1986), pp. 1186–1198

[41] L. Meegahapola, L. Udawatta, S. Witharana, The ocean thermal energy conversion strategies and analysis of current challenges, in: Proceedings of the International Conference on Industrial and Information Systems, August 2007, pp. 123–128

[42] The Hydrogen Economy: Opportunities, Costs, Barriers, and R&D Needs, Committee on Alternatives and Strategies for Future Hydrogen Production and Use, National Research Council, and National Academy of Engineering, National Academies Press, Washington, DC, 2004

[43] J. McCarthy, Hydrogen, Stanford University, 1995, Available online: http://www-formal. stanford.edu/jmc/progress/hydrogen.html.

[44] D. Palmer, Hydrogen in the Universe, NASA, 1997, Available online: http://imagine.gsfc. nasa.gov/.

[45] Hydrogen as an Energy Carrier, Technical Report by the Royal Belgian Academy Council of Applied Science, April 2006.

[46] F. Lau and C. E.G. Padro, Advances in Hydrogen Energy, Springer, New Orleans, 2000

[47] D. A.J. Rand and R. M. Dell, Hydrogen Energy: Challenges and Prospects, Royal Society of Chemistry, Cambridge, UK, 2008

[48] World Energy Intensity: Total Primary Energy Consumption per Dollar of Gross Domestic Product using Purchasing Power Parities: 1980-2004, Technical Report by the Energy Information Administration, Report released December 2008.

[49] All About Geothermal Energy, Geothermal Energy Association, Washington, DC, February 2007.

[50] The Future of Geothermal Energy, Impact of Enhanced Geothermal Systems (EGS) on the United States in the 21st Century, Technical Report by the Massachusetts Institute of Technology, 2006.

[51] J. Farigone, J. Hill, D. Tillman, S. Polasky, P. Hawthome, Land clearing and biofuel carbon debt, Science 319 (2008), pp. 1235–1238

[52] T. Searchinger, R. Heimlich, R. A. Houghton, F. Dong, A. Amani, J. Fabiosa, S. Tokgaz, D. Hayes, Use of U.S. croplands for biofuels increases greenhouse gases through emission from land-use change, Science 319 (2008), pp. 1238–1240

[53] R. E. Sonntag, C. Borgnakke, G. J. Van Wylen, Fundamentals of Thermodynamics, sixth ed., Wiley, Haboken, NJ, 2002

[54] M. J. Moran and H. N. Shapiro, Fundamentals of Engineering Thermodynamics, fifth ed., Wiley, Haboken, NJ, 2003

[55] F. Wicks, J. Maleszweski, C. Wright, J. Zarbnicky, Thermodynamic analysis of an enhanced gas and steam cycle, in: Proceedings of the 37th Intersociety Energy Conversion Engineering Conference, July 2004, pp. 456-459.

[56] L. Beltracchi, A direct manipulation for water-based Rankine cycle heat engines, IEEE Trans. Syst. Man Cybern. 17 (3) (1987), pp. 478–487

[57] T. J. Ratajczak and M. Shahidehpour, Emerging technologies for coal-fired generation, in: Proceedings of the Power Engineering Society General Meeting, Quebec, Canada, October 2006, pp. 1–9

[58] Net Generation by Energy Source by Type of Producer, Annual Electric Power Report of the Energy Information Administration, US Department of Energy, December 2008.

[59] Renewables, Global Status Report 2006 Update, Technical Report by the Renewable Energy Policy Network for the 21st Century, REN21, 2007.

[60] 21st Century Complete guide to Hydropower, Hydroelectric Power, Dams, Turbine, Safety, Environmental Impact, Microhydropower, Impoundment, Pumped Storagem Diversion, Run-of-River, by U.S. Government, Progressive Management, September 2006.

[61] A. J. Wood and B. F. Wollenberg, Power Generation, Operation, and Control, second ed., Wiley-Interscience, Haboken, NJ, 1996

[62] L. L. Grigsby, Electric Power Generation, Transmission, and Distribution, second ed., CRC Press, Boca Raton, FL, 2007

[63] R. Chapo, Solar Energy Overview, December 2008, http://Ezinearticles.com, Available online: http://ezinearticles.com.

[64] US Department of Energy, Energy Efficiency and Renewable Energy, Solar Energy Technologies Program, PV Physics, Available online: http://www1.eere.energy.gov/solar.

[65] US Department of Energy, Energy Efficiency and Renewable Energy, Solar Energy Technologies Program, PV Systems, Available online: http://www1.eere.energy.gov/solar.

[66] Missouri Department of Natural Resources, Missouri's Solar Energy Resource, Available online: http://www.dnr.mo.gov.

[67] US Department of Energy, Energy Efficiency and Renewable Energy, Solar Energy Technologies Program, Concentrating solar power, Available online: http://www1.eere.energy.gov.

[68] A. Lasnier and T. G. Ang, Photovoltaic Engineering Handbook, Adam Hilger, Bristol and New York, 1990; pp. 69-97

[69] US Department of Energy, Energy Efficiency and Renewable Energy, Solar Energy Technologies Program, Current-voltage Measurements, Available online: http://www1.eere.energy.gov/solar.

[70] M. A.S. Masoum, H. Dehbonei, E. F. Fuchs, Theoretical and experimental analyses of photovoltaic systems with voltage and current-based maximum power-point tracking, IEEE Trans. Energy Convers. 17 (4) (2002), pp. 514–522

[71] H. Koizumi and K. Kurokawa, A novel maximum power point tracking method for PV module integrated converter, in: IEEE 36th Power Electronics Specialists Conference, 2005, pp. 2081–2086

[72] Y. Goswami, F. Kreith, J. Kreder, Principles of Solar EngineeringFundamentals of Solar Engineering, Taylor&Francis, Philadelphia, 1999

[73] A. Canova, L. Giaccone, F. Spertino, Sun tracking for capture improvement, in: Proceedings 22nd European Photovoltaic Solar Energy Conference (EUPVSEC), WIP Renewable Energies, Milano, September 2007, pp. 3053–3058

[74] H. D. Mohring, F. Klotz, H. Gabler, Energy yield of PV tracking systems, in: Proceedings of the 21st European Photovoltaic Solar Energy Conference (EUPVSEC), WIP Renewable Energies, Dresden, September 2006, pp. 2691–2694

[75] C. Aracil, J. M. Quero, L. Castañer, R. Osuna, L. G. Franquelo, Tracking system for solar power plants, in: Proceedings of the IEEE 32nd Annual Conference in Industrial Electronics, November 2006, pp. 3024–3029

[76] M. Dominguez, I. Ameijeiras, L. Castaner, J.M. Wuero, A. Guerrero, L.G. Franquelo, A novel light source position sensor, Patent Number P9901375.

[77] D. A. Pritchard, Sun tracking by peak power positioning for photovoltaic concentrator arrays, IEEE Control Syst. Mag. 3 (3) (2003), pp. 2–8

[78] D. P. Hohm and M. E. Ropp, Comparative study of maximum power point tracking algorithms using an experimental, programmable, maximum power point tracking test bed, in: Proceedings of the 28th IEEE Photovoltaic Specialists Conference, 2000, pp. 1699–1702

[79] K. H. Hussein, Maximum photovoltaic power tracking: an algorithm for rapidly changing atmospheric conditions, IEEE Proc. Transm. Distrib. 142 (1) (1995), pp. 59–64

[80] D. O'Sullivan, H. Spruyt, A. Crausaz, PWM conductance control, in: Proceedings of the IEEE 19th Annual Power Electronics Specialists Conference, vol. 1, 1988, pp. 351–359

[81] L. Bangyin, D. Shanxu, L. Fei, X. Pengwei, Analysis and improvement of maximum power point tracking algorithm based on incremental conductance method for photovoltaic array, in: Proceedings of the IEEE 7th International Conference on Power Electronics and Drive Systems, November 2007, pp. 637–641

[82] D. P. Hohm and M. E. Ropp, Comparative study of maximum power point tracking algorithm, in: Proceedings of the 28th IEEE Photovoltaic Specialists Conference, September 2000, pp. 1699–1702

[83] T. Esram and P. L. Chapman, Comparison of photovoltaic array maximum power point tracking techniques, IEEE Trans. Energy Convers. 22 (2) (2007), pp. 439–449

[84] N. Femia, G. Petrone, G. Spagnuolo, M. Vitelli, Optimization of perturb and observe maximum power point tracking method, IEEE Trans. Power Electron. 20 (4) (2005), pp. 963–973

[85] J. H.R. Enslin and D. B. Snyman, Simplified feed-forward control of the maximum power point in PV installations, in: IEEE Industrial Electronics Conference, vol. 1, 1992, pp. 548–553

[86] A. S. Kislovski and R. Redl, Maximum-power-tracking using positive feedback, in: IEEE 25th Power Electronics Specialists Conference, vol. 2, 1994, pp. 1065–1068

[87] C. Y. Won, D. H. Kim, S. C. Kim, W. S. Kim, H. S. Kim, A new maximum power point tracker of photovoltaic arrays using fuzzy controller, in: IEEE Power Electronics Specialists Conference, 1994, pp. 396–403

[88] J. J. Schoeman and J. D. van Wyk, A simplified maximal power controller for terrestrial photovoltaic panel arrays, in: Proceedings of the 13th Annual IEEE Power Electronics Specialists Conference, 1982, pp. 361–367

[89] M. Buresch, Photovoltaic Energy Systems, McGraw-Hill, New York, 1983

[90] G. W. Hart, H. M. Branz, C. H. Cox, Experimental tests of open loop maximum power-point tracking techniques, Sol. Cells 13 (1984), pp. 185–195

[91] D. J. Patterson, Electrical system design for a solar powered vehicle, in: Proceedings of the 21st Annual IEEE Power Electronics Specialists Conference, 1990, pp. 618–622

[92] W. Xiao and W. G. Dunford, A modified adaptive hill climbing MPPT method for photo-voltaic power systems, in: Proceedings of the 35th Annual IEEE Power Electronics Specialists Conference, 2004, pp. 1957–1963

[93] H.-J. Noh, D.-Y. Lee, D.-S. Hyun, An improved MPPT converter with current compensation method for small scaled PV applications, in: Proceedings of the 28th Annual Conference on Industrial Electronics Society, 2002, pp. 1113-1118.

[94] K. Kobayashi, H. Matsuo, Y. Sekine, A novel optimum operating point tracker of the solar cell power supply system, in: Proceedings of the 35th Annual IEEE Power Electronics Specialists Conference, 2004, pp. 2147–2151

[95] B. Bekker and H. J. Beukes, Finding an optimal PV panel maximum power point tracking method, in: Proceedings of the 7th AFRICON Conference in Africa, 2004, pp. 1125–1129

[96] S. Yuvarajan and S. Xu, Photo-voltaic power converter with a simple maximum-power-point-tracker, in: Proceedings of the 2003 International Symposium on Circuits and Systems, 2003, pp. III-399–III-402

[97] R. M. Hilloowala and A. M. Sharaf, A rule-based fuzzy logic controller for a PWM inverter in photo-voltaic energy conversion scheme, in: Proceedings of the IEEE Industrial Application Society Annual Meeting, 1992, pp. 762–769

[98] C. -Y. Won, D. -H. Kim, S. -C. Kim, W. -S. Kim, H. -S. Kim, A new maximum power point tracker of photovoltaic arrays using fuzzy controller, in: Proceedings of the 25th Annual IEEE Power Electronics Specialists Conference, 1994, pp. 396–403

[99] T. Senjyu and K. Uezato, Maximum power point tracker using fuzzy control for photovoltaic arrays, in: Proceedings of the IEEE International Conference on Industrial Technologies, 1994, pp. 143–147

[100] G. -J. Yu, M. -W. Jung, J. Song, I. -S. Cha, I. -H. Hwang, Maximum power point tracking with temperature compensation of photovoltaic for air conditioning system with fuzzy controller, in: Proceedings of the IEEE Photovoltaic Specialists Conference, 1996, pp. 1429–1432

[101] M. G. Simoes, N. N. Franceschetti, M. Friedhofer, A fuzzy logic based photovoltaic peak power tracking control, in: Proceedings of the IEEE International Symposium on Industrial Electronics, 1998, pp. 300–305

[102] A. M.A. Mahmoud, H. M. Mashaly, S. A. Kandil, H. El Khashab, M. N.F. Nashed, Fuzzy logic implementation for photovoltaic maximum power tracking, in: Proceedings of the 9th IEEE International Workshop on Robot Human Interactive Communications, 2000, pp. 155–160

[103] N. Patcharaprakiti and S. Premrudeepreechacharn, Maximum power point tracking using adaptive fuzzy logic control for grid connected photovoltaic system, in: IEEE Power Engineering Society Winter Meeting, 2002, pp. 372–377

[104] B.M. Wilamowski, X. Li, Fuzzy system based maximum power point tracking for PV system, in: Proceedings of the 28th Annual Conference in IEEE Industrial Electronics Society, 2002, pp. 3280-3284.

[105] M. Veerachary, T. Senjyu, K. Uezato, Neural-network-based maximum-power-point tracking of coupled-inductor interleaved-boost converter-supplied PV system using fuzzy controller, IEEE Trans. Ind. Electron. 50 (4) (2003), pp. 749–758

[106] N. Khaehintung, K. Pramotung, B. Tuvirat, P. Sirisuk, RISC-microcontroller built-in fuzzy logic controller of maximum power point tracking for solar-powered light-flasher applications, in: Proceedings of the 30th Annual Conference of the IEEE Industrial Electronics Society, 2004, pp. 2673-2678.

[107] T. Hiyama, S. Kouzuma, T. Imakubo, Identification of optimal operating point of PV modules using neural network for real time maximum power tracking control, IEEE Trans. Energy Convers. 10 (2) (1995), pp. 360–367

[108] K. Ro and S. Rahman, Two-loop controller for maximizing performance of a grid-connected photovoltaic-fuel cell hybrid power plant, IEEE Trans. Energy Convers. 13 (3) (1998), pp. 276–281

[109] A. Hussein, K. Hirasawa, J. Hu, J. Murata, The dynamic performance of photovoltaic supplied dc motor fed from DC-DC converter and controlled by neural networks, in: Proceedings of the International Joint Conference on Neural Networks, 2002, pp. 607–612

[110] X. Sun, W. Wu, X. Li, Q. Zhao, A research on photovoltaic energy controlling system with maximum power point tracking, in: Proceedings of the Power Conversion Conference, 2002, pp. 822–826

[111] K. Samangkool and S. Premrudeepreechacharn, Maximum power point tracking using neural networks for grid-connected system, in: Proceedings of the International Conference on Future Power Systems, November 2005, pp. 1–4

[112] L. Zhang, Y. Bai, A. Al-Amoudi, GA-RBF neural network based maximum power point tracking for grid-connected photovoltaic systems, in: Proceedings of the International Conference on Power Electronics, Machines and Drives, 2002, pp. 18-23.

[113] P. Midya, P. T. Krein, R. J. Turnbull, R. Reppa, J. Kimball, Dynamic maximum power point tracker for photovoltaic applications, in: Proceedings of the 27th Annual IEEE Power Electronics Specialists Conference, 1996, pp. 1710–1716

[114] M. Bodur and M. Ermis, Maximum power point tracking for low power photovoltaic solar panels, in: Proceedings of the 7th Mediterranean Electrotechnical Conference, 1994, pp. 758–761

[115] T. Kitano, M. Matsui, D. -H. Xu, Power sensor-less MPPT control scheme utilizing power balance at DC link-system design to ensure stability and response, in: Proceedings of the 7th Annual Conference of IEEE Industrial Electronics Society, 2001, pp. 1309–1314

[116] M. Matsui, T. Kitano, D. -H. Xu, Z. -Q. Yang, A new maximum photovoltaic power tracking control scheme based on power equilibrium at DC link, in: Conference Record of the 1999 IEEE Industry Applications Conference, 1999, pp. 804–809

[117] S. B. Kjaer, J. K. Pedersen, F. Blaabjerg, Power inverter topologies for photovoltaic modules—a review, in: Proceedings of the IAS'02 Conference, vol. 2, 2002, pp. 782–788

[118] F. Blaabjerg, Z. Chen, S. B. Kjaer, Power electronics as efficient interface in dispersed power generation systems, IEEE Trans. Power Electron. 19 (5) (2004), pp. 1184–1194

[119] M. Meinhardt and G. Cramer, Past, present and future of grid connected photovoltaic- and hybrid-power-systems, in: Proceedings of the IEEE Power Engineering Society Summer Meeting, vol. 2, 2000, pp. 1283–1288

[120] T. Shimizu, M. Hirakata, T. Kamezawa, H. Watanabe, Generation control circuit for photo-voltaic modules", IEEE Trans. Power Electron. 16 (3) (2001), pp. 293–300

[121] M. Meinhardt and D. Wimmer, Multistring-converter: the next step in evolution of string-converter technology, in: Proceedings of the EPE'01 Conference, Graz, Austria, 2001

[122] B. K. Bose, P. M. Szczeny, R. L. Steigerwald, Microcomputer control of a residential photovoltaic power conditioning system, IEEE Trans. Ind. Appl. 21 (5) (1985), pp. 1182–1191

[123] B. Lindgren, Topology for decentralized solar energy inverters with a low voltage ac-bus, in: Proceedings of the European Conference on Power Electronics and Applications, 1999

[124] S. B. Kjaer and F. Blaabjerg, A novel single-stage inverter for AC-module with reduced low-frequency ripple penetration, in: Proceedings of the 10th European Conference on Power Electronics and Applications, 2003, pp. 2–4

[125] A. Lohner, T. Meyer, A. Nagel, A new panel-integratable inverter concept for grid-connected photovoltaic systems, in: Proceedings of the ISIE'96 Conference, vol. 2, 1996, pp. 827–831

[126] C. Dorofte, Comparative Analysis of Four dc/dc Converters for Photovoltaic Grid Intercon-nection & Design of a dc/dc Converter for Photovoltaic Grid Interconnection, Technical Report, Aalborg University, Aalborg, Denmark, 2001.

[127] S.J. Chiang, Design and Implementation of Multi-Functional Battery Energy Storage Sys-tems, Ph.D. Dissertation, Dep. Elect. Eng., National Tsing Hua University, Hsin-Chu, Taiwan, ROC, 1994.

[128] C. M. Liaw, T. H. Chen, S. J. Chiang, C. M. Lee, C. T. Wang, Small battery energy storage system, IEEE Proc. Electr. Power Appl. 140 (1) (1993), pp. 7–17

[129] H. Holttinen, P. Meibom, A. Orths, F. Van Hulle, C. Ensslin, L. Hofmann, J. McCann, J. Pierik, J. O. Tande, A. Estanquerio, L. Soder, G. Strbac, B. Parsons, J. C. Smith, B. Lemstrom, Design and operation of power systems with large amounts of wind power, first results of IEA collaboration, in: Proceedings of the Global Wind Power Conference, Adelaide, Australia, September 18-21, 2006

[130] World Wind Energy Association, Bonn, Germany, Press Release, February 21, 2008.

[131] Danish Wind Power Association, Available online: http://www.windpower.org/en.

[132] A. Betz, Introduction to the Theory of Machines, Oxford-Pergamon Press, Oxford, UK, 1966

[133] N. A. Ahmed and M. A. Miyateke, A Stand-alone hybrid generation system combining solar photovoltaic and wind turbine with simple maximum power point tracking control,

in: Proceedings of the 5th International Power Electronics and Motion Control Conference, vol. 1, August 2006, pp. 1–7

[134] J. R. Hendershot and T. J.E. Miller, Design of Brushless Permanent-Magnet Motors, Oxford Magna Physics Publications, England, 1994

[135] H.-Woo Lee, Advanced Control for Power Density Maximization of the Brushless DC generator, Ph.D. Dissertation, University of Arlington, Texas, 2003.

[136] M. A. Khan, P. Pillay, M. Malengret, Impact of direct-drive WEC systems on the design of a small PM wind generator, in: Proceedings of the IEEE Bologna Power Tech Conference, vol. 2, June 2003, p. 7

[137] L. Soderlund, J. -T. Eriksson, J. Salonen, H. Vihriala, R. Perala, A permanent-magnet generator for wind power applications, IEEE Trans. Magn. 32 (1996), pp. 2389–2392

[138] Tog Inge Reigstad: Direct Driven Permanent Magnet Synchronous Generators with Diode Rectifiers for Use in Offshore Wind Turbines, Ph.D. Dissertation, Norwegian University of Science and Technology Department of Electrical Power Engineering, June 2007.

[139] H. Li and Z. Chen, Optimal direct-drive permanent magnet wind generator systems for different rated wind speeds, in: Proceedings of the European Conference on Power Electronics and Applications, September 2007, pp. 1–10

[140] A. Grauers, Design of Direct-Driven Permanent-Magnet Generators for Wind Turbines, Ph.D. Dissertation, Chalmers University of Technology, Goteburg, 1996.

[141] J. Marques, H. Pinheiro, H.A. Gründling, J.R. Pinherio, H.L. Hey, A survey on variable-speed wind turbine system, in: Cientifico Greater Forum of Brazilian Electronics of Power, COBEP'03, Cortaleza, 2003, pp. 732-738.

[142] M. M. Neam, F. F.M. El-Sousy, M. A. Ghazy, M. A. Abo-Adma, The dynamic performance of an isolated self-excited induction generator driven by a variable-speed wind turbine, in: Proceedings of the International Conference on Clean Electric Power, May 2007, pp. 536–543

[143] G. S. Kumar and A. Kishore, Dynamic analysis and control of output voltage of a wind turbine driven isolated induction generator, in: Proceedings of the IEEE International Conference on Industrial Technology, December 2006, pp. 494–499

[144] M. Orabi, M. Z. Youssef, P. K. Jain, Investigation of self-excited induction generators for wind turbine applications, in: Proceedings of the Canadian Conference on Electrical and Computer Engineering, vol. 4, May 2004, pp. 1853–1856

[145] D. Seyoum, M. F. Rahman, C. Grantham, Inverter supplied voltage control system for an isolated induction generator driven by a wind turbine, in: Proceedings of the Industry Applications Conference (38th IAS Annual Meeting), vol. 1, October 2003, pp. 568–575

[146] E. Muljadi, J. Sallan, M. Sanz, C. P. Butterfield, Investigation of self-excited induction generators for wind turbine applications, in: Proceedings of the IEEE Industry Applications Conference (34th IAS Annual Meeting), vol. 1, October 1999, pp. 509–515

[147] T. Ahmed, O. Noro, K. Matsuo, Y. Shindo, M. Nakaoka, Wind turbine coupled three-phase self-excited induction generator voltage regulation scheme with static VAR compensator controlled by PI controller, in: Proceedings of the International Conference on Electrical Machines and Systems, vol. 1, November 2003, pp. 293–296

[148] T. Ahmed, O. Noro, E. Hiraki, M. Nakaoka, Terminal voltage regulation characteristics by static Var compensator for a three-phase self-excited induction generator, IEEE Trans. Ind. Appl. 40 (2004), pp. 978–988

[149] W. Qiao, G. K. Veneyagamoorthy, R. G. Harley, Real-time implementation of a STATCOM on a wind farm equipped with doubly fed induction generators, in: Proceedings of the IEEE Industry Applications Conference (41st IAS Annual Meeting), vol. 2, October 2006, pp. 1073–1080

[150] J. M. Elder, J. T. Boys, J. L. Woodward, The process of self-excitation in induction generators, IEEE Proc. B 130 (1983), pp. 103–108

[151] T. F. Chan, K. A. Nigim, L. L. Lai, "Voltage and frequency control of self-excited slip-ring induction generators," IEEE Trans. Energy Convers. 19 (2004), pp. 81–87.

[152] M.T. Abolhassani, H.A. Toloyat, P. Enjeti, Stator flux-oriented control of an integrated alternator/active filter for wind, in: Proceedings of the IEEE International Electric Machines and Drives Conference, vol. 1, June 2003, pp. 461-467.

[153] E.-H. Kim, S.-B. Oh, Y.-H. Kim, C.-S. Kim, Power control of a doubly fed induction machine without rotational transducers, in: Proceedings of the Power Electronics and Motion Control Conference, vol. 2, August 2000, pp. 951-955.

[154] H. Azaza, A. Masmoudi, On the dynamics and steady state performance of a vector controlled DFM drive systems, in: Proceedings of the IEEE International Conference on Man and Cybernetics, vol. 6, October 2002, p. 6.

[155] A. Tapia, G. Tapia, J. X. Ostolaza, J. R. Saenz, Modeling and control of a wind turbine driven DFIG, IEEE Trans. Energy Convers. 18 (2003), pp. 194–204

[156] L. Jiao, B. -T. Ooi, G. Joos, F. Zhou, Doubly-fed induction generator (DFIG) as a hybrid of asynchronous and synchronous machines, Electr. Power Syst. Res. 76 (2005), pp. 33–37

[157] S. Muller, M. Diecke, R. W. De Doncker, Doubly fed induction generator systems for wind turbines, IEEE Ind. Appl. Mag. 8 (2002), pp. 26–33

[158] R. Pena, J. C. Clare, G. M. Asher, A doubly fed induction generator using back-to-back PWM converters supplying an isolated load from a variable speed wind turbine, IEEE Proc. Electr. Power Appl. 143 (1996), pp. 380–387

[159] L. Xu, Y. Tang, Stator field oriented control of doubly-excited induction machine in wind power generation system, in: Proceedings of the 25th Mid West Symposium on Circuit and Systems, August 1992, pp. 1449-1466.

[160] L. Xu and W. Cheng, Torque and reactive power control of a doubly fed induction machine by position sensorless scheme, IEEE Trans. Ind. Appl. 31 (1995), pp. 636–642

[161] S. Doradla, S. Chakrovorty, K. Hole, A new slip power recovery scheme with improved supply power factor, IEEE Trans. Power Electron. 3 (1988), pp. 200–207

[162] R. Pena, J. Clare, G. Asher, Doubly fed induction generator using back-to-back PWM converters and its application to variable-speed wind-energy conversion, IEEE Proc. Electr. Power Appl. 143 (1996), pp. 231–241

[163] Y. Tang and L. Xu, A flexible active and reactive power control strategy for a variable speed constant frequency generating systems, IEEE Trans. Power Electron. 10 (1995), pp. 472–478

[164] A. Feijo, J. Cidrs, C. Carrillo, Third order model for the doubly-fed induction machine, Electr. Power Syst. Res. 56 (2000), pp. 121–127

[165] B. H. Chowdhury and S. Chellapilla, Double-fed induction generator control for variable speed wind power generation, Electr. Power Syst. Res. 76 (2006), pp. 786–800

[166] J. A. Sanchez, C. Veganzones, S. Martinez, F. Blazquez, N. Herrero, J. R. Wilhelmi, Dynamic model of wind energy conversion systems with variable speed synchronous generator and full-size power converter for large-scale power system stability studies, Renew. Energ. 33 (2008), pp. 1186–1198

[167] A. Bouscayrol, P. Delarue, X. Guillaud, Power strategies for maximum control structure of a wind energy conversion system with a synchronous machine, Renew. Energ. 30 (2005), pp. 2273–2288

[168] G. Raina and O. P. Malik, Variable speed wind energy conversion using synchronous machine, IEEE Trans. Aerosp. Electron. Syst. 21 (1985), pp. 100–105

[169] G. Raina and O. P. Malik, Wind energy conversion using a self-excited induction generator, IEEE Trans. Power App. Syst. 102 (1983), pp. 3933–3936

[170] B. Borowy and Z. Salameh, Dynamic response of a stand alone wind energy conversion system with battery energy storage to a wind gust, IEEE Trans. Energy Convers. 12 (1997), pp. 73–78

[171] Z. Chen and E. Spooner, Grid power quality with variable speed wind turbine, IEEE Trans. Energy Convers. 16 (2001), pp. 148–154

[172] J.A. Baroudi, V. Dinavahi, A.M. Knight, A review of power converter topologies for wind generators, in: IEEE International Conference on Electric Machines and Drives, May 2005, pp. 458-465.

[173] J.H.R. Enslin, J. Knijp, C.P.J. Jansen, P. Bauer, Integrated approach to network stability and wind energy technology for on-shore and offshore applications, in: Power Quality Conference, May 2003, pp. 185-192.

[174] L. Ran, J.R. Bumby, P.J. Tavner, Use of turbine inertia for power smoothing of wind turbines with a DFIG, in: Proceedings of the 11th International Conference on Harmonics and Quality Power, September 2004, pp. 106-111.

[175] K. Strunz, E.K. Brock, Hybrid plant of renewable stochastic source and multilevel storage for emission-free deterministic power generation, in: Proceedings of the Quality and Security Electric Power Delivery Systems CIGRE/IEEE PES International Symposium, October 8-10, 2003, pp. 214-218.

[176] J. P. Barton and D. G. Infield, Energy storage and its use with intermittent renewable energy, IEEE Trans. Energy Convers. 19 (2004), pp. 441–448

[177] R. Cardenas, R. Pena, G. Asher, J. Clare, Power smoothing in wind generation systems using a sensorless vector controlled induction machine driving a flywheel, IEEE Trans. Energy Convers. 19 (2004), pp. 206–216

[178] C. Abbey and G. Joos, Supercapacitor energy storage for wind energy applications, IEEE Trans. Ind. Appl. 43 (2007), pp. 769–776

[179] Ocean Energy: Technology Overview, Renewable Development Initiative, Available online: http://ebrdrenewables.com/sites/renew/ocean.aspx.

[180] W.J. Jones, M. Ruane, Alternative Electrical Energy Sources for Maine, Appendix I, Wave Energy Conversion by J. Mays, Report No. MIT-E1 77-010, MIT Energy Laboratory, July 1977.

[181] Ocean Energy, Report of the US Department of Interior Minerals Management Service, 2007.

[182] A. Wolfbrandt, Automated design of a linear generator for wave energy converters—a simplified model, IEEE Trans. Magn. 42 (7) (2007), pp. 1812–1819

[183] Ocean Energy: Technology Overview, Renewable Development Initiative, Available online: http://ebrdrenewables.com.

[184] D.A. Dixon, Fish and the Energy Industry, EPRI (Electric Power Research Institute) Research Report, ASMFC Energy Development Workshop, October 2006.

[185] Geothermal Energy, Power from the Depths, Technical Report by NREL for US Department of Energy, DOE/Gp-10097-518 FS18, December 8, 1997.

[186] S. Sheth, M. Shahidehpour, Geothermal energy in power systems, in: Proceedings of the IEEE Power Engineering Society General Meeting, vol. 2, June 2004.

[187] Renewable Energy into the Mainstream, International Energy Agency (IEA), Renewable Energy Working Party, Netherlands, October 2002.

[188] M.H. Dickson, M. Fanelli, Instituto di Geoscienze e Georisorse, Pisa, Italy, Available online: http://iga.igg.cnr.it/geothermal.php.

[189] G. W. Braun and H. K. McCluer, Geothermal power generation in United States, Proc. IEEE 81 (3) (1993), pp. 434–448

[190] Geothermal Energy Assessment, The World Bank Group, Available online: http://www.worldbank.org/html/fpd/energy/geothermal/assessment.

[191] Clean Energy Basics: Introduction to Geothermal Energy Production, NREL, http://www.nrel.gov/energy.

[192] W. P. Short, Trends in the American geothermal energy industry, Geotherm. Resour. Counc. Bull. 20 (9) (1991), pp. 245–259

[193] Geothermal Energy Facts, Advanced Level, Geothermal Education Office, Available online: http://www.geothermal.marin.org/geoenergy.

[194] DoE Fundamentals Handbook: Nuclear Physics and Reactor Theory, vol. 1-2, US Department of Energy, FSC-6010, January 1993.

[195] D. Bodansky, Nuclear Energy: Principles, Practices, and Prospects, second ed., Springer, Berlin, 2008

[196] Reactor Protection & Engineered Safety Feature Systems, The Nuclear Tourist, Available online: http://www.nucleartourist.com/systems/rp.htm.

[197] I. Hore-Lacy, Nuclear Energy in the 21st Century: World Nuclear University Press, first ed., Academic Press, London, UK, 2006

[198] C. Ma, F. von Hippel, Ending the production of highly enriched uranium for naval reactors, The Nonproliferation Review, James Martin Center for Nonproliferation, 2001.

[199] D. Hinds, C. Maslak, Next-Generation Nuclear Technology: The ESBWR, American Nuclear Society, Nuclear News, January 2006, pp. 35-40.

[200] J. Perkins, Fusion Energy: The Agony, the Ecstasy, and Alternatives, November 1997, http://PhysicsWorld.com, Available online: http://physicsworld.com/cws/article/print/1866.

[201] L. J.M.J. Blomen and M. N. Mugerwa, Fuel Cell Systems, Plenum, New York, 1993

[202] Fuel Cell Handbook, fifth ed., US Department of Energy, Office of Fossil Energy, National Energy Technology Laboratory, Morgantown, WV, 2000.

[203] R. Anahara, S. Yokokawa, M. Sakurai, Present status and future-prospects for fuel-cell power-systems, Proc. IEEE 81 (1993), pp. 399–408

[204] S. G. Meibuhr, Electrochim Acta 11 (1966), p. 1301

[205] J. Larminie, Fuel Cell Systems Explained, second ed., SAE International, West Sussex, England, 2003

[206] A.J. Appleby, Fuel cell, in: McGraw-Hill Encyclopedia of Science & Technology, ninth ed., v. 7, McGraw-Hill, New York, 2002, pp. 549-552.

[207] M. W. Ellis, M. R. Von Spakovsky, D. J. Nelson, Fuel cell systems: efficient, flexible energy conversion for the 21st century, Proc. IEEE 89 (12) (2001), pp. 1808–1818

[208] S. Rahman, Fuel cell as a distributed generation technology, in: Proceedings of the IEEE Power Engineering Society Meeting, vol. 1, June 2001, pp. 551-552.

[209] L. Gao, Z. Jiang, R. A. Dougal, An actively controlled fuel cell/battery hybrid to meet pulsed power demands, J. Power Sources 130 (1-2) (2004), pp. 202–207

[210] M. Uzunoglu and M. S. Alam, Dynamic modeling, design, and simulation of a combined PEM fuel cell and ultracapacitor systems for stand-alone residential applications, IEEE Trans. Energy Convers. 21 (3) (2006), pp. 767–775

[211] X. Yu, M. R. Starke, L. M. Tolbert, B. Ozpineci, Fuel cell power conditioning for electric power applications: a summary, IET Electr. Power Appl. 1 (5) (2007), pp. 643–656

[212] J. Midgett, Assessing a Hydrothermal Liquefaction Process using Biomass Feedstocks, M.Sc. Thesis, Department of Biological and Agricultural Engineering, Louisiana State

University, May 2008, Available online: http://etd.lsu.edu/docs/available/etd-01162008-150910/unrestricted/jasonmidgettthesis.pdf.

[213] E.D. Attanasi, R.F. Meyer, Natural Bitumen and Extra-Heavy Oil, Survey of Energy Resources, 22nd ed., World Energy Council, London, UK, 2010.

[214] M.R. Beychok, Process and Environmental Technology for Producing SNG and Liquid Fuels, U.S. EPA Report, EPA-660/2-75-011, May 1975.

[215] T. Kaneko, F. Derbyshire, E. Makino, D. Gray, M. Tamura, Coal Liquefaction, Ullmann's Encyclopedia of Industrial Chemistry, Wiley-VCH, Berlin, 2001

[216] P. Patel, A comparison of coal and biomass as feedstocks for synthetic fuel production, Alternative Energy Sources: An International Compendium, MIT Technology Review, 2007.

[217] P.R. Carlson, Power Generation Through Controlled Convection (Aeroelectric Power Generation), U.S. Patent, 3,894,393, issued on July 15, 2015.

[218] H. Alfvén, Existence of electromagnetic-hydrodynamic waves, Nature 150 (1942), p. 405

[219] F.J. Holler, D.A. Skoog, S.R. Crouch, Principles of Instrumental Analysis, Cengage Learning, Boston, MA, 2007 (Chapter 1).

Chapter 3

Photovoltaic System Conversion

Lana El Chaar

Electrical Engineering Department, The Petroleum Institute, Abu Dhabi, UAE

Chapter Outline

3.1 INTRODUCTION

For many years, fossil fuels have been the primary source of energy. However, due to the limited supply, the rate of deployment of fossil fuels is more rapid than their rate of production, and hence, fossil fuels will eventually run out. Moreover, the threat of global climate change caused by carbon dioxide (CO_2) emissions from fossil fuels is one of the main reasons for the increasing consensus to reduce the consumption of such fuels. This reduction can be achieved by switching to renewable energy for many energy-requiring applications, since it is "clean" and "green." Today, the global trend is to use nondepletable clean source of energy for a healthier and greener environment to save the future generation. The most efficient and harmless energy source is probably solar energy, which is so technically straightforward to use in many applications. Almost, all renewable energy sources, except nuclear and geothermal, are the energy forms originating from the solar energy.

Solar energy is considered one of the most promising energy sources due to its infinite power. Thus, modern solar technologies have been penetrating the market at faster rates, and photovoltaic (PV) technology that has the greatest impact, not because of the amount of electricity it produces but because photovoltaic cells – working silently, not polluting – can generate electricity wherever sun shines, even in places where no other form of electricity can be obtained [1]. PV is a combination of the Greek word for light and the name of the physicist Alessandro Volta [2]. PV is the direct conversion of sunlight into electricity by means of solar cells.

This chapter will highlight in brief how solar cells produce electricity and will discuss in detail the various techniques available to track the sun in order to maximize the output power generated by the PV array. Moreover, the various components required to operate PV systems efficiently will be described.

3.2 SOLAR CELL CHARACTERISTICS

Solar cells are composed of various semiconductor materials that become electrically conductive when supplied with heat or light. The majority of the first-generation solar cells produced are composed of silicon (Si), which exists in sufficient quantities. However, more than 95% of these cells have power conversion efficiency about 17% [4], whereas solar cells developed over the last decade in laboratory environment have efficiency as high as 31% [5]. All technologies related to capturing solar energy to be used as direct electricity generator are described as photovoltaic technology, which is subdivided into crystalline, thin film, and nanotechnology.

Doping technique is used to obtain excess of positive charge carriers (p-type) or a surplus of negative charge carriers (n-type). When two layers of different doping are in contact, then a p-n junction is formed on the boundary.

An internal electric field is built up causing the separation of charge carriers released by light, freeing electrons within the electric field proximity, which then pull the electrons from the p-side to the n-side (Fig. 3.1). The primary solar cell equivalent circuit (Fig. 3.2) contains a current source with a parallel diode, in addition to parasitic series (R_s; normally small) [6] and shunt (R_{sh}) resistances (relatively large) [7]. R_s is mainly affected by the factors such as the bulk resistance of the semiconductor material, metallic contacts, and interconnections, whereas R_{sh} is affected mainly by the p-n junction nonidealities and impurities near the junction [8].

A simplified equivalent circuit is shown in Fig. 3.3.

The diode current is given by the Shockley equation:

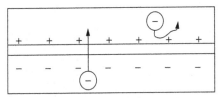

FIGURE 3.1 Effect of the Electric Field in a PV Cell [3].

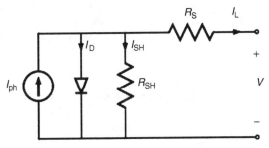

FIGURE 3.2 Solar cell equivalent circuit [8].

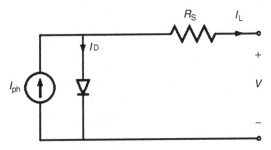

FIGURE 3.3 Model for a PV cell [9].

$$I_D = I_0 \left[\exp\left(\frac{qV}{nkT}\right) - 1 \right] \tag{3.1}$$

where I_0 is the reverse saturation current, q is the charge carrier, k is the Boltzman constant, T is the cell temperature, and n is the ideality factor.

The PV module has two limiting components (Fig. 3.3): open-circuit voltage (V_{oc}) and short-circuit current (I_{sc}). To determine I_{sc}, set $V = 0$ and $I_{sc} = I_{ph}$ Eq. (3.4), and this value changes proportionally to the cell irradiance. To determine V_{oc}, set the cell current $I_L = 0$, hence Eq. (3.3) leads to

$$V_{OC} = \frac{nkT}{q} \ln\left[\frac{I}{I_0}\right] \tag{3.2}$$

The PV module can also be characterized by the maximum point when the product (V_{mp} (voltage, where power is maximum) $\times I_{mp}$ (current, where power is maximum)) is at its maximum value. The maximum power output is derived by

$$\frac{d(V \times I)}{dt} = 0 \tag{3.3}$$

and

$$V_{mp} = V_{OC} - \frac{kT}{q} \ln \left[\frac{V_{mp}}{nkT/q} + 1 \right] \tag{3.4}$$

A PV module is normally rated using its W_p, which is normally $1 \, kW/m^2$ under standard test conditions (STC), which defines the PV performance at an incident sunlight of $1000 \, W/m^2$, a cell temperature of $25°C$ ($77°F$), and an air mass (AM) of 1.5. The product ($V_{mp} \times I_{mp}$) is related to the product generated by ($V_{OC} \times I_{SC}$) by a fill factor (FF) that is a measure of the junction quality and series resistance, and it is given by

$$FF = \frac{V_{mp} \times I_{mp}}{V_{OC} \times I_{SC}} \tag{3.5}$$

The closer the FF is to unity, the higher the quality of the PV module.

Finally, the last and most important factor of merit for a PV module is its efficiency (η), which is defined as

$$\eta = \frac{FF \times V_{OC} \times L_{OC}}{p_{in}} \tag{3.6}$$

P_{in} represents the incident power depending on the light spectrum incident on the PV cell.

To achieve the desired voltage and current levels, solar cells are connected in series (N_s) and parallel (N_p) combinations forming a PV module. The PV parameters are then affected as shown below [9]:

$$I_{phtotal} = N_p I_{ph} \tag{3.7}$$

$$I_{0total} = N_p I_0 \tag{3.8}$$

$$n_{total} = N_s n \tag{3.9}$$

$$R_{stotal} = \frac{N_s}{N_p} R_s \tag{3.10}$$

This model is shown in Figure 3.4.

In order to obtain the appropriate voltages and outputs for different applications, single solar cells are interconnected in series (for larger voltage) and in parallel (for larger current) to form the photovoltaic module. Then, several of these modules are connected to each other to form the photovoltaic array. This array is then fitted with aluminum or stainless steel frame and covered with transparent glass on the front side (Fig. 3.5).

The voltage generated by the array depends primarily on the design and materials of the cell, whereas the electric current depends primarily on the incident solar irradiance and the cell area. This current fluctuates since the path of the sun varies dramatically over the year, with winter and summer seasons being the two extreme excursions. The elevation angle of the sun ($\theta_{sun}^{elevation}$) is

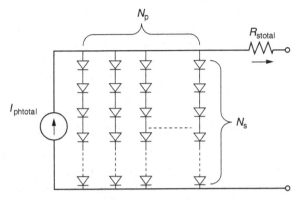

FIGURE 3.4 PV module circuit model.

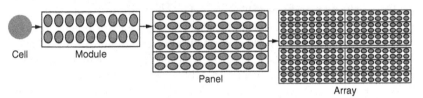

FIGURE 3.5 Photovoltaic cells, modules, panels, and array [10].

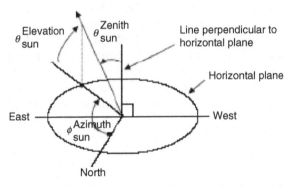

FIGURE 3.6 Azimuth, zenith, and elevation angles of a vector pointed toward the sun [11].

expressed in degrees above the horizon. Azimuth angle ($\phi_{sun}^{azimuth}$) of the sun is expressed in degrees from true north. Zenith angle (θ_{sun}^{zenith}) of the sun equals 90 degrees less than the elevation angle of the sun, or

$$\theta_{sun}^{zenith} = 90° - \theta_{sun}^{elevation} \tag{3.11}$$

Azimuth, zenith, and elevation angles are illustrated in Fig. 3.6

The output from a typical solar cell that is exposed to the sun, therefore, increases from zero at sunrise to a maximum at midday, and then falls again to zero at dusk. The radiation of the sun varies when reaching the surface of the earth due to absorption and scattering effect in the earth's atmosphere. PV system designers require the estimate of the insolation expected to fall on a randomly tilted surface, hence need a good evaluation of global radiation on a horizontal surface, horizontal direct and diffuse components, in order to estimate the amount of irradiation striking a tilted plane.

3.3 PHOTOVOLTAIC TECHNOLOGY OPERATION

Photovoltaic technology is used to produce electricity in areas where power lines do not reach. In developing countries, it helps improving living conditions in rural areas, especially in health care, education, and agriculture. In industrialized countries, such technology has been used extensively and integrated with the utility grid.

Photovoltaic arrays are usually mounted in a fixed position and tilted toward the south to optimize the noontime and the daily energy production. The orientation of fixed panel should be carefully chosen to capture the most energy for the season, or for a year. Photovoltaic arrays have an optimum operating point called the maximum power point (MPP) as shown in Fig. 3.7 [12].

It is noted that power increases as voltage increases, reaching a peak value and decreases as the resistance increases to a point where current drops off. According to the maximum power transfer theory, this is the point where the load is matched to the solar panel's resistance at a certain level of temperature and insolation. The I–V curve changes as the temperature and insolation levels change as shown in Fig. 3.8 and thus the MPP will vary accordingly [13].

It is shown that the open-circuit voltage increases logarithmically while the short-circuit current increases linearly as the insolation level increases [14]. Moreover, increasing the temperature of the cell decreases the open-circuit voltage and increases slightly the short-circuit current, causing reduction in the efficiency of the cell.

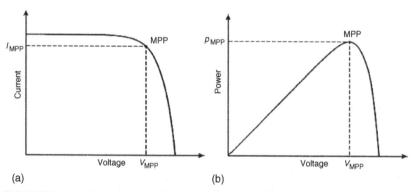

(a)   (b)

FIGURE 3.7 (a) I–V characteristic of a solar cell showing maximum power point (MPP); (b) P–V characteristic of a solar cell showing MPP.

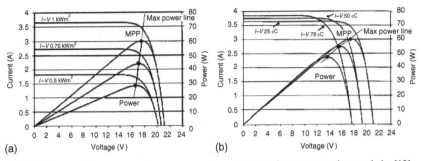

FIGURE 3.8   (a) PV panel insolation characteristics; (b) PV panel temperature characteristics [13].

The PV panels, usually mounted on the roof or at a near open area, are fixed to face the sun at an angle matching the country's latitude. If possible, seasonal adjustment of the module's direction toward the sun is done manually. Since solar power technology is relatively expensive, it is important to operate panels at their maximum power conditions. However, to collect as much solar radiation as possible, it is more convenient and efficient to use a sun tracking mechanism causing the module's surface to track the sun throughout the day.

The tracking can be along either one axis or two axes, whereby double axes tracking provides higher power output. The energy yield can be thus increased by about 20% to 30% depending on the seasonal climate and geographical location [15–17]. Although some claim that a fixed system costs less and requires almost no maintenance [18], different tracking mechanisms utilized to control the orientation of the PV panels have proved their superiority over fixed systems in terms of converted power efficiency.

To get maximum power from the PV panel at the prevailing temperature and insolation conditions, either the operating voltage or current should be controlled by a maximum power point tracker (MPPT) that should meet the following conditions [19]:

- Operate the PV system close to the MPP irrespective of the atmospheric changes.
- Have low cost and high conversion efficiency.
- Provide an output interface compatible with the battery-charging requirement.

3.4   MAXIMUM POWER POINT TRACKING COMPONENTS

The MPPT increases the energy that can be transferred from the array to an electrical system. The main function is to adjust the panel's output voltage to supply the maximum energy to the load. Most current designs consist of three basic components: switch-mode dc–dc converter, control system, and tracking component.

The switch-mode converter is the core of the entire supply because the energy drawn, stored as magnetic energy, is released at different potential levels. By setting up the switch-mode section in various topologies such as buck or boost converter, voltage converters are designed providing a fixed input voltage or current, which correspond to the maximum power point, allowing the output resistance to match the battery. To achieve the above-stated mechanism, a controller is essential to continuously monitor the PV system and ensure its operation at the PV maximum power point by tracking this MPP. The controller's aim is to continuously measure the voltage and current values generated from the PV, and compare them to certain treshhold values in order to apply either voltage controlled method or power feedback control [20].

3.4.1 Voltage Feedback Control

With the PV array terminal voltage being the controlled variable, voltage feedback controller forces the PV array to operate at its MPP by changing the array terminal voltage and neglecting the variation in the temperature and insolation level [20,21].

3.4.2 Power Feedback Control

In this method, power delivered to the load is the controlled variable. To achieve maximum power, dp/dv should be zero. This control scheme is not affected by the characteristics of the PV array, yet it increases power to the load and not power from the PV array [20,21]. Factors such as fast shadows may cause trackers to lose the MPP momentarily. It is very critical to ensure that the time lost in seeking MPP again, which equates the energy lost while the array is off power point, is very short. On the other hand, if lighting conditions do change, the tracker needs to respond within a short amount of time to the change avoiding energy loss. Therefore, the controller's most important feature is its capability to quickly adjust the system to operate back at the MPPT.

3.5 MPPT CONTROLLING ALGORITHMS

Several proposed algorithms to accomplish MPPT are described in the following sections.

3.5.1 Perturb and Observe (PAO)

The PAO method has a simple feedback structure and few measured parameters. It operates by periodically perturbing (i.e. incrementing or decrementing) the duty cycle while controlling the array current as shown in Fig. 3.9 and comparing

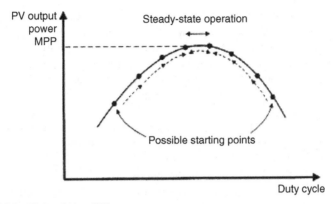

FIGURE 3.9 PAO technique [19].

the PV output power with that of the previous perturbation cycle. It measures the derivative of power Δp and the derivative of voltage Δv to determine the movement of the operating point. If the perturbation leads to an increase (or decrease) in array power, the subsequent perturbation is made in the same (or opposite) direction. This cost-effective technique can be easily implemented and is characterized by continuously tracking and very efficiently extracting the maximum power from PV. However, such method may fail under rapidly changing atmospheric conditions due to its slow tracking speed.

3.5.2 Incremental Conductance Technique (ICT)

The ICT process based on the fact that the derivative of the power with respect to the voltage (dp/dv) vanishes at the MPP because it is the maximum point on the curve as shown in Fig. 3.10.

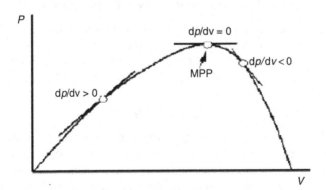

FIGURE 3.10 The slope "conductance" of the P–V curve [22].

The ICT algorithm detects the MPP by comparing di/dv against $\frac{-I}{V}$ till it attains the voltage operating point at which the incremental conductance is equal to the source conductance [23,24]. The Reference [23] describes in detail the ICT algorithm used for maximum power point tracking. The algorithm starts by measuring the present values of the I and V, then uses the corresponding stored value (I_b and V_b) measured during the preceding cycle to calculate the incremental changes as: $dI = I - I_b$ and $dV = V - V_b$. Based on the result obtained, the control reference signal V_{ref} will be adjusted in order to move the array voltage toward the MPP voltage. At the MPP, $di/dv = \frac{-I}{V}$, no control action is needed; therefore, the adjustment stage will be bypassed and the algorithm will update the stored parameters at the end of the cycle. In order to detect any changes in weather conditions, the algorithm detects whether a control action took place when the array was operating at the previous cycle MPP ($dv = 0$). This technique is accurate and well suited for rapid changes in atmospheric conditions; however, because the increment size approach is used to determine how fast the system is responding, ICT requires precise calculations of both instantaneous and increasing conductance.

3.5.3 Constant Reference

One very common MPPT technique is to compare the PV array voltage (or current) with a constant reference voltage (or current), which corresponds to the PV voltage (or current) at the maximum power point, under specific atmospheric conditions. The resulting difference signal (error signal) is used to drive a power conditioner, which interfaces the PV array to the load. Although the implementation of this method is simple, the method itself is not very accurate because it does not consider the effects of temperature and irradiation variations in addition to the difficulty in choosing the optimum point [19].

3.5.4 Current-Based Maximum Power Point Tracker

Current-based maximum power point tracker (**CMPPT**) is another MPPT technique that exists [22]. Employed numerical methods show a linear dependence between the "cell currents corresponding to maximum power" and the "cell short-circuit currents." The current I_{MPP} operating at the MPP is calculated using the following equation:

$$I_{MPP} = M_C I_{SC} \qquad (3.12)$$

where M_C is the "current factor" that differs from one panel to another and is affected by the panel surface conditions, especially if partial shading covers the panel [25]. Although this method is easy to implement, additional switch is added to the power converter to periodically short the PV array, increase the cost, and reduce the output power. This method also suffers from a major drawback due to periodic tuning requirement.

3.5.5   Voltage-Based Maximum Power Point Tracker

Similar to the above-mentioned method, voltage-based maximum power tracking (VMPPT) technique can also be applied [22]. The MPP operating voltage is calculated directly from V_{OC}

$$V_{MPP} = M_V V_{OC} \tag{3.13}$$

where M_V is the "voltage factor." The open-circuit voltage V_{OC} is sampled by an analogue sampler, and then V_{MPP} is calculated by Eq. (3.13). This operating V_{MPP} voltage is the reference voltage for the voltage control loop as shown in Fig. 3.11. This method always "results in a considerable power error because the output voltage of the PV module only follows the unchanged reference voltage during one sampling period" [9]. Albeit the implementation of this procedure is simple, it endures several disadvantages such as momentarily power converter shutdown causing power loss. Furthermore, such process depends greatly on the I–V characteristics and requires periodic tuning.

Other researchers argue that these two practices are considered to be "fast, practical, and powerful methods for MPP estimation of PV generators under all insolation and temperature conditions" [27].

3.5.6   Other Methods

Automated techniques such as Fibonacci line search, ripple correlation control method, neural network, and fuzzy logic have also been introduced for MPPT. In order to generate a clear understanding in determining the advantages and disadvantages of each algorithm, a comprehensive experimental comparison between different MPPT algorithms was made and run for the same PV setup at South Dakota State University [28], and results showed that the ICT method has the highest efficiency of 98% in terms of power extracted from the PV array, the PAO technique has the efficiency of 96.5%, and finally, the constant voltage method has the efficiency of 88%.

The ICT method provided good performance under rapidly changing weather conditions and provided the highest tracking efficiency, although four sensors were required to perform the measurements for computations and decision making [23]. If the system required more conversion time in tracking the MPP, a large amount of power loss would occur [20]. On the contrary, under

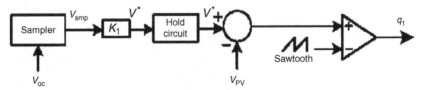

FIGURE 3.11   The conventional MPPT controller using open-circuit voltage V_{oc} [26].

perturb and observe method, losses are reduced if the sampling and execution speed were increased. The main benefit of this procedure is that only two sensors are required, which resulted in the reduction of hardware requirements and cost.

3.6 PHOTOVOLTAIC SYSTEMS' COMPONENTS

Once the PV array is controlled to perform efficiently, a number of other components are required to control, convert, distribute, and store the energy produced by the array. Such components may vary depending on the functional and operational requirements of the system. They may require battery banks and controller, dc–ac inverters, in addition to other components such as overcurrent, surge protection and, other processing equipment. Fig. 3.12 shows a basic diagram of a photovoltaic system and the relationship with each component.

Photovoltaic systems are classified into two major classes: grid-connected photovoltaic systems and stand-alone photovoltaic systems.

3.6.1 Grid-Connected Photovoltaic System

Grid-connected photovoltaic systems are composed of PV arrays connected to the grid through a power conditioning unit and are designed to operate in parallel with the electric utility grid as shown in Fig. 3.13. The power conditioning unit may include the MPPT, the inverter, the grid interface as well as the control system needed for efficient system performance [29] There are two general types of electrical designs for PV power systems: systems that interact with the utility power grid as shown in Fig. 3.13a and have no battery backup capability, and systems that interact and include battery backup as well as shown in Fig. 3.13b. The latter type of system incorporates energy storage in the form of a battery to keep "critical load" circuits operating during utility outage. When an outage occurs, the unit disconnects from the utility and powers specific circuits of the

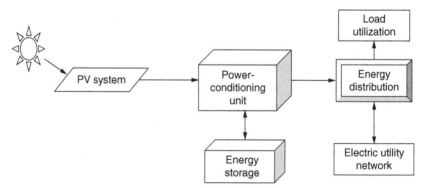

FIGURE 3.12   Major photovoltaic system components [8].

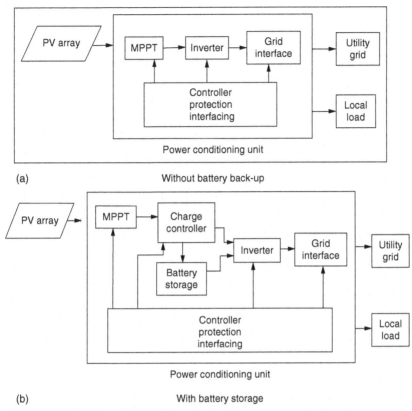

(a)   Without battery back-up

(b)   With battery storage

FIGURE 3.13   Grid-Connected PV system.

load. If the outage occurs in daylight, the PV array is able to assist the load in supplying the loads.

The major component in both systems is the DC-AC inverter or also called the power conditioning unit (PCU). The inverter is the key to the successful operation of the system, but it is also the most complex hardware. The inverter requirements include operation over a wide range of voltages and currents and regulated output voltage and frequency while providing AC power with good power quality which includes low total harmonic distortion and high power factor, in addition to highest possible efficiency for all solar irradiance levels. Several interconnection circuits have been described in [30,31]. Inverters can be used in a centralized connection (Fig 3.14a for the whole array of PV or each PV module string is connected to a single inverter (Fig. 3.14b [29]. The second proposed procedure is more efficient since it minimizes the losses due to voltage/current mismatching as well as it enhances it modularity capability. Moreover, the inverter may contain protective devices that monitor the grid and islands the grid from the PV system in case of fault occurrence [32].

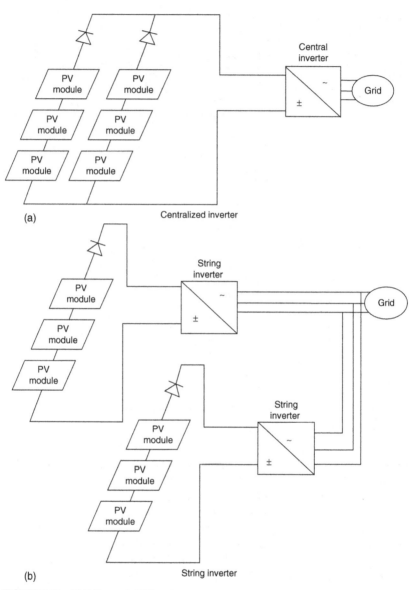

(a)   Centralized inverter

(b)   String inverter

FIGURE 3.14   Grid-Connected PV system.

For the last twenty years, researchers have been working on developing different inverter topologies that satisfy the above listed requirements. The evolution of solid state devices such as Metal Oxide semiconductor Field Effect Transistors (MOSFETs), Insulated Gate Bipolar Transistors (IGBTs), microprocessors, PWM integrated circuits have allowed improvements on the inverter. However, more research is being carried to ensure quality control, reliability and lower cost since inverters are the key for a sustainable photovoltaic market.

The main advantage of PV systems is their flexibility to be implemented in remote locations where grid connection is either impossible or very expensive to execute. Such systems are called stand-alone PV systems and are described in the following section.

3.6.2 Stand-Alone Photovoltaic Systems

Stand-alone photovoltaic systems are usually a utility power alternate. They generally include solar charging modules, storage batteries, and controls or regulators as shown in Fig. 3.15. Ground or roof-mounted systems will require a mounting structure, and if ac power is desired, an inverter is also required. In many stand-alone PV systems, batteries are used for energy storage as they may account for up to 40% of the overall stand-alone PV system cost over its lifetime [33].

These batteries cause losses in the PV system due to limited availability of time and energy to recharge the battery in addition to the insufficient battery maintenance. Hence, a charge controller is then used to control the system and prevent the battery from overcharging and overdischarging. Overcharging shortens the battery life and may cause gassing while undercharging may lead to sulphation and stratification, which result in the reduction in battery effectiveness and lifetime [34–37].

Batteries are often used in PV systems for storing energy produced by the PV array during daytime and supplying it to electrical loads as needed

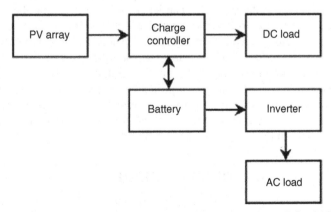

FIGURE 3.15   Diagram of stand-alone PV system with battery storage power DC and AC loads [8].

(during nighttime or cloudy weather). Moreover, batteries are also needed in the tracker systems to operate at MPP in order to provide electrical loads with stable voltages. Nearly, most of the batteries used in PV systems are deep cycle lead-acid batteries [38]. These batteries have thicker lead plates that make them tolerate deep discharges. The thicker the lead plates, the longer the life span of the batteries. The heavier the battery for a given group size, the thicker the plates and the better the battery will tolerate deep discharges [39].

All deep cycle batteries are rated in ampere-hour (AH) capacity, a quantity of the amount of usable energy it can store at nominal voltage [40]. A good charge rate is approximately 10% of the total capacity of the battery per hour. This will reduce the electrolyte losses and the damage to the plates [38]. A PV system may have to be sized to store a sufficient amount of power in the batteries to meet power demand during several days of cloudy weather, known as "days of autonomy." The Institute of Electrical and Electronics Engineers (IEEE) has set several guidelines and standards for sizing lead-acid batteries (IEEE Std 1013–1990) [41], for selecting, charging, and testing in stand-alone PV systems (IEEE Std 1361–2003) [42], and for installing and maintaining them (IEEE Std 937–2007) [43].

Nickel–cadmium batteries are also used for PV stand-alone systems but are often expensive and "may have voltage compatibility issues with certain inverters and charge controls" [44]. However, their main advantage is they are not affected by temperature as other battery types, hence mostly recommended for industrial or commercial applications in cold locations. IEEE has also drafted some guidelines for installation and maintenance (IEEE Std 1145–1999) [45].

To extend battery's lifetime and for efficient system's operation, a charge controller is needed to regulate the flow of electricity from the PV modules to the battery and the load. The controller keeps the battery fully charged without overcharging it. Many controllers have the ability to sense the excess of electricity drawn from batteries to the load and stop the flow until sufficient charge is restored to the batteries. The latter can greatly extend the battery's lifetime. However, controllers in stand-alone photovoltaic system are more complex devices that depend on battery state of charge, which in turn depends on many factors and is difficult to measure. The controller must be sized to handle the maximum current produced. Several characteristics should be considered before selecting a controller such as adjustable set-points including high-voltage and low-voltage disconnects, temperature compensation, low-voltage warning, and reverse current protection. Moreover, the controller should ensure that no current flows from the battery to the array at night.

3.7   FACTORS AFFECTING PV OUTPUT

PV systems produce power in proportion to the intensity of sunlight striking the solar array surface. Thus, there are some factors that affect the overall output of the PV system and are discussed below.

3.7.1   Temperature

Output power of a PV system decreases as the module temperature increases. For crystalline modules, a representative temperature reduction factor suggested by the California Energy Commission (CEC) is 89% in the middle of spring or in a fall day, under full-light conditions.

3.7.2   Dirt and Dust

Dirt and dust can accumulate on the solar module surface, blocking some of the sunlight and reducing the output. A typical annual dust reduction factor to use is 93%. Sand and dust can cause erosion of the PV surface, which affects the system's running performance by decreasing the output power to more than 10% [46–49].

3.7.3   DC–AC Conversion

Because the power from the PV array is converted back to ac as shown earlier, some power is being lost in the conversion process, in addition to losses in the wiring. Common inverters used have peak efficiencies of about 88–90%.

3.8   PV SYSTEM DESIGN

The goal for a solar direct electricity generation system or photovoltaic system is to provide high-quality, reliable, and green electrical power.

3.8.1   Criteria for a Quality PV System

The criteria for quality PV system are as follows:

- Be properly sized and oriented to provide electrical power and energy
- Good control circuit to reduce electrical losses, overcurrent protection, switches, and inverters
- Good charge controller and battery management system, should the system contain batteries

3.8.2   Design Procedures

The first task in designing a PV system is to estimate the system's load. This is achieved by defining the power demand of all loads, the number of hours used per day, and the operating voltage [50]. From the load ampere-hours and the given operating voltage for each load, the power demand is calculated. For a stand-alone system, the system voltage is the potential required by the largest load. When ac loads dominate, the dc system voltage should be chosen to be compatible with the inverter input.

3.8.3 Power-Conditioning Unit

The choice of the PCU has a great impact on the performance and economics of the system. It depends on the type of waveform produced, which in turn depends on the method used for conversion, as well as the filtering techniques of unwanted frequencies. Several factors must be considered when selecting or designing the inverter:

- The power conversion efficiency
- Rated power
- Duty rating, the amount of time the inverter can supply maximum load
- Input voltage
- Voltage regulation
- Voltage protection
- Frequency requirement
- Power factor

3.8.4 Battery Sizing

The amount of battery storage needed depends on the load energy demand and on weather patterns at the site. There is always a trade-off between keeping cost low and meeting energy demand.

SUMMARY

This chapter discussed the conversion of solar energy into electricity using photovoltaic system. There are two types of PV systems: the grid-connected PV system and the stand-alone PV system. All major components for such systems have been discussed. Maximum power point tracking is the most important factor in PV systems to provide the maximum power. For this reason, several tracking systems have been described and compared. Factors affecting the output of such systems have been defined and steps for a good and reliable design have been considered.

REFERENCES

[1] www.worldenergy.org/.
[2] Photovoltaics: Solar Electricity and Solar Cells in Theory and Practice, www.solarserver.de/ wissen/photovoltaic-e.html.
[3] J. Toothman, S. Alsous, How Solar Cells Work, http://science.howstuffworks.com/ environmental/energy/solar-cell.htm (accessed October 2010).
[4] H. Moller, Semiconductors for Solar Cells, Artech House, Inc., London, 1993.
[5] Berkeley Lab, http://www.lbl.gov/msd/pis/walukiewicz/02$\delimiter"026E30F$02_08_full_ solar_spectrum.html.
[6] D. Sera, R. Teodorescu, P. Rodriguez, PV panel model based on datasheet values, in: Industrial Electronics, ISIE 2007, IEEE International Symposium Volume, 2007, pp. 2392–2396.

[7] C. Chu, C. Chen, Robust maximum power point tracking method for photovoltaic cells: a sliding mode control approach, Sol. Energy 83.8 (2009) 1370–1378.

[8] EE362L, Power Electronics, Solar Power, I-V Characteristics, Version October 14, 2005, http://www.ece.utexas.edu/~grady/EE362L_Solar.pdf.

[9] D. Lee, H. Noh, D. Hyun, I. Choy, An improved MPPT converter using current compensation method for small scaled PV-Applications, in: Applied Power Electronics Conference and Exposition (APEC'03), vol. 1, February 2003, pp. 540–545.

[10] Photovoltaic Fundamentals, http://www.fsec.ucf.edu/PVT/pvbasics/index.htm.

[11] The Solar Sprint PV Panel, http://chuck-wright.com/SolarSprintPV/SolarSprintPV.html.

[12] M.A. Serhan, Maximum Power Point Tracking system: An Adaptive Algorithm for Solar Panels, Thesis, American University of Beirut, January 2005.

[13] J.H.R. Enslin, M.S. Wolf, D.B. Snyman, W. Swiegers, Integrated photovoltaic maximum power point tracking converter, IEEE Trans. Ind. Electron. 44(6) (1997) 769–773.

[14] A.D. Hansen et al., Models for a Stand Alone PV System, Technical Report Riso National Laboratory, Roskilde, Norway, December 2000, http://www.risoe.dk/solenergu/rapporter/pdf/sec-r-12.pdf.

[15] K.S. Karimov, J.A. Chattha, M.M. Ahmed, et al., Journal of references, Acad. Sci. Tajikistan XLV(9) (2002) 75–83.

[16] A.A. Khalil, M. El-Singaby, Position control of sun tracking system; circuits and systems, in: Proceedings of the 46th IEEE International Midwest Symposium, vol. 3, December 27–30, 2003, pp. 1134–1137.

[17] W. Dankoff, Glossary of Solar Water Pumping Terms and Related Components, 2009, Available at: www.conergy.us/Desktopdefault.aspx/tabid-332/449_read-3816/.

[18] R.A. Bentley, Global oil depletion—methodologies and results, in: Proceedings of the 3rd International Workshop on Oil and Gas Depletion, Australian Association for the Study of Peak Oil and Gas (ASPO), Berlin, Germany, 2004, pp. 25–26.

[19] E. Kroutoulis, K. Kalaitzakis, N.C. Voulgaris, Development of a microcontroller-based, photovoltaic maximum power tracking control system, IEEE Trans. Power Electron. 16(1) (2001) 46–54.

[20] C. Hua, C. Shen, Comparative study of peak power tracking techniques for solar storage system, in: IEEE Applied Power Electronics Conference and Exposition, vol. 2, February 1998, pp. 679–685.

[21] C. Hua, J. Lin, DSP-based controller in battery storage of photovoltaic system, in: IEEE IECON 22nd International Conference on Industrial Electronics, Control and Instrumentation, vol. 3, 1996, pp. 1705–1710.

[22] Liu Shengyi, Maximum Power Point Tracker Model, Control Model, University of South Carolina, May 2000, Available from: http://vtb.engr.sc.edu/modellibrary_old.

[23] K.H. Hussein, I. Mutta, T. Hoshino, M. Osakada, Maximum photovoltaic power tracking: an algorithm for rapidly changing atmospheric conditions, IEE Proc. Gen. Transm. Distrib. 142(1) (1995) 59–64.

[24] K.K. Tse, H.S.H. Chung, S.Y.R. Hui, M.T. Ho, Novel maximum power point tracking technique for PV panels, in: IEEE Power Electronics Specialists Conference, vol. 4, June 2001, pp. 1970–1975.

[25] T. Noguchi, S. Togashi, R. Nakamoto, Short-current pulse-based maximum power point tracking method for multiple photovoltaic-and-converter module system, IEEE Trans. Ind. Electron. 49(1) (2002) 217–223.

[26] HowStuffworks, http://science.howstuffworks.com/solar-cell5.html.

[27] M. Masoum, H. Dehbonei, E. Fuschs, Theoretical and experimental analyses of photovoltaic systems with voltage-and current-based maximum power point tracking, IEEE Trans. Convers. 17(4) (2002) 514–522.

[28] D. Hohm, M. Ropp, Comparative study of maximum power point tracking algorithms using an experimental, programmable, maximum point test bed, in: IEEE Photovoltaic Specialists Conference, September 2000, pp. 1699–1702.

[29] G. Chicco, R. Napoli, F. Spertino, Experimental evaluation of the performance of grid-connected photovoltaic systems, in: IEEE Melecon, Dubrovnik, Croatia, 2004, pp. 1011–1016.

[30] N. Mohan, T. Undeland, W. Robbins, Power Electronics: Converter, Applications and Design, third ed., John Wiley and Sons, New York, 2003.

[31] M.H. Rashid, Power Electronics: Circuits, Devices, and Applications, Prentice Hall, 2004.

[32] R. Carbone, Grid-connected photovoltaic systems with energy storage, in: International Conference on Clean Electrical Power, Capri, Italy, 2009, pp. 760–767.

[33] S. Duryea, S. Islam, W. Lawrance, A battery management system for stand-alone photovoltaic energy systems, IEEE Ind. Appl. Mag. 7(3) (1999) 67–72.

[34] E. Lorenzo, L. Narvarte, K. Preiser, R. Zilles, A field experience with automotive batteries in SHS's, in: Proceeding of the 2nd World Conference and Exhibition on Photovoltaic Solar Energy Conversion, Vienna, 1998, pp. 3266–3268.

[35] P. Diaz, M.A. Egido, Experimental analysis of battery charge regulation in photovoltaic systems, Prog. Photovolt. Res. Appl. 11(7) (2003) 481–493.

[36] H. Yang, H. Wang, G. Chen, G. Wu, Influence of the charge regulator strategy on state of charge and lifetime of VRLA battery in household photovoltaic systems, Sol. Energy 80 (2006) 281–287.

[37] J. Garche, A. Jossen, H. Doring, The influence of different operating conditions, especially over-discharge, on the lifetime and performance of lead acid batteries for photovoltaic systems, J. Power Sources 67 (1997) 201–212.

[38] J. Enslin, D. Snyman, Combined low cost, high efficient inverter, peak power tracker and regulator for PV applications, IEEE Trans. Power Electron. 6(1) (1991) 73–82.

[39] D. Linden, Handbook of Batteries, McGraw-Hill, New York, 1995.

[40] W. Jian, L. Jianzheng, Z. Zhengming, Optimal control of solar energy combined with MPPT and battery charging, in: Proceedings of IEEE International Conference on Electrical Machines and Systems, vol. 1, November 2003, pp. 285–288.

[41] IEEE Recommended Practice for Sizing Lead-Acid Batteries for Stand-Alone Photovoltaic (PV) systems, E-ISBN: 0-7381-2990-9, ISBN: 1-55937-068-8, http://ieeexplore.ieee.org/stamp/stamp.jsp?tp=&arnumber=210970.

[42] IEEE Guide for Selection, Charging, Test, and Evaluation of Lead-Acid Batteries Used in Stand-Alone Photovoltaic (PV) Systems, E-ISBN: 0-7381-3581-X, PDF: ISBN: 0-7381-3581-X SS95086, http://ieeexplore.ieee.org/stamp/stamp.jsp?tp=&arnumber=1263341.

[43] IEEE Recommended Practice for Installation and Maintenance of Lead-Acid Batteries for Photovoltaic (PV) Systems, E-ISBN: 978-0-7381-5592-0, ISBN: 978-0-7381-5591-3, http://ieeexplore.ieee.org/stamp/stamp.jsp?tp=&arnumber=4238866.

[44] Solar Energy International, Photovoltaics Design and Installations, New Society Publishers, 2008.

[45] EEE Recommended Practice for Installation and Maintenance of Nickel-Cadmium Batteries for Photovoltaic (PV) System, E-ISBN: 0-7381-1088-4, ISBN: 1-55937-072-6, http://ieeexplore.ieee.org/stamp/stamp.jsp?tp=&arnumber=89824.

[46] J.P. Thornton, The Effect of Sand-Storm on Photovoltaic Array and Components, in: Solar Energy Conference, 1992.
[47] L. Chaar, A. Jamaleddine, F. Ajmal, H. Khan, Effect of wind blown sand and dust on PV arrays especially in UAE, in: Power Systems Conference (PSC), South Carolina, USA, March 2008.
[48] B. Mohandes, L. El-Chaar, L. Lamont, Application study of 500 W photovoltaic (PV) system in the UAE, Appl. Sol. Energ. J. 45(4) (2009) 242–247.
[49] T. Al Hanai, R. Bani Hashim, L. El-Chaar, L. Lamont, Study of a 900 W, thin-film, amorphous silicon PV system in a dusty environment, in: International Conference on Renewable Energy: Generation and Applications—ICREGA'10, Al-Ain, UAE, March 2010.
[50] Sandia National Laboratories: Stand Alone Photovoltaic Systems: A Handbook of Recommended Design Practices, National Technical Information Service, Springfield, VA, 1988.

Chapter 4

Wind Turbine Applications

Juan M. Carrasco, Eduardo Galván, and Ramón Portillo
Department of Electronic Engineering, Engineering School, Seville University, Spain

Chapter Outline

4.1 WIND ENERGY CONVERSION SYSTEMS

Wind energy has matured to a level of development where it is ready to become a generally accepted utility generation technology. Wind turbine technology has undergone a dramatic transformation during the last 15 years, developing from a fringe science in the 1970s to the wind turbine of the 2000s using the latest in power electronics, aerodynamics, and mechanical drive train designs [1,2].

Most countries have plans for increasing their share of energy produced by wind power. The increased share of wind power in the electric power system makes it necessary to have grid-friendly interfaces between the wind turbines and the grid in order to maintain power quality.

In addition, power electronics is undergoing a fast evolution, mainly due to two factors. The first one is the development of fast semiconductor switches, which are capable of switching quickly and handling high powers. The second factor is the control area, where the introduction of the computer as a real-time controller has made it possible to adapt advanced and complex control algorithms. These factors together make it possible to have cost-effective and grid-friendly converters connected to the grid [3,4].

4.1.1 Horizontal-axis Wind Turbine

A horizontal-axis wind turbine is the most extensively used method for wind energy extraction. The power rating varies from a few watts to megawatts on large grid-connected wind turbines.

In relation to the position of the rotor regarding the tower, the rotors are classified as leeward (rotor downstream the tower) or windward (rotor upstream the tower), this last configuration being the most widely used.

These turbines consist of a rotor, a gearbox, and a generator. The group is completed with a nacelle that includes the mechanisms, as well as a tower holding the whole system and hydraulic subsystems, electronic control devices, and electric infrastructure as it is shown in Fig. 4.1 [1]. A photograph of a real horizontal-axis wind turbine is shown in Fig. 4.2. We will briefly explain the above-mentioned devices.

4.1.1.1 The Rotor

The rotor is the part of the wind turbine that transforms the energy from the wind into mechanical energy [1]. The area swept by the rotor is the area that captures the energy from the wind. The parameter measuring the influence of the size of the capturing area is the ratio area/rated power. Thus, for the same installed power, more energy will be delivered if this ratio is greater, and so, more equivalent hours (kWh/kW). Values for this ratio close to 2.2 m^2/kW are found today in locations with high average wind speed (>7 m/s), but there is a trend to elevate this ratio above 2.5 m^2/kW for certain locations of medium and low potential. In this case, the technical limits are the high tangential speed at the tip

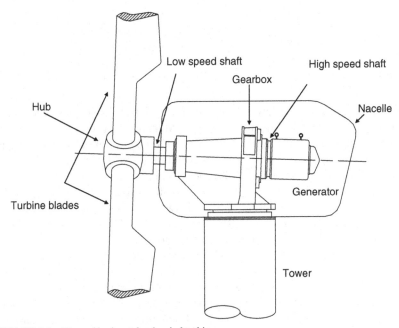

FIGURE 4.1 View of horizontal-axis wind turbine.

FIGURE 4.2   Wind turbine in Monteahumada (Spain). Made S.A. AE-41PV.

of the blade, that force to lower the speed of the rotors, hence the variable speed and the technology used are most important. Making a bigger rotor for a certain wind turbine involves the possibility of using it for a lower wind speed location by compensating wind loss with a bigger capturing area. The rotor consists of a shaft, blades, and a hub, which holds the fastening system of the blades to the shaft. The rotor and the gearbox form the so-called drive train.

A basic classification of the rotors is between constant pitch and variable-pitch machines, according to whether the type of tie of the blade to the hub is constant or whether it allows rotation to the rotor axis.

The pitch control of a wind turbine makes it possible to regulate energy extraction at high speed wind condition. On the other hand, the use of variable speed makes the systems more expensive to build and maintain.

The use of variable-speed generators (other than 50 Hz of the grid), allows the reduction of sudden load surges. This condition differentiates between constant speed and variable speed generators. The hub includes the blade pitch controller in case of variable pitch, and the hydraulic brake system in case of constant pitch. The axis to which the hub is tied to the so-called low speed shaft is usually hollow which allows for the hydraulic conduction for regulation of the power by varying the blade pitch or by acting on the aerodynamic brakes in case of constant pitch.

4.1.1.2   The Gearbox

The function of the gearbox, shown in Fig. 4.1, is to adapt a low rotation speed of the rotor axis to a higher one in the electric generator [1,2]. The gearbox may have parallel or planetary axis. It consists of a system of gears that connect the low speed shaft to the high speed shaft connected to the electric generator by a coupler. In some cases, using multi-pole, the gearbox is not necessary.

4.1.1.3   The Generator

The main objective of the generator is to transform the mechanical energy captured by the rotor of the wind turbine into electrical energy that will be injected into the utility grid.

Asynchronous generators are commonly used in wind turbine applications with fixed speed or variable speed control strategies. Also, in large power wind turbine applications synchronous machines are used. In the asynchronous generator, the electric energy is produced in the stator when the rotating speed of the rotor is higher than the speed of the rotary field of the stator. The asynchronous generator needs to take energy from the grid to create the rotary field of the stator. Because of this, the power factor is decreased and so a capacitor bank is needed. The synchronous generator with an excitation system includes electromagnets in the rotor that generate the rotating field. The rotor electromagnets are fed back with a DC current by rectifying part of the electricity generated. Another kind of generator recently used is the permanent magnet [5]. This type of machine does not need an excitation system, and it is used mainly for low power wind turbine applications. The advantages of using an asynchronous generator are low cost, robustness, simplicity, and easier coupling to the grid, yet its main disadvantage is the necessity of a power factor compensator and a lower efficiency.

4.1.1.3.1   Induction Machine

The induction generator, as can be deducted from its torque/speed characteristic, has a nearly constant speed in a wide working torque range, as they are positive (working as a motor) or negative (working as a generator). This characteristic curve is very useful for machines with constant speed, as the machine is auto-regulated to keep the synchronous frequency. But the situation is very different when we proceed to change the speed of the generator. It is then necessary to use power converters in order to adapt the generator frequency to the frequency of the grid [3,6,7]. The general principles applicable to change the speed of an induction generator can be deduced from the following equation:

$$N_r = N_1 \cdot (1 - s) = \frac{60 \cdot f_1}{p} \cdot (1 - s) \tag{4.1}$$

where N_r is the generator speed (rpm), N_1 is the generator synchronous speed, s is the induction generator slip, p is the pole pair number, and f_1 is the excitation stator frequency (Hz).

From Eq. (4.1), it is immediately inferred that the speed can be controlled in either ways; one way is changing the synchronous speed and the other is changing the slip. The speed is deduced from the number of pole pairs p and the supplying frequency into the machine f_1. The slip can be easily changed when modifying the torque/slip characteristic curve. This modification can be achieved as follows: first, by changing the input voltage of the generator; second, by changing the resistance of the rotor circuit; and third, by injecting a voltage into the rotor so that it has the same frequency as the electromotive force induced in it and an arbitrary magnitude and phase. The techniques used to vary the supplying frequency permit a wide range of variation of the speed, from 0 to 100% or even greater than the synchronous speed. Another variable-speed technique is achieved by changing the number of poles which permit a regulation of the speed in discrete steps. If we proceed to vary the slip, then the range of variation of the speed is within a narrow margin of regulation.

Among all these techniques, only the variation of the voltage can be actually implemented using a squirrel cage machine with a short-circuited rotor. The rest are implemented by means of a wound-rotor machine.

The stator voltage can be varied by means of a power converter [4,8,9]. This converter should be connected in series to the generator and to the grid. Since it is only necessary to vary the voltage of the generator and not its frequency, an AC–AC converter can be used. Furthermore, the power converter bears all the power of the generator so it deals with all the disadvantages of the other wide-range of control techniques.

For the speed to be varied by changing the slip, it is necessary to work with wound rotor induction machines.

4.1.1.3.2 Synchronous Machine with Excitation System

As it is well-known, the general principle to change the speed of a synchronous machine is summarized in the following equation [3,6,7]:

$$N_r = N_1 = \frac{60 \cdot f_1}{p} \tag{4.2}$$

The only way to control the speed is by changing the number of pole pairs or by supplying frequency into the machine, f_1. Therefore, wide range or discrete steps are permitted. The synchronous machine will always be controlled in a wide range of the rotor speed, ω_r. In this kind of system, the excitation current permits an easier torque and power control.

4.1.1.3.3   Permanent Magnet Synchronous Machine

As with the synchronous generator with excitation system, the permanent magnet synchronous machine can be controlled in a wide range of rotor speeds ω_r. In this case, a magnetic field control has to be made from the power converter. The advantage of this machine is better performance and less complexity [3,6,7,10].

4.1.1.4   Power Electronic Conditioner

The power electronic conditioner is a converter that is mainly used in variable speed applications. This converter is connected between the generator machine and the utility grid by an isolating transformer and permits different frequency and voltage levels in its input and output. The power converter is connected to the stator voltage or to the rotor of a wound rotor machine. This system includes large power switches that can be GTOs, Thyristors, IGCTs, or IGBTs arranged in different topologies.

4.1.2   Simplified Model of a Wind Turbine

The mechanical power P_m in the low speed shaft can be expressed as a function of the available power in the wind P_v by the Eq. (4.3):

$$P_m = C_p\,(\lambda, \beta) \cdot P_w \tag{4.3}$$

where $C_p\,(\lambda, \beta)$ is the power coefficient, which is a function of the blade angle β and the dimensionless variable $\lambda = \omega_L R / v_w$ (where ω_L is the angular speed on the low speed shaft, R is the turbine radius, and v_w the wind speed). In Fig. 4.3 an analytical approximation of the power coefficient $C_p(\lambda, \beta)$ is shown.

In Fig. 4.4 the power characteristic of a wind turbine P_m is shown.

The power of the wind can be expressed by the following equations [1,2]:

$$P_w = \frac{1}{2}\rho\pi R^2 v_w^3 \tag{4.4}$$

where ρ is the air density.

Substitution of Eq. (4.4) in Eq. (4.3) and including λ in such expression, the following can be obtained:

$$Q_L = \frac{C_p}{2\lambda^3}\rho\pi R^5 \omega_L^2 \tag{4.5}$$

where Q_L is the torque in the low speed shaft that the wind turbine draws from the wind.

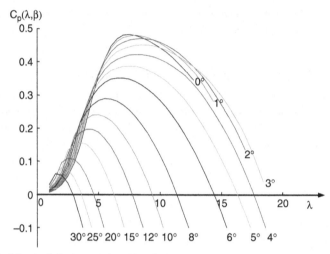

FIGURE 4.3 Analytical approximation of $C_p(\lambda, \beta)$ characteristic (blade pitch angle β as parameter).

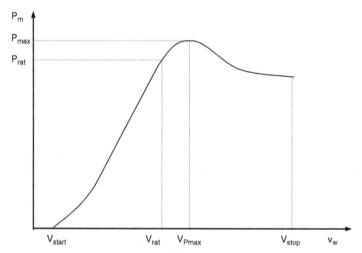

FIGURE 4.4 Power characteristic of the wind turbine.

This Eq. (4.5) is represented in Fig. 4.5.

Neglecting mechanical losses, the total torque on the high speed shaft, Q_t is equal to the torque in the low speed shaft, Q_L, divided by the gearbox ratio, G.

$$Q_t = \frac{Q_L}{G} \tag{4.6}$$

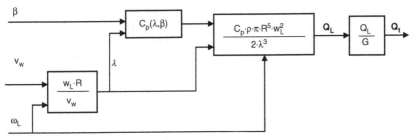

FIGURE 4.5   Torque calculation block diagram.

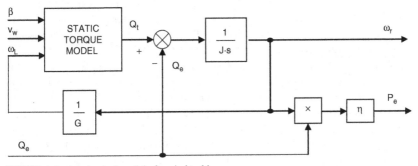

FIGURE 4.6   Mechanical model of a wind turbine.

Equation (4.7) shows the differential equation for the dynamics of the rotational speed that depends on the difference of load and generator torque.

$$Q_t - Q_e = J\frac{d\omega_r}{dt} \tag{4.7}$$

In Eq. (4.7) J is the total inertia of the system referred to the high speed shaft.

Figure 4.6 shows the block diagram of the simplified mechanical model of a wind turbine. Also, it has been represented by the electrical power P_e, obtained by multiplying the electrical torque Q_e by the rotor speed ω_r and the electrical performance η.

4.1.3   Control of Wind Turbines

Many horizontal axes, grid-connected, and medium- to large-scale wind turbines are regulated by pitch control, and most of the wind turbines built so far have practically constant speed, since they use an AC generator, directly connected to the distribution grid, which determines its speed of rotation.

In the last few years, variable speed control has been added to pitch-angle control design in order to improve the performance of the system [11]. Variable speed operation of a wind turbine has important advantages vs the constant speed ones. The main advantages of variable speed wind tunnel are the reduction of electric power fluctuations by changes in kinetic energy of the rotor, the potential reduction of stress loads on the blades and the mechanical transmissions, and the possibility to tune the turbine to local conditions by adjusting the control parameters. On the other hand, variable speed control is normally used with fixed pitch angle and very few applications using both controls have been reported [12,13].

In short, four different wind turbine types are provided depending on the controller [14]:

- No control. The generator is directly connected to a constant frequency grid, and the aerodynamics of the blade is used to regulate power in high winds.
- Fixed speed pitch regulated. In this case, the generator is also directly connected to a constant frequency grid, and pitch control is used to regulate power in high winds.
- Variable speed stall regulated. A frequency converter decouples the generator from the grid, allowing the rotor speed to be varied by controlling the generator reaction torque. In high winds, this speed control capability is used to slow the rotor down until aerodynamic stall limits the power to a desired level [15].
- Variable speed pitch regulated. A frequency converter decouples the generator from the grid, allowing the rotor speed to be varied by controlling the generator reaction torque. In high winds, the torque is held at a rated level, and pitch control is used to regulate the rotor speed, and hence, also the power [13].

A power converter will be mainly used in variable speed applications. In fixed speed control, a power converter could be used for a better system performance, for example, smooth transition during turn on, harmonics, and flicker reduction, etc. Next, the operation of the most general controller, namely, the variable speed pitch regulator controller is explained. Another controller can be obtained from this control scheme, but will not be presented here.

4.1.3.1 Variable Speed Variable Pitch Wind Turbine

Objectives of variable speed control systems are summarized by the following general goals [12,16,17]:

- to regulate and smooth the power generated
- to maximize the energy captured
- to alleviate the transient loads throughout the wind turbine
- to achieve unity power factor on the line side with no low frequency harmonics current injection
- to reduce the machine rotor flux at light load reducing core losses

Objectives for the pitch-angle control are similar to the variable speed. If pitch-angle control is used together with variable speed, better performance in the system is obtained. For instance, to permit starting, blade pitch angle differs from the operation pitch angle, allowing an easier starting and optimum running. Moreover, the power and speed can be limited through rotor pitch regulation.

The control diagram is shown schematically in Fig. 4.7. The generator torque Q_e and the pitch angle β control the wind turbine. The control system acquires the actual generated electric power P_e and the generator speed, ω_r, and calculates the reference generator torque Q_e^{ref} and the reference pitch angle β^{ref}, using two control loops [14].

In low winds it is possible to maximize the energy captured by following a constant tip speed ratio λ load line which corresponds to operating at the maximum power coefficient. This load line is a quadratic curve in the torque-speed plane as it is shown in Fig. 4.8. During that time, the pitch angle is adjusted to a constant value, the maximum power pitch angle.

If there is a minimum allowed operating speed, then it is not possible to follow this curve in very low winds, and the turbine is then operated at a constant speed N_{min} shown in Fig. 4.8. On the other hand, in high wind speed, it is necessary to limit the torque Q_{rate} or power P_{rate} of the generator to a constant value.

The control parameters are: the minimum speed ω_r^{min}, the maximum speed in constant tip speed ratio mode ω_r^{max}, the nominal steady-state operating speed ω_r^{rat}, and the parameter K_λ which defines the constant tip speed ratio line $Q_e = K_\lambda \omega_r^2$. K_λ is given by (4.8):

$$K_\lambda = \frac{\pi \rho R^5 C_p\,(\lambda, \beta)}{2\lambda^3 G^3} \tag{4.8}$$

When the generator torque demand is set to $K_\lambda \omega_r^2$ where ω_r is the measured generator speed, this ensures that in the steady state the turbine will maintain the tip speed ratio λ_{opt} and the corresponding maximum power coefficient $C_p(\lambda, \beta)$.

Figure 4.9 shows the simplified control loops used to generate pitch and torque demands. When operating below rated power, the torque controller is

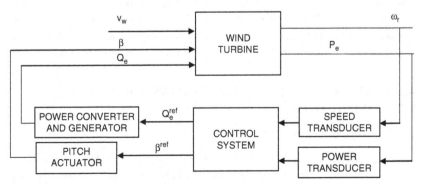

FIGURE 4.7 Block control diagram of the variable speed pitch regulated wind turbine.

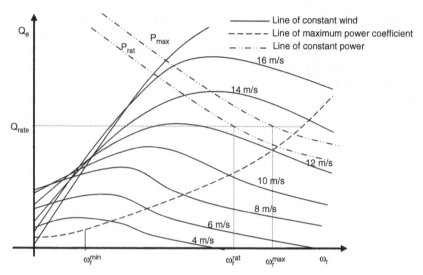

FIGURE 4.8 Variable speed pitch regulated operating curve.

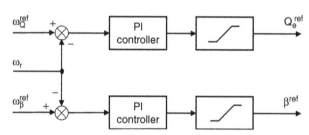

FIGURE 4.9 Pitch regulated variable speed control loops.

active, and the pitch demand loop is active when operating above rated power. Below rated, the speed set-point, ω_Q^{ref}, is the optimum speed given by the optimal tip speed ratio curve and pitch angle is held at zero. Above rated, the reference generator torque is hold rated constant value Q_{rate} and the pitch angle controller is achieving the reference nominal speed ω_β^{ref}. During this control interval, the captured power is constant because the reference torque is maintained at a rated torque of the machine and the rotor speed is controlled to maintain a rated value.

4.2 POWER ELECTRONIC CONVERTERS FOR VARIABLE SPEED WIND TURBINES

4.2.1 Introduction

Power electronic converters can operate the stator of synchronous or asynchronous machines. In other applications, the power converter can be connected to the rotor of a wound rotor induction machine. In the first case, the converter handles the overall power of the machine and it operates in a wide speed range. In the wound rotor machine case, the converter handles a fraction of the rated power but it does not allow a very low speed to obtain higher energy from the wind. So, the advantage is that the power converter is smaller and cheaper than the stator converter.

4.2.2 Full Power Conditioner System for Variable Speed Turbines

In this section, the different topologies of power electronic converters that are currently used for wide range speed control of generators will be presented. A variable speed wind turbine control method within a wide range has the following advantage and disadvantage compared to those for narrow-range speed control. The advantage of a variable speed wind turbine control method in a wind range is that it allows for very low speed to obtain higher energy from the wind. On the other hand, the disadvantage is that the power converter must be rated to the 100% of the nominal generation power.

The power conditioners, next to be considered in this section, may be used for synchronous as well as asynchronous generators. For both cases the control block to be employed is defined. The main objective of power converters to be used for wind energy applications is to handle the energy captured from the wind and the injection of this energy into the grid. The characteristics of the generator to be connected to the grid and where to inject the electric energy are decisive when designing the power converter. To attain this design, it is necessary to consider the type of semiconductor to be used, components and subsystems.

By using cycloconverters (AC/AC) or frequency converters based on double frequency conversion, normally AC/DC–DC/AC, and connected by a DC link, a rapid control of the active and reactive power can be accomplished along with a low incidence in the distribution electric grid. The commutation frequency of the power semiconductors is also an important factor for the control of the wind turbine because it allows not only to maximize the energy captured from the wind but also to improve the quality of the energy injected into the electrical grid. Because of this, the semiconductors required are those that have a high power limit and allow a high commutation frequency. The insulated gate bipolar transistor (IGBTs) are commonly used because of their high breakdown voltage and because they can bear commutation frequencies within the range of 3–25 kHz, depending on the power handled by the device. Other semiconductors such as gate turn-off thyristors (GTOs) are used for high power applications allowing lower commutation frequencies, and thus, worsening not only the control of the generator but also the quality of the energy injected into the electric grid. [4,8–10,18].

The different topologies used for a wide-range rotor speed control are described next. Advantages and disadvantages for using these topologies, as power electronics is concerned, are:

Advantages:

- Wide-range speed control
- Simple generator-side converter and control
- Generated power and voltage increased with speed
- VAR-reactive power control possible

Disadvantages:

- One or two full-power converter in series
- Line-side inductance of 10–15% of the generated power
- Power loss up to 2–3% of the generated power
- Large DC link capacitors

4.2.2.1   Double Three Phase Voltage Source Converter Connected by a DC-link

Figure 4.10 shows the scheme of a power condition for a wind turbine. The three phase inverter on the left side of the power converter works as a driver controlling the torque generator by using a vectorial control strategy. The three phase inverter on the right side of the figure permits the injection of the energy extracted from the wind into the grid, allowing a control of the active and reactive power injected into the grid. It also keeps the total harmonic distortion coefficient as low as possible improving the quality of the energy injected into the public grid. The objective of the DC-link is to act as an energy storage, so that the captured energy from the wind is stored as a charge in the capacitors and is instantaneously injected into the grid. The control signal is set to maintain a constant reference to the voltage of the capacitors battery V_{dc}. The control strategy for the connection to the grid will be described in Section 4.2.3.

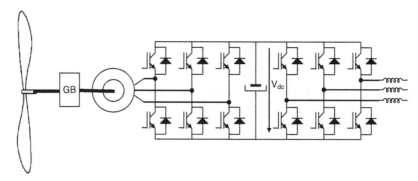

FIGURE 4.10   Double three phase voltage source inverter connected by a DC link used in wind turbine applications.

The power converter shown in Fig. 4.10 can be used for a variable speed control in generators of wind turbines, either for synchronous or asynchronous generators.

4.2.2.1.1 Asynchronous Generator

Next to be considered is the case of an asynchronous generator connected to a wind turbine. The control of a variable speed generator requires a torque control, so that for low speed winds the control is required with optimal tip speed ratio, λ_{opt}, to allow maximum captured wind energy from low speed winds. The generator speed is adjusted to the optimal tip speed ratio λ_{opt} by setting a reference speed. For high speed winds the pitch or stall regulation of the blade limits the maximum power generated by the wind turbine. For low winds it is necessary to develop a control strategy, mentioned in Section 4.2.

The adopted control strategy is an algorithm for indirect vector control of an induction machine [3,6,7], which is described next and shown in Fig. 4.11.

A reference speed, ω_Q^{ref}, has been obtained from the control strategy used in order to achieve optimal speed ratio working conditions of the wind turbine to capture the maximum energy from the wind. In Fig. 4.11, the calculation block to obtain ω_Q^{ref} is shown, that is fed by the actual rotor speed, ω_r, and the electrical power generated, P_e, by the asynchronous generator. Using this ω_Q^{ref} and the actual rotor speed, ω_r, which is measured by the machine, the reference for the

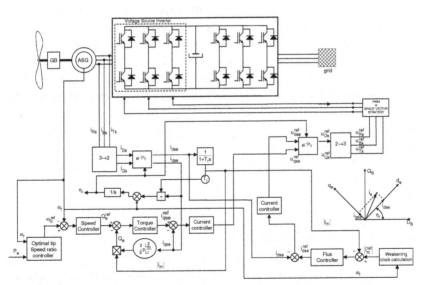

FIGURE 4.11 Schematic of the rotor flux-oriented of a squirrel cage induction generator used in a variable speed wind turbine.

electric torque, Q_e^{ref}, is obtained from the speed regulator which is necessary to set the reference torque in the machine shaft in order to achieve the control objectives.

In Fig. 4.11 the induction generator is driven by a voltage-source pulse width moducation (PWM) inverter, which is connected by a second voltage-source PWM inverter to the public grid through a DC link battery capacitors. The output voltage of the inverter is controlled by a PWM technique in order to follow the voltage references, u_{Rs}^{ref}, u_{Ss}^{ref}, u_{Ts}^{ref}, provided by the control algorithm in each phase. There are many types of modulation techniques which are not discussed in detail here.

A flux model has been used to obtain the angular speed of the rotor flux and the modulus of magnetizing current, $|i_m|$, that has also been used to calculate the electromagnetic torque, Q_e, as shown in Fig. 4.11, using the Eq. (4.9).

$$Q_e = \frac{3}{2} \cdot p \cdot \frac{L_m^2}{L_r} \cdot |i_m| \cdot i_{qse} \qquad (4.9)$$

The speed controller provides the reference of the torque, Q_e^{ref}, and the torque controller gives the reference value of the quadrature-axis stator current in the rotor flux-oriented reference frame i_{qse}^{ref}.

In Fig. 4.11 the field weakening block, used to obtain the reference value of the modulus of the rotor magnetizing current space phasor $\left|\vec{i}_m^{ref}\right|$, is rotor speed-dependent. This reference signal is then compared with the actual value of the rotor magnetizing current, $|i_m|$, and the error generated is used as input to the flux controller. The output of this controller is the direct-axis stator current reference expressed in the rotor-flux-oriented reference frame i_{dse}^{ref}.

The difference in values of the direct and quadrature-axis stator current references $(i_{dse}^{ref}, i_{qse}^{ref})$ and their actual values (i_{dse}, i_{qse}) are given as inputs to the respective PI current controllers. The outputs of these PI controllers are values of the direct and quadrature-axis stator voltage reference expressed in the rotor-flux-oriented reference frame $(u_{dse}^{ref}, u_{qse}^{ref})$. After this they are transformed into the steady reference frame $(u_{Ds}^{ref}, u_{Qs}^{ref})$, using the $e^{-j\eta r}$ transformation. This is followed by the $2 \rightarrow 3$ block and finally, the reference values of the three phase stator voltage $(u_{Rs}^{ref}, u_{Ss}^{ref}, u_{Ts}^{ref})$ are obtained. These signals are used to control the pulse-width modulator, which transforms these reference signals into appropriate on–off switching signals to command the inverter phase.

4.2.2.1.2 Synchronous Generator

When the generator, used to transform the mechanical energy into electrical energy in the wind turbine, is a salient-pole synchronous machine with an electrically excited rotor, the control criteria are the same as the one applied in the induction generator case, so as to minimize the angular speed error in order to obtain an optimum tip speed ratio performance of the wind turbine. In this section, a drive control based on magnetizing field-oriented control is described and applied to a wind turbine with a salient-pole synchronous machine. This control method can be applied to the synchronous generator using a voltage-source inverter or a cycloconverter as it is shown in Figs. 4.10 and 4.13 respectively, using the same block control diagram.

In both cases, a controllable three-phase rectifier supplies the excitation winding on the rotor of the synchronous machine. As it is well-known, the cycloconverter is a frequency converter which converts power directly from a fixed frequency to a lower frequency. Each of the motor phases is supplied through a three-phase transformer, an antiparallel thyristor bridge, and the field winding is supplied by another three-phase transformer and a three-phase-rectifier using the bridge connection.

In the control block described in Fig. 4.12, the rotor speed and monitored current have been used in order to control the relationship between the magnetizing flux and currents of the machines by modifying the voltages of the power converter and the excitation current in the field winding.

The reference rotor speed, ω_Q^{ref}, and the measured rotor speed, ω_r, are compared and the error is introduced into the speed controller. The output voltage is proportional to the electromagnetic torque PI, of the synchronous machine and the reference torque, Q_e^{ref}, is obtained. Dividing the reference

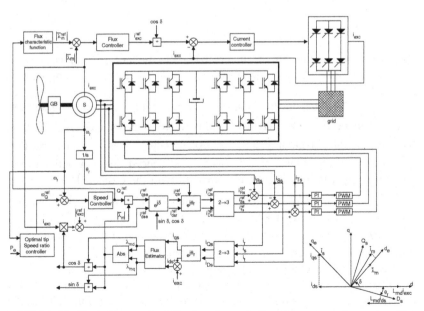

FIGURE 4.12   Block diagram of a field-oriented control of a salient-pole synchronous machine used in wind turbine applications.

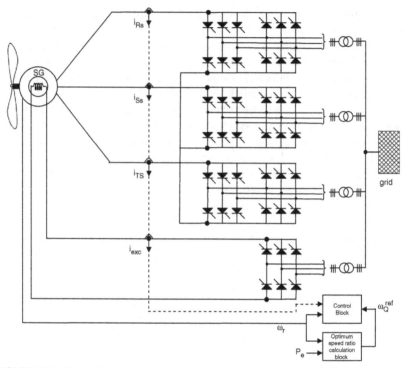

FIGURE 4.13 Schematic of the cycloconverter synchronous generator used in variable speed wind turbines.

torque by the modulus of the magnetizing flux-linkage space phasor $|\lambda_m|$, the reference value of the torque-producing stator current, $|\lambda_m|$, component is obtained.

Using a characteristic function of the magnetizing flux reference and the actual rotor speed, ω_r, the magnetizing flux reference is obtained, $\left|\lambda_m^{ref}\right|$. Below base speed, this function yields a constant value of the magnetizing flux reference, λ_m^{ref}; above base speed, this flux is reduced. The magnetizing flux controller, $|\bar{\lambda}_m|$, is introduced and compared with the estimated magnetizing flux of the synchronous machine, $\left|\lambda_m^{ref}\right|$, obtaining the error, which is fed into the flux controller as shown in Fig. 4.12. The flux controller maintains the magnetizing linkage flux to a pre-set value independent of the load. As an output of this controller the reference excitation current, i_{exc}^{ref} is obtained.

The value of the reference magnetizing stator current, i_{exc}^{ref}, is obtained using a steady state in which there is no reactive current drawn from the stator [3]. In this case the power factor is unity and the stator current value is the optimal. The zero reactive power condition can be fulfilled by controlling the magnetizing current stator component that is shown in Fig. 4.12.

The stator current components (i_{dsc}^{ref}, i_{qse}^{ref}) are first transformed into the stator current components established in the rotor reference frame (i_{dsr}^{ref}, i_{qsr}^{ref}). After these components are transformed into the stationary-axes current components by a similar transformation, but taking into account that the phase displacement between the stator direct axis of the rotor is θ_r. The obtained two-axis stator current references (i_{Qs}^{ref}, i_{Ds}^{ref}), are transformed into the three phase stator current references (i_{Rs}^{ref}, i_{Ss}^{ref}, i_{Ts}^{ref}) by the application of the three phase to two phase transformation. The reference stator currents are compared with their respective measured currents, and their errors are fed into the respective stator current controllers.

4.2.2.2   Step-up Converter and Full Power Converter

An alternative for the power conditioning system of a wind turbine is to choose a synchronous generator and a three phase diode rectifier, as shown in Fig. 4.14. Such a choice is based on low cost compared with an induction generator connected to a voltage source inverter used as a rectifier. When the speed of the synchronous generator alters, the voltage value on the DC-side of the diode rectifier will change. A step-up chopper is used to adapt the rectifier voltage to the DC-link voltage of the inverter, see Fig. 4.14. When the inverter system is analyzed, the generator/rectifier system can be modeled as an ideal current source. The step-up chopper used as a rectifier utilizes a high switching frequency so the bandwidth of these components is much higher than the bandwidth of the generator. The generator/rectifier current is denoted as i_{dc} and is independent of the value of the DC-link voltage. The inductor of the step-up converter is denoted as LDC. The capacitor of the DC-link has the value C_{dc}. The inductors on the AC-side of the inverter have the inductance L$_{AC}$. The three-phase voltage system of the grid has phase voltages e_R, e_S, and e_T and the phase currents are denoted as i_{Rg}, i_{Sg}, and i_{Tg}. The DC-link voltage value is denoted as V_{dc}.

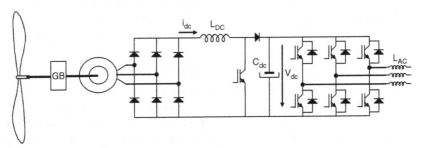

FIGURE 4.14   Step-up converter in the rectifier circuit and full power inverter topology used in wind turbine applications.

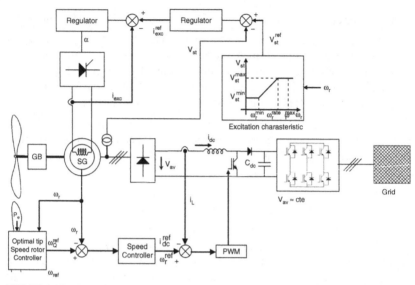

FIGURE 4.15   Control block diagram of a step-up converter in the rectifier circuit and full power inverter used in wind turbine applications.

The control system, shown in Fig. 4.15, is based on the measurement of the rotor speed ω_r of the synchronous generator by means of a speed transducer. This value is compared with the reference rotor speed obtained by the control algorithm of the variable speed wind turbine used in the application in order to achieve optimal speed ratio, and therefore, to capture the maximum energy from the wind.

The objective of the synchronous machine excitation system is to keep the stator voltage V_{st} following the excitation characteristic of the generator, shown in Fig. 4.15. This excitation characteristic is linear in the range of the minimum of the rotor speed ω_r^{min} and the rated value of the rotor speed ω_Q^{rate}. Outside this range, the stator voltage is saturated to V_{st}^{min} or V_{st}^{max}. For rotor speeds ω_r in the range between ω_r^{min} and ω_r^{rate}, the control is carried out using an inductor current i_{dc} proportional to the shaft torque of the generator. Above the rated rotor speed, this current is proportional to the power because the stator voltage V_{st} is constant.

4.2.2.3   Grid Connection Conditioning System

Injection into the public grid is accomplished by means of PWM voltage source interver. This requires the control of the current of each phase of the inverter, as shown in Fig. 4.16. There are several methods of generating the reference current to be injected into any of the phases of the inverter. A very useful method to calculate this current was proposed by Professor Akagi [19], that is applied to the active power filters referred to as "Instantaneous Reactive Power Theory."

The control block shown in Fig. 4.16 is based on the comparison of the actual capacitor array voltage V_{dc} with a reference voltage V_{dc}^{ref}. Subtracting the actual capacitor array voltage V_{dc}^{ref} from the reference voltage V_{dc}^{ref}, the error signal voltage is obtained. This error signal is fed into a compensator, usually a PI compensator, that transforms the output to an active instant power signal that, after being injected into the electric grid, is responsible for the error of the regulator of the voltage in the capacitor array to be zero.

Transformation of the phase voltage e_{rg}, e_{sg}, and e_{tg} into the α–β orthogonal coordinates is given by the following expression:

$$\begin{bmatrix} e_\alpha \\ e_\beta \end{bmatrix} = \sqrt{\frac{2}{3}} \cdot \begin{bmatrix} 1 & -1/2 & -1/2 \\ 0 & \sqrt{3}/2 & -\sqrt{3}/2 \end{bmatrix} \cdot \begin{bmatrix} e_{rg} \\ e_{sg} \\ e_{tg} \end{bmatrix} \qquad (4.10)$$

Using the inverse transformation equations from the control algorithm [19] and the reference of real power of the capacitor array, it is possible to derive the phase sinusoidal reference current to be injected by the converter into the grid.

$$\begin{bmatrix} i_{rg}^{ref} \\ i_{sg}^{ref} \\ i_{tg}^{ref} \end{bmatrix} = \sqrt{\frac{2}{3}} \cdot \begin{bmatrix} 1 & 0 \\ -1/2 & \sqrt{3}/2 \\ -1/2 & -\sqrt{3}/2 \end{bmatrix} \cdot \begin{bmatrix} e_\alpha & e_\beta \\ -e_\beta & e_\alpha \end{bmatrix}^{-1} \cdot \begin{bmatrix} p^{ref} \\ q^{ref} \end{bmatrix} \qquad (4.11)$$

As can be seen in the above equation, the reference of the reactive power appears which is normally set at zero so that the current is being injected into the public grid with unity power factor.

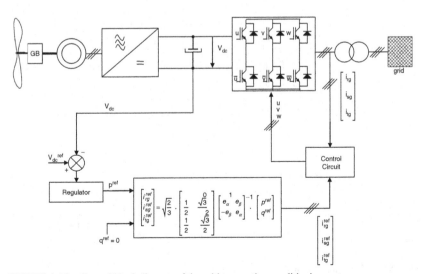

FIGURE 4.16 Control block diagram of the grid connection conditioning system.

There is another type of wind energy extraction system where the objective is to compensate the reactive power generated by non-linear loads of the grid. In this case, the reactive reference power is set to the corresponding value, for either reactive or capacitive power factor "leading or lagging."

4.2.3 Rotor Connected Power Conditioner for Variable Speed Wind Turbines

As it was introduced before, variable speed can also be obtained by slip change of induction generator using a wound rotor machine.

Since it is necessary to have an electric connection to the rotor winding, rotation is achieved by using a slip ring. The power delivered by the rotor through the slip rings is equal to the product of the slip by the electrical power that flows into the stator P_S, Eq. (4.12). This is the so-called "slip power" P_{slip}.

$$P_{slip} = s \cdot P_S \tag{4.12}$$

The slip power can be handled as follows:

- It can be dissipated in a resistor (Fig. 4.17).
- Using a single doubly fed scheme, the slip power is returned to the electrical grid or to the machine stator (Fig. 4.18).
- Using a cascaded scheme. This is accomplished by connecting a second machine. Part of the power is transferred as mechanical power through the shaft, and the other part is transferred to the grid by a power converter (Fig. 4.19).

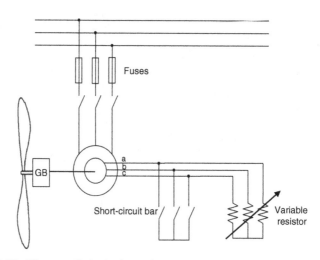

FIGURE 4.17 Slip power dissipation in a resistor.

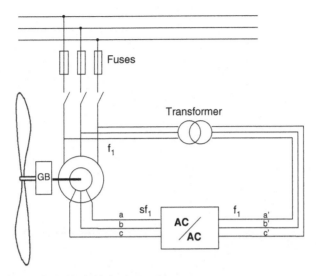

FIGURE 4.18 Single doubly fed induction machine.

- A single frame cascaded or brushless doubly fed induction machine [20] can be used in the same way as before, but using only one machine instead of two (Fig. 4.20).

4.2.3.1 Slip Power Dissipation

Figure 4.17 shows a system in which the power delivered by the rotor is dissipated in a resistor.

The variable resistor can be substituted by the power converter in Fig. 4.21. The power converter controls the power delivered to the resistor using an uncontrolled rectifier and a parallel DC–DC chopper. This design has the disadvantage of the current having a higher harmonic content. This is caused by the rectangular rotor current waveform in the case of a three phase uncontrolled rectifier. This disadvantage can be avoided using a six IGBT's controlled rectifier, however this topology increases the cost significantly.

The variation of the rotor resistance is not a recommended technique due to the high copper losses in the regulation resistance and so, the generator system efficiency is lower. It only can be efficient within a very narrow range of the rotor speed. Another disadvantage is that this technique is applicable only to wound-rotor machines and so, slip rings and brushes are needed. In order to solve this problem some brushless schemes are proposed. A solution is to use a rotor auxiliary winding which couples the power to an external variable resistor. The scheme can be observed in Fig. 4.22.

Another solution is to dissipate the energy within a resistor placed in the rotor as it is shown in Fig. 4.23. This method is currently used in generators for wind

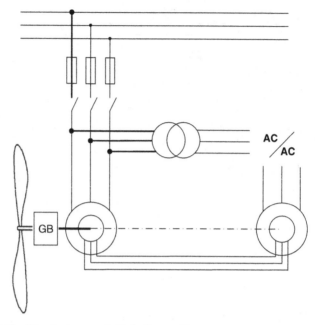

FIGURE 4.19 Wound rotor cascaded induction machines.

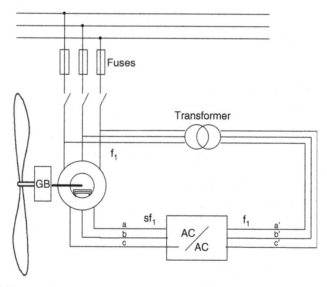

FIGURE 4.20 Brushless doubly fed induction machine.

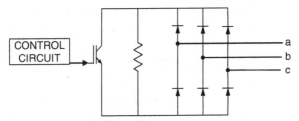

FIGURE 4.21 Variable resistor using a power converter.

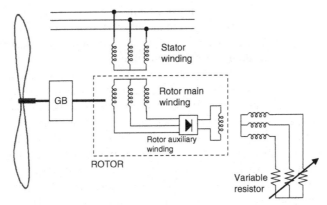

FIGURE 4.22 Slip power dissipation in an external resistor using brushless machine.

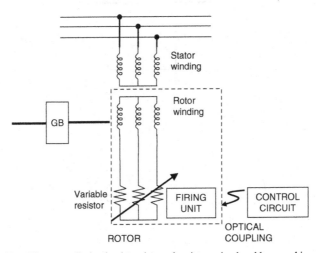

FIGURE 4.23 Slip power dissipation in an internal resistor using brushless machine.

conversion systems, but as the efficiency of the system decreases with increasing the slip, the speed control is limited to a narrow margin. This scheme includes the power converter and the resistors in the rotor. Trigger signals to the power switches are accomplished by optical coupling.

4.2.3.2   Single Doubly Fed Induction Machine

In Fig. 4.18 the connection scheme shows that slip power is injected into the public grid by a power converter and a transformer. The power converter changes the frequency and controls the slip power. In some cases a transformer is used due to public grid voltages which can be higher than rotor voltages.

Disregarding losses, the simplified scheme of Fig. 4.24 shows real power flux in all different connection points of the diagram. In this figure, the electrical power in the stator machine P_S, the mechanical power $(1 - s) \cdot P_S$, and the slip power and power converter $s \cdot P_S$ are represented.

In generation mode, the power is positive when the arrow direction shown in Fig. 4.24 is considered. The power handled on the power converter depends on the sign of the machine slip. When this slip is positive, i.e. subsynchronous mode of operation, the slip power goes through the converter from the grid to the rotor of the machine. On the other hand, when the slip is negative, i.e. hypersynchronous mode of operation, the slip power comes out of the rotor to the power converter. Since the slip power is the real power through the converter, this power is determined directly by the maximum slip or by the speed range of the machine. For instance, if the speed range used is 20% of the synchronous speed, the power rating of the converter is 20% of the main power.

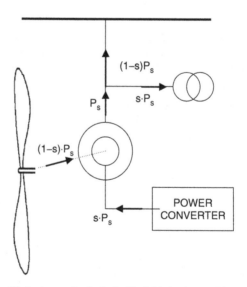

FIGURE 4.24   Simplified scheme of a single doubly fed induction machine.

4.2.3.3 Power Converter in Wound-rotor Machines

The power converter used in wound-rotor machines can be a force-commutated converter connected to a line-commutated converter by an inductor as shown in Fig. 4.25. In this case, the power can only flow from the rotor to the grid, and the induction generator works above synchronous speed. Main disadvantages of this scheme are a low power factor and a high content of low frequency harmonics whose frequency depends on the speed.

Since the feeding frequency of the rotor is much lower than the grid's, a cycloconverter, as shown in Fig. 4.26, can be used. In this case the controllability of the system is greatly improved. The cycloconverter is an AC–AC converter based on the use of two three-phase thyristor bridges connected in parallel, one for each phase. This scheme allows working with speed above and below the synchronous speed.

The two schemes mentioned before (Figs. 4.25 and Fig. 4.26) are based on line switched converters. A disadvantage of this scheme is that voltages in the rotor decrease when the machine is working in frequencies close to the synchronous frequency. This fact makes the line-commutated converter not to commute satisfactorily, and we need to use a forced-commutated converter. Also, when a forced-switched converter is used, quality of the voltage and current injected into the public grid is improved. The forced-switched power converter scheme is shown in Fig. 4.27. The converter includes two three-phase AC–DC converters linked by a DC capacitor battery. This scheme allows, on one hand, a vector control of the active and reactive power of the machine, and on

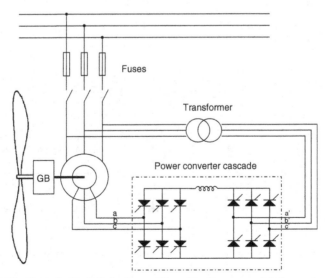

FIGURE 4.25 Single doubly fed induction machine with a force-commutated converter connected to a line-commutated converter.

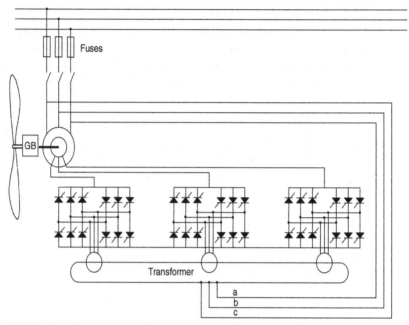

FIGURE 4.26 Single doubly fed induction machine with a cycloconverter.

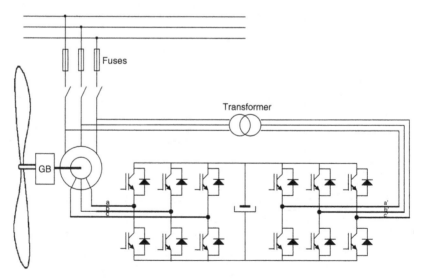

FIGURE 4.27 Single doubly fed induction machine with two fully controlled AC–DC power converters.

the other hand, a decrease by a high percentage of the harmonic content injected into the grid by the power converter.

4.2.3.4 Control of Wound-rotor Machines

The vector control of the rotor flux can be accomplished very easily in a wound-rotor machine. Power converters that can be used for vector control applications are either a controlled rectifier in series with an inverter or a cycloconverter.

In this system, the slip power can flow in both directions, from the rotor to the grid (subsynchronous) or from the grid to the rotor (hyper-synchronous). In both working modes, the machine must be working as a generator. When the speed is hyper-synchronous, the converter connected to the rotor will work as a rectifier and the converter connected to the grid as an inverter, as was deduced before from Fig. 4.24.

The wound-rotor induction machine can be modeled as follows (see nomenclature page):

$$u_S = R_S \cdot i_S + \frac{d\lambda_S}{dt} \tag{4.13}$$

$$u'_r = n^2 \cdot R_r \cdot i'_r + \frac{d\lambda'_r}{dt} - j \cdot \omega_r \cdot \lambda'_r \tag{4.14}$$

$$\lambda_S = L_S \cdot i_S + n \cdot L_m \cdot i'_r \tag{4.15}$$

$$\lambda'_r = n^2 \cdot L_S \cdot i'_r + n \cdot L_m \cdot i_S \tag{4.16}$$

$$Q_e = \frac{3}{2} \cdot \frac{\rho}{2} \cdot Im\left\{\lambda_r \cdot i_r^*\right\} \tag{4.17}$$

$$P_S = \frac{3}{2} \cdot Re\left\{u_S \cdot i_s^*\right\} \tag{4.18}$$

$$Q_S = \frac{3}{2} \cdot Im\left\{u_S \cdot i_S^*\right\} \tag{4.19}$$

In the induction machine model, rotor magnitudes, the rotor voltage u_r, the rotor flux λ_r, and the rotor current i_r are referred to the stator, and so, the rotor voltage u'_r, the rotor flux λ'_r, and the rotor current i'_r are defined as:

$$u'_r = n \cdot u_r \cdot e^{j \cdot \theta_r} \tag{4.20}$$

$$\lambda'_r = n \cdot \lambda_r \cdot e^{j \cdot \theta_r} \tag{4.21}$$

$$i'_r = \frac{i_r \cdot e^{j \cdot \theta_r}}{n} \tag{4.22}$$

The stator magnetizing current i_m is defined as:

$$i_m = \frac{\lambda_S}{n \cdot L_m} = \frac{L_S}{n \cdot L_m} \cdot i_S + i'_r \tag{4.23}$$

Figure 4.28 shows the control block of a wound-rotor induction machine. The wound-rotor induction machine is controlled by the rotor using a power

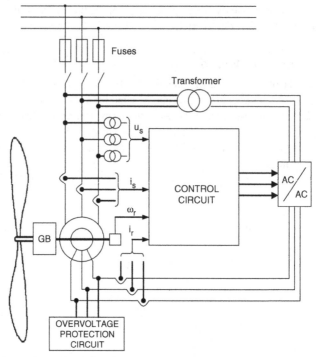

FIGURE 4.28 Wind turbine control block.

converter that controls the rotor current i_r' by changing the rotor voltage u_r'. The control of the stator current via the rotor current makes sense only if the converter power is kept lower than the rated power of the machine. The AC stator voltage generates a rotating magnetic field with angular frequency ω_e. Relative to the rotor, this magnetic field rotates only with the angular slip frequency. The frequency of voltages induced in the rotor is low, so voltages of the power converter are low too. Active and reactive power of the induction generator, or a certain percentage of them, can be controlled by the rotor current.

The machine model can be referred to the reference axes that move with the magnetizing current. This system of coordinates rotates with an angle θ_e relative to the stator. In these axes $i_{qm} = 0$ as shown in Fig. 4.29.

Equations of the rotor and stator voltage become:

$$u_S = R_S \cdot i_S + \frac{d\lambda_S}{dt} + j \cdot \omega_e \cdot \lambda_S \tag{4.24}$$

$$u_r' = n^2 \cdot R_r \cdot i_r' + \frac{d\lambda_r'}{dt} + j \cdot (\omega_e - \omega_r) \cdot \lambda_r' \tag{4.25}$$

$$\omega_e = \frac{d\theta_e}{dt} \tag{4.26}$$

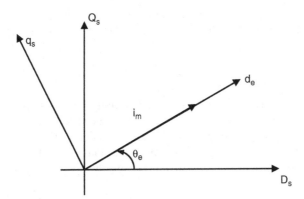

FIGURE 4.29   Stator and rotor reference frames.

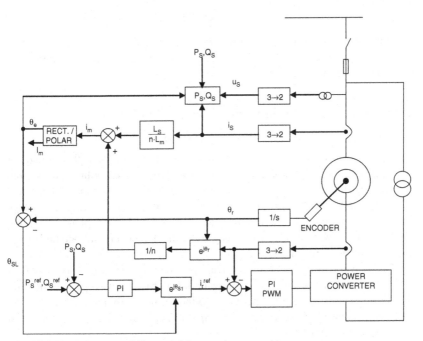

FIGURE 4.30   Power control diagram of the wound-rotor machine.

Supposing steady-state conditions and disregarding the resistors in the stator and the rotor, because the voltage drop is very low in comparison to the stator voltage, the stator and rotor voltages can be determined as:

$$u_S = j\omega_e \lambda_S \tag{4.27}$$

$$u'_r = j\omega_{s1} \lambda'_r \tag{4.28}$$

The flux is determined from the stator voltage u_S and the angular frequency ω_e of the AC system. Since both are constant, the flux linkage and magnetizing current are constant too.

$$i_m = \frac{\lambda_S}{nL_m} = \frac{u_{ds} + ju_{qs}}{j\omega_e nL_m} \qquad (4.29)$$

As i_m is constant, the stator current can be controlled at any time, by means of controlling the rotor current that can be deduced from Eq. (4.23). From Eq. (4.29) we can also deduce that the direct component of the stator voltage u_{ds} is zero due to the quadrature component of the stator magnetizing current i_{qm} is zero and so, the active and reactive power can be obtained by the following equations:

$$P_S = \frac{3}{2} u_{ds} i_{dsm} \qquad (4.30)$$

$$Q_S = \frac{3}{2} u_{qs} i_{qsm} \qquad (4.31)$$

Figure 4.30 shows a block diagram of a vector control of active and reactive power for a wound-rotor machine. Vector control of cascaded doubly fed machine is presented in [21].

4.2.4 Grid Connection Standards for Wind Farms

4.2.4.1 Voltage Dip Ride-through Capability of Wind Turbines

As wind capacity increases, network operators have to ensure that consumer power quality is not compromised. To enable large-scale application of wind energy without compromising power system stability, the turbines should stay connected and contribute to the grid in case of a disturbance such as a voltage dip. Wind farms should generate similar to conventional power plants supplying active and reactive power for frequency and voltage recovery, immediately after the fault has been produced.

Thus, several utilities have introduced special grid connection codes for wind farm developers, covering reactive power control, frequency response, and fault ride-through, especially in places where wind turbines provide for a significant part of the total power. Examples are Spain [22], Denmark [23], and part of Northern Germany [24].

The correct interpretation of these codes is crucial for wind farm developers, manufacturers, and network operators. They define the operational boundary of a wind turbine connected to the network in terms of frequency range, voltage tolerance, power factor, and fault ride-through. Among all these requirements, fault ride-through is regarded as the main challenge to the wind turbine manufacturers. Though the definition of fault ride-through varies, the E.ON (German Transmission and Distribution Utility) regulation is likely to set the standard [24]. This stipulates that a wind turbine should remain stable and

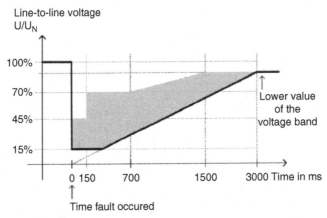

Line-to-line voltage
U/U_N

FIGURE 4.31   E. On Netz requirements for wind farm behavior during faults.

connected during the fault, while voltage at the point of connection drops to 15% of nominal (i.e. a drop of 85%) for a period of 150 ms: see Fig. 4.31.

Only when the grid voltage drops below the curve, the turbine is allowed to disconnect from the grid. When the voltage is in the shaded area, the turbine should also supply reactive power to the grid in order to support grid voltage restoration.

A major drawback of variable-speed wind turbines, especially for turbines with doubly fed induction generators (DFIGs), is their operation during grid faults [25,26]. Faults in the power system, even far away from the location of the turbine, can cause a voltage dip at the connection point of the wind turbine. The dip in the grid voltage will result in an increase of the current in the stator windings of the DFIG. Because of the magnetic coupling between the stator and rotor, this current will also flow in the rotor circuit and the power electronic converter. This can lead to the permanent damage of the converter. It is possible to try to limit the current by current-control on the rotor side of the converter; however, this will lead to high voltages at the converter terminals, which might also lead to the destruction of the converter. A possible solution that is sometimes used is to short-circuit the rotor windings of the generator with the so-called crowbars.

The key of the protection technique is to limit the high currents and to provide a bypass for it in the rotor circuit via a set of resistors that are connected to the rotor windings (Fig. 4.32). This should be done without disconnecting the converter from the rotor or from the grid. Thyristors can be used to connect the resistors to the rotor circuit. Because the generator and converter stay connected, the synchronism of operation remains established during and after the fault. The impedance of the bypass resistors is of importance but not critical. They should be sufficiently low to avoid excess voltage on the converter terminals. On the other hand, they should be high enough to limit the current. A range of values can be found that satisfies both conditions. When the fault in the grid

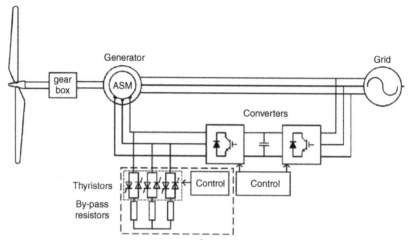

FIGURE 4.32 DFIG bypass resistors in the rotor circuit.

is cleared, the wind turbine is still connected to the grid. The resistors can be disconnected by inhibiting the gating signals and the generator resumes normal operation.

4.2.4.2 Power Quality Requirements for Grid-connected Wind Turbines

The grid interaction and grid impact of wind turbines has been focused in the past few years. The reason behind this interest is that wind turbines are among utilities considered to be potential sources of bad power quality. Measurements show that the power quality impact of wind turbines has been improved in recent years. Especially variable-speed wind turbines have some advantages concerning flicker. But a new problem is faced with variable-speed wind turbines. Modern forced-commutated inverters used in variable-speed wind turbines produce not only harmonics but also inter-harmonics.

The IEC initiated the standardization on power quality for wind turbines in 1995 as a part of the wind turbine standardization in TC88. In 1998, the IEC issued a draft IEC-61400-21 standard for "Power Quality Requirements for Grid Connected Wind Turbines" [27]. The methodology of that IEC standard consists on three analyses. The first one is the flicker analysis. IEC-61400-21 specifies a method that uses current and voltage time series measured at the wind turbine terminals to simulate the voltage fluctuations on a fictitious grid with no source of voltage fluctuations other than the wind turbine switching operation. The second one is switching operations. Voltage and current transients are measured during the switching operations of the wind turbine (start-up at cut wind speed and start-up at rated wind speed). The last one is the harmonic analysis which is carried out by the fast fourier transform (FFT) algorithm.

Rectangular windows of eight cycles of fundamental frequency width, with no gap and no overlapping between successive windows are applied. Furthermore, the current total harmonic distortion (THD) is calculated up to 50th harmonic order [28,29].

Recently, high frequency harmonics and inter-harmonics are treated in the IEC 61000-4-7 and IEC 61000-3-6 [30,31]. The methods for summing harmonics and inter-harmonics in the IECE61000-3-6 are applicable to wind turbines. In order to obtain a correct magnitude of the frequency components, the use of a well-defined window width, according to the IEC 61000-4-7, Amendment 1 is of a great importance, as it has been reported in ref. [32]

Wind turbines not only produce harmonics, they also produce inter-harmonics, i.e. harmonics which are not a multiple of 50 Hz. Since the switching frequency of the inverter is not constant but varies, the harmonics will also vary. Consequently, since the switching frequency is arbitrary, the harmonics are also arbitrary. Sometimes they are a multiple of 50 Hz and sometimes they are not. Figure 4.33 shows the total harmonics spectrum from a variable-speed wind turbine. As can be seen in the figure, at lower frequencies there are only pure harmonics but at higher frequencies there are a whole range of harmonics and inter-harmonics. This whole range of harmonics and inter-harmonics represents variations in the switching frequency of that wind turbine.

4.3 MULTILEVEL CONVERTER FOR VERY HIGH POWER WIND TURBINES

4.3.1 Multilevel Topologies

In 1980s, power electronics concerns were focused on the increase of the power converters by increasing the voltage or current to fulfill the requirements of the emerging applications. There were technological drawbacks, that endure nowadays, which make impossible to increase the voltage or current in the individual power devices, so researchers were developing new topologies based on series and parallel association of individual power devices in order to manage higher levels of current and voltage, respectively. Due to the higher number of individual power devices on such topologies it is possible to obtain more than the classical two levels of voltage at the output of the converter hence the multilevel denomination for this converter.

4.3.2 Diode Clamp Converter (DCC)

In 1981, Nabae *et al.* presented a new neutral-point-clamped PWM inverter (NPC-PWM) [33]. This converter was based on a modification of the two-level converter topology. In the two-level case, each power switch must support at the most a voltage equal to DC-link total voltage so the switches should be dimensioned to support such voltage.

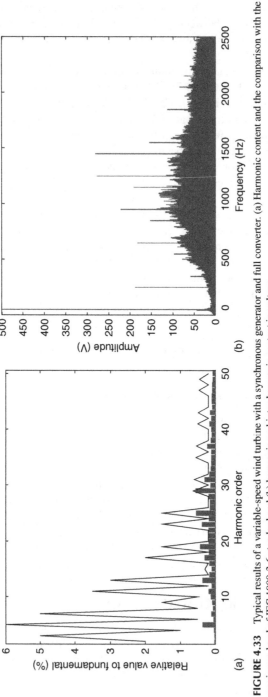

FIGURE 4.33 Typical results of a variable-speed wind turbine with a synchronous generator and full converter. (a) Harmonic content and the comparison with the maximum level of IEC 1000-3-6 standard and (b) harmonic and inter-harmonic content in voltage.

The proposed modification adds two new switches and two clamp diodes in each phase. In this converter each transistor support at the most a half of the total DC-link voltage; hence, if the used power devices have the same characteristics of those used in the two-level case, the DC-link can be doubled and hence, the power which the converter can manage. Figure 4.34 shows one phase of a three-level DCC with the capacitors voltage divider and the additional switches and diodes.

The analysis of the DCC converter states shows that there are three different switching configurations. These possible configurations are shown in Table 4.1.

When transistors S_3 and S_4 are switched on, the phase is connected to the lowest voltage in the DC-link. In the same manner, when the transistors S_1 and S_2 are switched on, the phase is connected to the highest voltage in the DC-link, and when the transistors S_2 and S_3 are switched on, the phase is connected to the mid DC-link voltage through one of the transistors and clamping diodes.

4.3.3 Full Converter for Wind Turbine Based on Multilevel Topology

In order to decrease the cost per megawatt and to increase the efficiency of the wind energy conversion, the nominal power of wind turbines has been continuously growing in the last years. The limitations of the two-level converters power ratings versus three-level ones and the capacity of this to reduce the harmonic distortion and electromagnetic interferences (EMI) make the multilevel converters suitable for modern high power wind turbine applications.

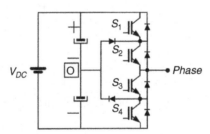

FIGURE 4.34 Three-level DCC.

TABLE 4.1 Switching Configurations for the Three-level DCC

State	S_1	S_2	S_3	S_4	Phase-O voltage
0	Off	Off	On	On	$-V_{DC}/2$
1	Off	On	On	Off	0
2	On	On	Off	Off	$V_{DC}/2$

Figure 4.35 shows the diagram of high power wind turbine directly connected to the utility grid, with a full converter based on two coupled three-level DCC. The converter connected to the generator acts like an AC–DC converter and its main function is to extract the energy from the generator and to deliver it to the DC-link. The converter connected to the grid acts like a DC–AC converter and its main function is to collect the energy at the DC-link and to deliver it to the utility grid.

4.3.4 Modeling

The use of multilevel converters is limited by the following drawbacks: typically very complex, control and voltage imbalance problems at the DC-link capacitors. An analytical model of the whole system is necessary to study this dynamic and to develop control algorithms that meet with the design specifications.

In [34], a general modeling strategy is proposed to obtain the equations that describe the dynamics of the currents and the capacitors voltages as functions of the control signals that represent the voltage in each phase. Based on the nomenclature that can be seen in Fig. 4.36, this modeling strategy yields in the next mathematical model for the currents dynamics Eqs. (4.32), (4.33):

$$\begin{bmatrix} v_{sr1} \\ v_{sr2} \\ v_{sr3} \end{bmatrix} = L_r \begin{bmatrix} di_{r1}/dt \\ di_{r2}/dt \\ di_{r3}/dt \end{bmatrix} + \frac{1}{3} \begin{bmatrix} 2 & -1 & -1 \\ -1 & 2 & -1 \\ -1 & -1 & 2 \end{bmatrix} \begin{bmatrix} v_{r1} \\ v_{r2} \\ v_{r3} \end{bmatrix} \qquad (4.32)$$

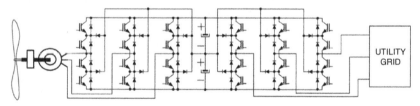

FIGURE 4.35 Diagram of a high power wind turbine with a full converter directly connected to the utility grid.

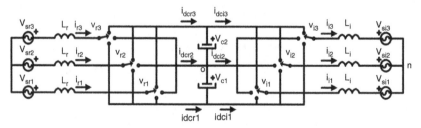

FIGURE 4.36 Nomenclature criterion for the modeling of the full DCC converter.

$$\begin{bmatrix} v_{si1} \\ v_{si2} \\ v_{si3} \end{bmatrix} = -L_i \begin{bmatrix} di_{i1}/dt \\ di_{i2}/dt \\ di_{i3}/dt \end{bmatrix} + \frac{1}{3} \begin{bmatrix} 2 & -1 & -1 \\ -1 & 2 & -1 \\ -1 & -1 & 2 \end{bmatrix} \begin{bmatrix} v_{i1} \\ v_{i2} \\ v_{i3} \end{bmatrix} \qquad (4.33)$$

And the next ones for capacitors voltages dynamics Eq. (4.34):

$$2C\frac{dx_1}{dt} = (\delta_{r1}i_{r1} + \delta_{r2}i_{r2} + \delta_{r3}i_{r3}) - (\delta_{i1}i_{i1} + \delta_{i2}i_{i2} + \delta_{i3}i_{i3})$$

$$x_1 = \frac{v_{c1} + v_{c2}}{2}$$

$$2C\frac{dx_2}{dt} = \left(\delta_{r1}^2 i_{r1} + \delta_{r2}^2 i_{r2} + \delta_{r3}^2 i_{r3}\right) - \left(\delta_{i1}^2 i_{i1} + \delta_{i2}^2 i_{i2} + \delta_{i3}^2 i_{i3}\right)$$

$$x_2 = \frac{v_{c2} - v_{c1}}{2}$$

$$(4.34)$$

where, x_1 and x_2 are chosen as variables to facilitate the controller design and represent the dynamics of the sum and the difference of the capacitors voltages, respectively.

As indicated in [34], it is useful to represent the system in $\alpha\beta\gamma$-coordinates because after the transformation appears the γ control signal as a third freedom degree of the control, moreover this transformation shows the direct relation between this control signal and the capacitors voltage balance. To change to the $\alpha\beta\gamma$-coordinates, an invariant power transformation has been used. The voltages and currents, which are vectors originally in abc-coordinates, are transformed into $\alpha\beta\gamma$-coordinates according to the following matrix transformation shown in Eq. (4.35):

$$T = \sqrt{\frac{2}{3}} \begin{bmatrix} 1 & -1/2 & -1/2 \\ 0 & \sqrt{3}/2 & -\sqrt{3}/2 \\ 1/\sqrt{2} & 1/\sqrt{2} & 1/\sqrt{2} \end{bmatrix} \qquad (4.35)$$

The transformed equations are:

$$L_r \begin{bmatrix} \dfrac{di_{r\alpha}}{dt} \\ \dfrac{di_{r\beta}}{dt} \end{bmatrix} = \begin{bmatrix} v_{sr\alpha} \\ v_{sr\beta} \end{bmatrix} - x_1 \begin{bmatrix} \delta_{r\alpha} \\ \delta_{r\beta} \end{bmatrix}$$

$$- x_2 \begin{bmatrix} \dfrac{\delta_{r\alpha}^2 - \delta_{r\beta}^2}{\sqrt{6}} + \dfrac{2\delta_{r\alpha}\delta_{r\gamma}}{\sqrt{3}} \\ -\sqrt{\frac{2}{3}}\delta_{r\alpha}\delta_{r\beta} + \frac{2}{\sqrt{3}}\delta_{r\beta}\delta_{r\gamma} \end{bmatrix} \qquad (4.36)$$

$$L_i \begin{bmatrix} \dfrac{di_{i\alpha}}{dt} \\ \dfrac{di_{i\beta}}{dt} \end{bmatrix} = - \begin{bmatrix} v_{si\alpha} \\ v_{si\beta} \end{bmatrix} + x_1 \begin{bmatrix} \delta_{i\alpha} \\ \delta_{i\beta} \end{bmatrix}$$

$$+ x_2 \left[\begin{array}{c} \dfrac{\delta_{i\alpha}^2 - \delta_{i\beta}^2}{\sqrt{6}} + \dfrac{2\delta_{i\alpha}\delta_{i\gamma}}{\sqrt{3}} \\ -\sqrt{\dfrac{2}{3}}\delta_{i\alpha}\delta_{i\beta} + \dfrac{2}{\sqrt{3}}\delta_{i\beta}\delta_{i\gamma} \end{array} \right] \tag{4.37}$$

$$2C\frac{dx_1}{dt} = \left[\delta_{r\alpha} \quad \delta_{r\beta} \right]\left[\begin{array}{c} i_{r\alpha} \\ i_{r\beta} \end{array} \right] - \left[\delta_{i\alpha} \quad \delta_{i\beta} \right]\left[\begin{array}{c} i_{i\alpha} \\ i_{i\beta} \end{array} \right]$$

$$2C\frac{dx_2}{dt} = \frac{2}{\sqrt{3}}\left[\delta_{r\alpha} \quad \delta_{r\beta} \right]\left[\begin{array}{c} i_{r\alpha} \\ i_{r\beta} \end{array} \right]\delta_{r\gamma}$$

$$+ \left[\dfrac{\delta_{r\alpha}^2 - \delta_{r\beta}^2}{\sqrt{6}}, -\sqrt{\dfrac{2}{3}}\delta_{r\alpha}\delta_{r\beta} \right]\left[\begin{array}{c} i_{r\alpha} \\ i_{r\beta} \end{array} \right] \cdots$$

$$- \frac{2}{\sqrt{3}}\left[\delta_{i\alpha} \quad \delta_{i\beta} \right]\left[\begin{array}{c} i_{i\alpha} \\ i_{i\beta} \end{array} \right]\delta_{i\gamma}$$

$$- \left[\dfrac{\delta_{i\alpha}^2 - \delta_{i\beta}^2}{\sqrt{6}}, -\sqrt{\dfrac{2}{3}}\delta_{i\alpha}\delta_{i\beta} \right]\left[\begin{array}{c} i_{i\alpha} \\ i_{i\beta} \end{array} \right] \tag{4.38}$$

In these final equations, it is important to point out the relation between the γ control signal and the input and output power of the DCC full converter.

4.3.5 Control

As it can be observed in Eqs. (4.36) and (4.37) the rectifier and inverter currents $i_{\alpha\beta}^r$, $i_{\alpha\beta}^i$ can be controlled separately due to the decoupling of these equations. Also it can be seen in Eq. (4.38) that the control objective on x_1 can be achieved using the normalized voltage references $\delta_{\alpha\beta}^r$ or $\delta_{\alpha\beta}^i$, and x_2 can be controlled using δ_γ^r or δ_γ^i. The implemented control consists basically of independently controlling the inverter and the rectifier. The inverter controls the voltage balance in the DC-link, whereas the rectifier controls the active and reactive power extracted from the generator.

4.3.5.1 Rectifier Control

Figure 4.37 shows the control scheme proposed for the rectifier. The objective of this controller is to make the currents of the generator such that the active and reactive power achieve the reference ones.

It is necessary to notice that the rectifier γ component of the normalized voltage is imposed to be equal to zero $\delta_\gamma^r = 0$ for not affecting on voltage balance, because this balancing is implemented on the inverter control.

4.3.5.2 Inverter Control

Inverter control is divided in two parts. The first part controls the sum of the capacitor voltages x_1, while the second part makes the difference between the capacitor voltages x_2 as small as possible.

4.3.5.3 Sum of the Capacitor Voltages Control

The controller scheme, which can be seen in Fig. 4.38, has been described before in [35], and it is appropriated for this application due to the similarities found in the equations.

The main objective of the controller is to achieve a desired value of the total DC-link voltage. Additionally, the controller can take a reactive power reference to control the power factor of the energy delivered to the utility grid.

4.3.5.4 Difference of the Capacitor Voltages Control

Avoiding the quadratic terms in δ from the equation of the difference of the capacitor voltages in Eq. (4.38), expression (4.39) is obtained:

$$\frac{dx_2}{dt} = K \cdot P^i_{ref} \cdot \delta^i_\gamma \qquad (4.39)$$

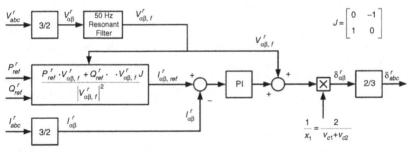

FIGURE 4.37 Control diagram of the rectifier.

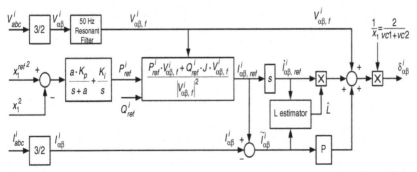

FIGURE 4.38 Inverter control scheme for the sum of the capacitors voltages.

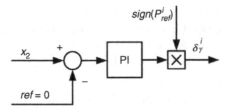

FIGURE 4.39   Proposed capacitors voltages balancing control.

where K is a constant. With this equation, the following control scheme (Fig. 4.39) is proposed:

The objective of the controller is to add a voltage reference in γ direction that depends on the sign of the power and the imbalance of the capacitors voltages.

4.3.5.5 Modulation

Finally, the normalized voltage references $\delta^r_{\alpha\beta\gamma}$ and $\delta^i_{\alpha\beta\gamma}$, obtained from the whole controllers, are translated to *abc*-coordinates and the 3D-space vector modulation algorithm [36] is used to generate the duty cycles and the switching times of power semiconductors.

4.3.6   Application Example

As it was explained before, the standards on energy quality related to renewable energy are focusing to request the plants to contribute to the general stability of the electrical system. To show that the exposed modeling strategy and control scheme can be used to meet the design specification, the electrical system of a wind turbine has been modeled. It consists of an asynchronous induction motor

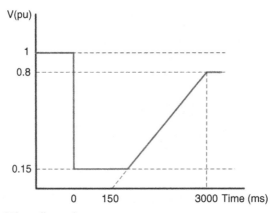

FIGURE 4.40   Voltage dip envelope.

connected to the utility grid through a full DCC converter. The parameters of the example are: nominal power: 3 MW, switching frequency: 2.5 kHz, DC-link nominal voltage: 5 kV, and utility grid line voltage: 2.6 kV. The experiment consists of studying the behavior of the system when there is a voltage dip in the utility grid due to a short-circuit. Figure 4.40 shows the envelope of the voltage dip that has been used to carry out the results.

Figure 4.41 shows the results obtained under the voltage dip condition. Good behavior of the currents on both the sides of the full DCC converter, DC-link voltage, and energy extracted from the generator illustrates the suitability of the control scheme and the model to study the system.

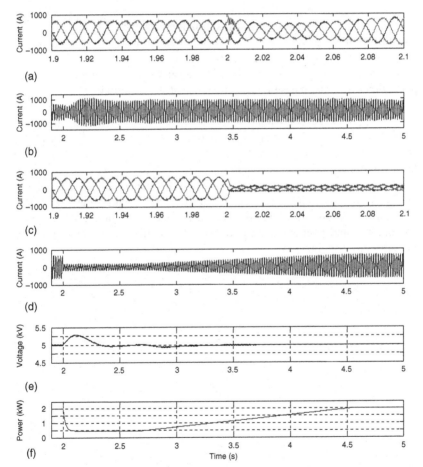

FIGURE 4.41 Response of the system to a voltage dip: (a) inverter side currents (detail); (b) inverter one phase current; (c) rectifier side currents (detail); (d) rectifier one phase current; (e) DC-link voltage; and (f) active power extracted from the generator.

4.4 ELECTRICAL SYSTEM OF A WIND FARM

4.4.1 Electrical Schematic of a Wind Farm

A wind farm is integrated by wind turbines and the substation that connect the farm to the utility grid to evacuate the electrical energy. The wind farm is arranged by string of wind turbines. Figure 4.42 shows a string compounded by several aerogenerators. These wind turbines are connected by manual switch breakers which isolate a wind turbine or it isolates the whole string. In variable-speed applications, an AC/AC power converter is used. This power converter is connected by a manual switch to the machine. The power converter includes a remote controlled switch breaker which isolates from the power transformer. The switch breaker is used for automatic reconnection after a fault. Figure 4.42 shows the transformer connection.

A schematic diagram of a typical substation is shown in Fig. 4.43. A large transformer, depicted in the figure, or several transformer connected in parallel, changes from the medium voltage to a higher voltage level. A typical voltage levels in Europe could be 20 kV/320 kV. The substation also incorporates bus bar, protections systems, measurement instrumentation, and auxiliary services circuit. Bus bar voltage measurement is made by voltage transformer. Each branch current, including several wind turbines, is measured by current transformer.

Some farms with lower rated power or connected to an isolated grid, e.g. wind-diesel systems, do not use this large transformer. The schematic of an isolated wind-diesel installation is represented in Fig. 4.44. Every power generator and load are connected to a medium voltage bus bar, in the typical range of 10–20 kV. The transformers are protected by circuit breakers that connect the lines directly to ground when open. A measurement system is used for power consumption and electrical quality control. Also auxiliaries' power supply feeds the substation equipment.

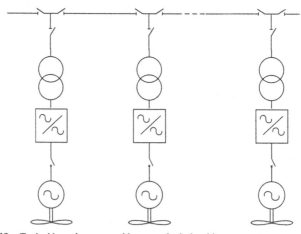

FIGURE 4.42 Typical branch composed by several wind turbines.

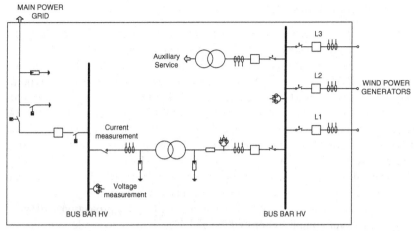

FIGURE 4.43 Schematic diagram of a typical wind farm substation.

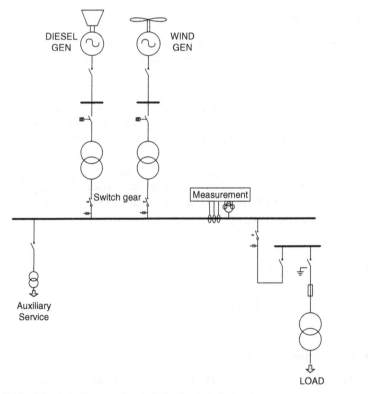

FIGURE 4.44 Schematic diagram of a wind–diesel generation system.

4.4.2   Protection System

Protection of wind power systems requires an understanding of system faults and their detection, as well as their safe disconnection. The protection system of a wind farm is mainly included in the substation. Circuit breaker and switchgear [37] are extensively used for overcurrent protection. New type of relay has been designed for the protection of wind farms that incorporate fixed-speed induction generators as described in [38]. A protection relay can be installed in the medium-voltage collecting line at the common point connection to the utility grid. This relay provides short-circuit protection for the collecting line and the medium-voltage (MV) and low-voltage (LV) circuits. Consequently, the relay allows wind farms to be constructed and adequately protected without the need to include fuses on the MV side of each generator–transformer.

The variable speed generator also includes digital relay protection and can be programmed for complex coordination and selectivity. This modern protection system can be used for voltage gap or sag function protection. Moreover, they can implement modern stabilization programs [39].

4.4.3   Electrical System Safety: Hazards and Safeguards

It is important to understand the hazards of electricity at the power system supply level. The safety of wind farm includes a good knowledge of electrical blast, electrocution, short circuits, overloads, ground faults, fires, lifting and pinching injuries. It is also recommended to review the principles, governmental regulations, work practices, and specialized equipment relating to electrical safety. Installers and maintenance personnel have to know the different types of "Personal Protective Equipment" through demonstrations of locking and tagging devices, protective clothing, and specialized equipment. The isolation and "Lockout Practices Procedures" for the lockout and isolation of electrical equipment can also be implemented into the existing site regulations and policies. A common practise is to use isolation transformer and grounding circuit breaker as they are being operated.

4.5   FUTURE TRENDS

Future trends relating to power electronics used in wind turbine applications can be summarized in the following points:

4.5.1   Semiconductors

Improvements in the performance of power electronics variable frequency drives for wind turbine applications have been directly related to the availability of power semiconductor devices with better electrical characteristics and lower prices because the device performance determines the size, weight, and cost of the entire power electronics used as interfaces in wind turbines [8,9,18].

The thyristor is the component that started power electronics. It is an old device with decreasing use in medium power applications, which was replaced by turn-off components like insulated gate bipolar transistor (IGBTs). The IGBT, which can be considered as an MOS bipolar Darlington, is now the main component for power electronics, and also for wind turbine applications. They are now mature technology turn-on components adapted to very high power (6 kV–1.2 kA), and they are in competition with gate turn-off thyristor (GTO) for high power applications [40].

Recently, the integrated gated control thyristor (IGCT) has been developed, consisting of the mechanical integration of a GTO plus a delicate hard drive circuit that transforms the GTO into a modern high performance component with a large safe operation area (SOA), lower switching losses, and a short storage time [41–43].

The comparison between IGCT and IGBT for frequency converters, used especially in wind turbines is explained below:

- IGBTs have higher switching frequency than IGCTs so they introduced less distortion in the grid. Accordingly if we use two three-phase systems in parallel, it is possible to double the resulting switching frequency without increasing the power loss, hence it is possible to have a total harmonic distortion (THD) of less than 2% without special harmonic filters.
- IGCTs are made like disk devices. They have to be cooled with a cooling plate by electrical contact on the high voltage side. This is a problem because high electromagnetic emission will occur. Another point of view is the number of allowed load cycles. Heating and cooling the device will always bring mechanical stress to the silicon chip and can be destroyed. This is a serious problem, especially in wind turbine applications. On the other hand, IGBTs are built like module devices. The silicon is isolated to the cooling plate and can be connected to ground for low electromagnetic emission even with higher switching frequency. The base plate of this module is made of a special material which has exactly the same thermal behavior as silicon, so nearly no thermal stress occurs. This increases the lifetime of the device by 10-fold approximately.
- The main advantage of IGCTs versus IGBTs is that they have a lower on state voltage drop, which is about 3.0 V for a 4500 V device. In this case, the power dissipation due to a voltage drop for a 1500 kW converter will be 2400 W per phase. On the other hand, in the case of IGBT, the voltage drop is higher than IGCTs. For a 1700 V device having a drop of 5 V, and in this case the power dissipation due to the voltage drop for a 1500 kW condition will be 5 kW per phase.

In conclusion, with the present semiconductor technology, IGBTs present better characteristics, for frequency converters in general and especially for wind turbine applications.

4.5.2   Power Converters

The technology of power electronic interfaces for variable-speed wind turbines is focused on the following points:

- Development of high efficiency/high quality voltage source AC/DC/AC converter for a main connection of variable wind turbines, operating with either a permanent magnet, a synchronous or an asynchronous generator.
- Operation at a power factor around one with higher-harmonic voltage distortion less than international standards.
- The power quality of the electrical output of the wind farms may be improved by the use of advanced static var compensators STATCOM or active power filters using power semiconductors like IGBTs, IGCTs, or GTOs. These kind of power conditioning systems are a new emerging family of FACTS (flexible AC transmission system) converters, which allow improved utilization of the power network. These systems will allow wind farms to reduce voltage drops and electrical losses in the network without the possibility of transient over voltage at islanding due to self-excitation of wind generators. Moreover, power conditioning systems equipment with different control algorithm can be used to control the network voltage, which will fluctuate in response to the wind farm output if the distribution network is weak [44,45].
- For large power wind turbine applications where it is necessary to increase the voltage level of the semiconductor of the power electronic interface, multilevel power converter technology is emerging as a new breed of power converter options for high-power applications. The general structure of the multilevel converter is to synthesize a sinusoidal voltage from several levels of voltages, typically obtained from capacitor voltage sources. Additionally, these converters have better performance and controllability because they use more than two voltage levels [46–48].

4.5.3   Control Algorithms

A variable pitch and speed wind turbine is a very complex non-lineal system. The control problem is more difficult to solve because some performance objectives, such as maximum power captured, minimum mechanical stress, constant speed, and power constant counteract each other. To solve this problem, a Fuzzy Logic control has recently been proposed. The Fuzzy Logic controller implements a rule-based structure [49] that can be easily adapted in order to optimize performance control objectives and has been widely used in introduction motor control applications [50–53]. In [54], a Fuzzy Logic controller is used to optimize the power captured using maximum power tracking algorithms. Other original structures have been proposed in [55]. The structure implements different control strategies depending on the rotor speed and generates a current torque control action. Presently, more complex control structures are being researched.

4.5.4 Offshore and Onshore Wind Turbines

One of the main trends in wind turbine technology is offshore installations. There are great wind resources at sea for installing wind turbines in many areas where the depth of the sea is relatively shallow. There are several demonstration plants that have had extremely positive results, so interest has increased in installing offshore wind farms, because of the development of large commercial power MW wind turbines. Offshore wind turbines may have slightly more favorable energy balance than onshore turbines, depending on local wind conditions. In places where onshore wind turbines are typically placed on flat terrain, offshore wind turbines will generally yield some 50% more energy than a turbine placed on a nearby onshore site. The reason is the low roughness of the sea surface. On the other hand, the construction and installation of a foundation requires 50% more energy than onshore turbines. It should be remembered, however, that offshore wind turbines have a longer life expectancy, around 25 to 30 years, than onshore turbines. The reason is that the low turbulence at sea gives lower fatigue loads on the wind turbine.

From a power electronics point of view, offshore wind turbines are interesting because, under certain circumstances, they become desirable to transmit the generated power to the load center over DC transmission lines (HVDC). This alternative becomes economically attractive versus AC transmission when a large amount of power is to be transmitted over a long distance from a remote wind farm to the load center [8]. Moreover, the transient stability and the dynamic damping of the electrical system oscillations can be improved by HVDC transmissions.

NOMENCLATURE

C_{dc}	DC link capacitor.
$C_p(\lambda, \beta)$	Power coefficient at tip speed ratio λ and pitch β.
e_{rg}, e_{sg}, e_{tg}	Instantaneous values of grid voltages.
e_α, e_β	Grid voltages expressed in an orthogonal reference frame.
f_1	Excitation frequency (the same as the grid frequency) in hertz.
f_2	Frequency of the voltage supplied to the rotor of machine 2 in hertz.
G	Gearbox ratio.
i_{dc}	DC inductor current of the step-up converter.
i_{dc}^{ref}	Desired DC inductor current of the step-up converter.
i_{dse}, i_{qse}	Instantaneous values of the direct and quadrature-axis stator current components, respectively, and expressed in rotor-flux-oriented reference frame.
$i_{dse}^{ref}, i_{qse}^{ref}$	Desired instantaneous values of the direct and quadrature-axis stator current components, respectively, and expressed in rotor-flux-oriented reference frame.

i_{dsr}, i_{qsr}	Stator current components established in the rotor reference frame.
$i_{dsr}^{ref}, i_{qsr}^{ref}$	Desired stator current components established in the rotor reference frame.
i_{ds}, i_{qs}	Instantaneous values of the direct and quadrature-axis stator current components, respectively, and expressed in the rotor reference phase.
i_{Ds}, i_{Qs}	Instantaneous values of the direct and quadrature-axis stator current components, respectively, and expressed in the stator reference frame.
$i_{Ds}^{ref}, i_{Qs}^{ref}$	Desired direct and quadrature-axis stator current components expressed in the stator reference frame.
i_{exc}	Synchronous generator excitation current.
i_{exc}^{ref}	Desired synchronous generator excitation current.
$i_{R_g}^{ref}, i_{S_g}^{ref}, i_{T_g}^{ref}$	Desired stator current.
$i_{R_r}, i_{S_r}, i_{T_r}$	Rotor current.
$i_{R_s}, i_{S_s}, i_{T_s}$	Stator current.
$i_{R_s}^{ref}, i_{S_s}^{ref}, i_{T_s}^{ref}$	Desired stator current.
i_{rg}, i_{sg}, i_{tg}	Public grid phase currents.
i_m^{ref}	Desired magnetizing current.
i_m	Magnetizing current.
i_{dsm}, i_{qsm}	Instantaneous values of the direct and quadrature-axis magnetizing stator current components, respectively, and expressed in the magnetizing current-oriented reference frame.
J	Total inertia of the system referred to the high speed shaft.
k_{ws}, k_{wr}	Winding factors of rotor and stator.
L_m	Coupled inductance.
L_{AC}	Inductance of inductors at the AC side of the inverter.
L_S, L_r	Stator and rotor windings inductances.
L_{DC}	Inductance of the step-up chopper.
n	Turns ratio of the machine. $n = k_{ws} \cdot n_s / k_{wr} \cdot n_r$
n_s, n_r	Number of turns of each rotor and stator phase.
N_1	Angular speed of the magnetic field (synchronous speed) expressed in rpm.
N_r	Angular speed of the generator rotor expressed in rpm.
p	Number of pole pairs.
p_1, p_2	Number of pole pairs of machine number 1 and 2.
P_1	Electrical power in the stator of principal machine number 1.
P_2	Electrical power in the stator of auxiliary machine number 2.
P_e	Electrical generated power.
P_m	Mechanical power in the low speed shaft.
P_{max}	Generated maximum power.
P_{rate}	Generator rate power.
P_S, Q_S	Active and reactive power through the stator.
P_S^{ref}, Q_S^{ref}	Desired active and reactive power through the stator.
P_{slip}	Slip power.
P_w	Available wind power.
p^{ref}, q^{ref}	Desired real power and desired reactive power on the grid side.
Q_e	Electromagnetic torque of the machine.
Q_e^{ref}	Desired electromagnetic torque of the machine.
Q_L	Torque in the low speed shaft.

Q_{rate}	Generator rate torque.
Q_t	Torque in the high speed shaft.
R	Rotor radius.
R_S, R_r	Stator and rotor winding resistors.
s	Slip.
$s_1 = \omega_1 - \omega_r / \omega_1$	Slips of machine principal number 1.
$s_2 = \omega_2 - \omega_r / \omega_2$	Slips of machine principal number 2.
T_s, T_r	Stator and rotor time constants, respectively.
$u_{dse}^{ref}, u_{qse}^{ref}$	Desired direct and quadrature-axis stator voltage expressed in the rotor-flux-oriented reference frame.
$u_{Ds}^{ref}, u_{Qs}^{ref}$	Desired direct and quadrature-axis stator voltage expressed in the stator reference frame.
$u_{R_r}, u_{S_r}, u_{T_r}$	Rotor voltage.
$u_{R_s}, u_{S_s}, u_{T_s}$	Stator voltage.
$u_{R_s}^{ref}, u_{S_s}^{ref}, u_{T_s}^{ref}$	Desired stator voltage.
u_r, i_r, λ_r	Rotor voltage, current, and flux, respectively, referred to a reference frame that rotates with the rotor.
u_r', i_r', λ_r'	Rotor voltage, current, and flux, respectively, referred to a reference frame fixed with the stator.
u_s, i_s, λ_s	Stator voltage, current, and flux, respectively, referred to a reference frame fixed with the stator.
u_{ds}, u_{qs}	Direct and quadrature-axis stator voltage expressed in the magnetizing current reference frame.
v_{rat}	Rated wind speed.
v_{pmax}	Maximum power wind speed.
v_{start}	Start wind speed.
v_{stop}	Stop wind speed.
v_w	Wind speed.
V_{dc}	DC-Link capacitor voltage.
V_{dc}^{ref}	Desired DC-Link capacitor voltage.
V_{st}^{max}	Maximum RMS stator voltage of the synchronous generator.
V_{st}^{min}	Maximum RMS stator voltage of the synchronous generator.
β	Pitch angle.
β^{ref}	Desired pitch angle.
δ	Load angle.
$\lambda = \omega \cdot R / V_v$	Tip speed ratio.
λ_{opt}	Optimal tip speed ratio.
λ_m	Magnetizing flux linkage vector.
$\lambda_{md}, \lambda_{mq}$	Instantaneous values of the direct and quadrature axis magnetizing flux linkage components expressed in the rotor reference frame.
$\|\lambda_m\|$	Estimated modulus of magnetizing flux linkage vector.
$\left\|\lambda_m^{ref}\right\|$	Desired modulus of magnetizing flux linkage vector.
η	Electrical performance.
η_r	Phase angle of the rotor flux linkage space phasor with respect to the direct axis of the stator reference frame.
θ_e	Magnetizing current angle.

θ_{sl}	Angle corresponding to the angular slip frequency.
θ_r	Rotor angle.
ρ	Air density.
$\omega_1 = 2\pi f_1/p_1$	Angular speed of the rotating magnetic flux produced in the stator of machine 1 relative to the stator.
$\omega_2 = 2\pi f_2/p_2$	Angular speed of the rotating magnetic flux produced in the rotor of machine 2 relative to the rotor.
ω_L	Low-speed shaft angular speed.
ω_r	Angular speed of the generator rotor.
ω_r^{ref}	Reference rotor speed.
ω_r^{max}	Maximum angular speed of the synchronous generator.
ω_r^{min}	Minimum angular speed of the synchronous generator.
ω_Q^{ref}	Desired angular speed for the torque controller.
ω_β^{ref}	Desired angular speed for the pitch controller.
ω_r^{rate}	Rated value of the rotor speed.
ω_e	Electrical angular speed of the magnetizing current reference frame.

REFERENCES

[1] S. Heier, Grid Integration of Wind Energy Conversion Systems, John Wiley & Sons, Chichester, Sussex, UK, 1998.

[2] G.L. Johnson, Wind Energy Systems, Prentice-Hall, Inc., Englewood Cliffs, NJ, 1985.

[3] P. Vas, Vector Control of AC Machines, Oxford Clarendon Press, NY, USA, 1990.

[4] V. Subrahmanyam, Electric Drives: Concepts and Applications, MacGraw-Hill, NY, USA, 1996.

[5] M. Alatalo, M. Sc, T. Svensson, Variable speed direct-driven PM-generator with a PWM controlled current source inverter, in: European Community Wind Energy Conference, Lùbeck-Travemùnde, Germany, March 1993.

[6] P. Vas, Sensorless Vector and Direct Torque Control, Oxford University Press, NY, USA, 1998.

[7] C.-M. Ong, Dynamic Simulation of Electric Machinery Using Matlab/Simulink, Prentice-Hall PTR, Upper Saddle, NJ, 1998.

[8] N. Mohan, T.M. Undeland, W.P. Robbins, Power Electronics, Converters, Applications, and Design, second ed., John Wiley & Sons, Inc., New York: NY, 1995.

[9] R.E. Tarter, Solid-State Power Conversion Handbook, John Wiley & Sons, Inc., New York: NY, 1993.

[10] B.K. Bose, Power Electronics and Variable Frequency Drives: Technology and Applications, IEEE Press, Piscataway, New Jersey, 1997.

[11] S.A. Papathanassiou, M.P. Papadopoulos, A comparison of variable speed wind turbine configurations, in: Wind Energy Conference, France, March 1999.

[12] D.S. Zinger, E. Muljadi, Annualized wind energy improvement using variable speeds, IEEE Trans. Ind. Appl. 33(6) (1997) 80-83.

[13] K. Pierce, Control method for improved energy capture below rated power, in: ASME/JSME Joint Fluids Engineering Conference, third ed., San Francisco, CA, July 1999.

[14] E.A Bossanyi, Bladed for Windows, Garrad Hassan and Partners Limited, September 1997 (Theory Manual) Document No 282/BR/009.

[15] E. Muljadi, K. Pierce, P. Migliore, Control strategy for variable-speed, stall-regulated wind turbines, in: American Controls Conference, Philadelphia, NREL, June 1998.

[16] A.D. Simmons, L.L. Freris, J.A.M. Bleijs, Comparison of energy capture and structural implementations of various policies of controlling wind turbines, in: Wind Energy: Technology and Implementation, 1991 (Amsterdam EWEC'91).

[17] W.E. Leithead, S. de la Salle, D. Reardon, Wind turbine control objectives and design, in: European Community Wind Energy Conference, Madrid, Spain, September 1990.

[18] M.H. Rashid, Power Electronics. Circuits, Devices, and Applications, second ed., Prentice-Hall, Englewood Cliffs, NJ, 1993.

[19] H. Akagi, A. Nabae, S. Atoh, Control strategy of active power filters using multiple voltage-source PWM converters, IEEE Trans. Ind. Appl. IA-22(3) (1986) 460-465.

[20] S. Bhowmik, R. Spée, J.H.R. Enslin, Performance optimization for doubly fed wind power generation systems, IEEE Trans. Ind. Appl. 35(4) (1999) 949-958.

[21] B. Hopfensperger, D.J. Atkinson, R.A. Lakin, Application of vector control to the cascaded induction machine for wind power generation schemes, in: 7th IEE European on Power Electronics, EPE'97, Trondheim, Norway, September 1997.

[22] P.O. 12.3 Propuesta sobre requisitos de respuesta frente a huecos de tensión de las instalaciones eólica, Red Eléctrica de España SA, October 2005.

[23] C. Rasmussen, P. Jorgensen, J. Havsager, Integration of wind power in the grid in Eastern Denmark, in: Proceedings of the 4th International Workshop on Large-scale Integration of Wind Power and Transmission Network for Offshore Wind Farm, Billund, Denmark, October 20-21, 2003.

[24] E.ON Netz Grid Code, E.ON Netz GmbH, Bayreuth, Germany, August 1, 2003.

[25] M. Johan, W.H. de Hann Sjoerd, Ride-through of wind turbines with doubly-fed induction generator during a voltage dip, IEEE Trans. Energy Convers. 20(2) (2005) 4246-4254.

[26] X. Bing, F. Brendan, F. Damian, Study of fault ride through for DFIG wind turbines, IEEE Trans. Energy Convers. 20(2) (2005) 411-416.

[27] International Electrotechnical Commission, Draft IEC 61400-21: Power Quality Requirements for Grid Connected Wind Turbines, Committee Draft (CD), December 1998.

[28] D. Foussekis, F. Kokkalidis, S. Tentzevakis, D. Agoris, Power quality measurement on different type of wind turbines operating in the same wind farm, in: EWEC, 2003.

[29] S. Poul, G. Gert, S. Fritz, R. Niel, D. Willie, K. Maria, M. Evangelis, L. Ake, Standards for measurements and testing of wind turbine power quality, in: EWEC, 1999.

[30] International Electrotechnical Commission, IEC Standard, Amendment 1 to Publication 61000-4-7, Electromagnetic Compatibility, General Guide on Harmonics and Inter-Harmonics Measurements and Instrumentation, 1997.

[31] International Electrotechnical Commission, IEC Standard, Publication 61000-3-6, Electromagnetic Compatibility, Assessment of Emission Limits for Distorting Loads in MV and HV Power Systems, 1996.

[32] L. Ake, S. Poul, S. Fritz, Grid impact of variable speed wind turbines, in: EWEC'99, 1999.

[33] A. Nabae, H. Akagi, I. Takahashi, A new neutral-point-clamped PWM inverter, IEEE Trans. Ind. Appl. IA-17(5) (1981) 518-523.

[34] G. Escobar, J. Leyva, J.M. Carrasco, E. Galvan, R. Portillo, M.M. Prats, L.G. Franquelo, Modeling of a three level converter used in a synchronous rectifier application, in: Proceedings of the Power Electronics Specialists Conference, PESC'04, Aachen, Germany, vol. 6, 2004, pp. 4306-4311.

[35] G. Escobar, J. Leyva-Ramos, J.M. Carrasco, E. Galvan, R. Portillo, M.M. Prats, L.G. Franquelo, Control of a three level converter used as a synchronous rectifier, in: 2004 IEEE 35th Annual Power Electronics Specialists Conference, PESC'04, vol. 5, June 20-25, 2004, pp. 3458-3464.

[36] M.M. Prats, L.G. Franquelo, R. Portillo, J.I. León, E. Galván, J.M. Carrasco, A three dimensional space vector modulation generalized algorithm for multilevel converters, IEEE Power Electron. Lett. 1 (2003) 110-114.

[37] W.D. Goodwin, High-voltage auxiliary switchgear for power stations, Power Eng. J. [see also Power Engineer] 3(3) (1989) 145-154.

[38] S.J. Haslam, P.A. Crossley, N. Jenkins, Design and evaluation of a wind farm protection relay, IEE Proc. Gen. Transm. Distrib. 146(1) (1999) 37-44, **doi:10.1049/ip-gtd:19990045.

[39] M.P. Palsson, T. Toftevaag, K. Uhlen, J.O.G. Tande, Large-scale wind power integration and voltage stability limits in regional networks, in: Power Engineering Society Summer Meeting, IEEE Proceedings, vol. 2, 2002, pp. 762-769, doi:10.1109/PESS.2002.1043417.

[40] J.M. Peter, Main future trends for power semiconductors from the state of the art to future trends, in: Power Conversion Intelligent Motion (PCIM'99), Nürnberg, June 1999.

[41] H. Grüning, B.Ødegård, J. Rees, A. Weber, E. Carroll, S. Eicher, High power hard-driven GTO module for 4.5 kV/3 kA snubberless operations, in: PCI Europe Proceedings, Nürnberg, 1996.

[42] A. Jaecklin, Integration of power components—state of the art and trends, in: European Power Electronic Conference EPE'97, Trondheim, 1997.

[43] H.R. Zeller, High power components from the state of the art to future trends, in: PCIM Europe Proceedings, Nürnberg, 1998.

[44] J.M. Carrasco, M. Perales, B. Ruiz, E. Galván, L.G. Franquelo, S. Gutiérrez, E. Gonzalez DSP control of an active power line conditioning system, in: 7th IEE European Conference on Power Electronics, EPE'97, Trondheim, Norway, September 1997.

[45] J. Balcells, M. Lamich, D. González, Parallel active filter based on a three level inverter, in: European Power Electronic Conference EPE'99, Laussane, 1999.

[46] R. Teodorescu, F. Blaabjerg, J.K. Pedersen, E. Cengelci, S.U. Sulistijo, B.O. Woo, P. Enjeti, Multilevel converters—a survey, in: European Power Electronic Conference EPE'99, Lausanne, September 1999.

[47] K. Oguchi, T. Karaki, N. Hoshi, Space vector of output voltages of reactor-coupled three phase multilevel voltage-source inverters, in: European Power Electronic Conference EPE'99, Lausanne, September 1999.

[48] J.-S. Lai, F.Z. Peng, Multilevel converters—a new breed of power converters, IEEE Trans. Ind. Appl. 32(3) (1996) 509-517.

[49] M. Sugeno, An introductory survey of fuzzy control, Inform. Sci. 36 (1985) 59-83.

[50] E. Galván, A. Torralba, F. Barrero, M.A. Aguirre, L.G. Franquelo, Fuzzy-logic based control of an induction motor, in: Industrial Fuzzy Control and Applications, Tarrasa, Spain, March 1993.

[51] E. Galván, A. Torralba, F. Barrero, M.A. Aguirre, L.G. Franquelo, A robust speed control of AC motor drives based on fuzzy reasoning, in: Industry Application Society (IAS), Toronto, Canada, October 1993.

[52] F. Barrero, E. Galván, A. Torralba, L.G. Franquelo, Fuzzy self tuning system for induction motor controllers, in: IEE European Conference on Power Electronics, EPE'95, Seville, Spain, September 1995.

[53] F. Barrero, A. Torralba, E. Galván, L.G. Franquelo, A switching fuzzy controller for induction motors with self-tuning capability, in: International Conference on Industrial Electronics, IECON'95, Orlando, FL, USA, November 1995.

[54] M. Godoy Simoes, B.K. Bose, R.J. Spiegel, Fuzzy logic based intelligent control of a variable speed cage machine wind generation system, IEEE Trans. Power Electron. 12(1) (1997) 87-95.

[55] M. Perales, J. Pérez, F. Barrero, J.L. Mora, E. Galván, J.M Carrasco, L.G. Franquelo, D. de la Cruz, L. Fernández, A. Zazo, A new fuzzy based approach for a variable speed, variable pitch wind turbine, in: 8th International Fuzzy Systems Association World Congress, IFSA'99, 1999.

Chapter 5

High-Frequency-Link Power-Conversion Systems for Next-Generation Smart and Micro Grid

S.K. Mazumder, Sr.
Department of Electrical and Computer Engineering, Laboratory for Energy and Switching-Electronics Systems (LESES), University of Illinois, Chicago, IL, USA

Chapter Outline

Alternative Energy in Power Electronics

5.1 INTRODUCTION

Photovoltaic (PV), wind, and fuel-cell (FC) energy are the front-runner renewable- and alternate-energy solutions to address and alleviate the imminent and critical problems of existing fossil-fuel-energy systems: environmental pollution as a result of high emission level and rapid depletion of fossil fuel. The framework for integrating these "zero-emission" alternate-energy sources to the existing energy infrastructure has been provided by the concept of distributed generation (DG) based on distributed energy resources (DERs), which provides an additional advantage: reduced reliance on existing and new centralized power generation, thereby saving significant capital cost. DERs are parallel and standalone electric generation units that are located within the electric distribution system near the end user. DER, if properly integrated, can be beneficial to electricity consumers and energy utilities, providing energy independence and increased energy security. Each home and commercial unit with DER equipment can be a micropower station, generating much of the electricity it needs on-site and sell the excess power to the national grid. The projected worldwide market is anticipated to be $\$50 \times 10^9$ billion by 2015.

A key aspect of these renewable- or alternative-energy systems is an inverter (note: for wind, a front-end rectifier is needed) that feeds the energy available from the energy source to application load and/or grid. Such power electronics for next-generation renewable- or alternative-energy systems have to address several features including (1) cost, (2) reliability, (3) efficiency, and (4) power density. Conventional approach to inverter design is typically based on the architecture illustrated in Fig. 5.1a. A problematic feature of such an approach is the need for a line-frequency transformer (for isolation and voltage step-up), which is bulky, takes large footprint space, and is becoming progressively more expensive because of the increasing cost of copper. As such, recently, there has been significant interest in high-frequency (HF) transformer-based inverter approach to address some or all of the above-referenced design objectives. In such an approach, a HF transformer (instead of a line-frequency transformer) is used for galvanic isolation and voltage scaling, resulting in a compact and low-footprint design. As shown in Fig. 5.1b,c, the HF transformer can be inserted in the dc–dc or dc–ac converter stages for multistage power conversion. For single-stage power conversion, the HF transformer is incorporated into the integrated structure. In the subsequent sections, based on HF architectures, we describe several high-frequency-link (HFL) topologies [1–8], being developed at the University of Illinois at Chicago, which have applications encompassing photovoltaics, wind, and fuel cells. Some have applicability for energy storage as well.

S.K. Mazumder

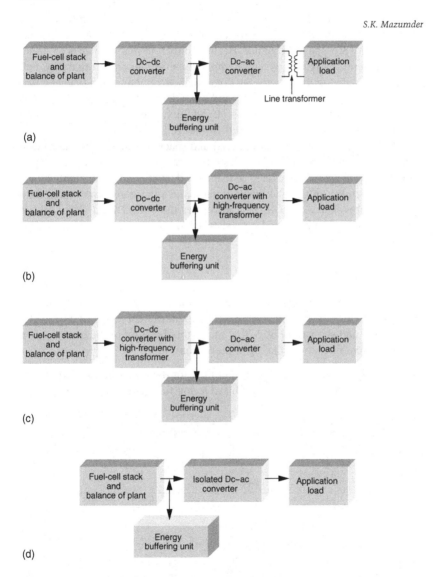

FIGURE 5.1   Inverter power-conditioning schemes [1] with (a) line-frequency transformer; (b) HF transformer in the dc–ac stage; (c) HF transformer in the dc–dc stage; and (d) single-stage isolated dc–ac converter.

5.2 LOW-COST SINGLE-STAGE INVERTER [2]

Low-cost inverter that converts a renewable- or alternative-energy source's low-voltage output into a commercial ac output is critical for success, especially for the low-power applications ($\leq$5 kW). Figure 5.2 shows one such single-stage isolated inverter, which was originally proposed in [10] as a "push–pull amplifier." It achieves direct power conversion by connecting load differentially across two bidirectional dc–dc Ĉuk converters and modulating them, sinusoidally, with 180° phase difference. Because only four main switches are used, it would potentially reduce system complexity, costs, improve reliability, and increase efficiency. Furthermore, the common source connection between two devices both at primary (Q_a and Q_c) and secondary sides (Q_b and Q_d) makes the gate-drive circuit relatively simple. In addition, the possibility of coupling inductors or integrated magnetics will further reduce the overall volume and weight, thereby achieving lesser material and space usage. Another advantage of this inverter is the reduction of turns ratio of the step-up transformer, which is usually required to achieve rated ac from low dc voltage. The inherent voltage boosting capability of the Ĉuk inverter can reduce the transformer turns-ratio requirement by at least half. Low transformer turns-ratio yields less leakage inductance and secondary winding resistance, which reduces the loss of duty cycle and secondary copper losses, respectively.

5.2.1 Operating Modes

In order to understand how the current flows and energy transfers during the switching and to help select the device rating, four different modes of the inverter are analyzed and shown in Fig. 5.3. It shows the direction of the current when the load current flows from the top to the bottom.

Mode 1: Figure 5.3a shows the current flow for the case when switch Q_a, Q_d are ON and Q_b, Q_c are OFF. During this time, the current flowing

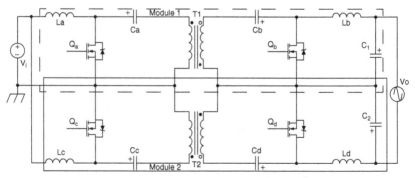

FIGURE 5.2 Schematic of the single-stage dc–ac differential-isolated Ĉuk inverter [2].

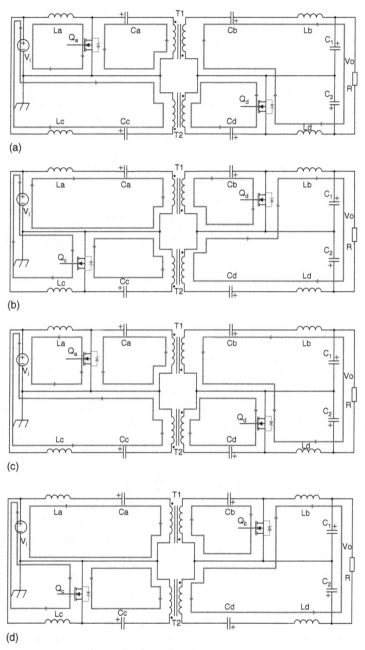

(a)

(b)

(c)

(d)

FIGURE 5.3 Direction of the current flow [2]: (a) and (b) for positive load current and (c) and (d) for negative load current. (a) Mode 1, when Q_a, Q_d are ON and Q_b, Q_c are OFF; (b) Mode 2, when Q_a, Q_d are OFF and Q_b, Q_c are ON; and (c) and (d) are Modes 3 and 4 corresponding to negative load current.

through the input inductor La increases and the inductor stores the energy. At the same time, the capacitor Ca discharges through Q_a, and thus, there is transfer of energy from the primary side to the secondary side through the transformer T1. The capacitor Cb is discharged to the circuit formed by Lb, C_1, and the load R. Meanwhile, the inductor Ld stores energy, and its current increases. The capacitor Cd discharges through Q_d. The power flows in opposite direction in the Module 2 from the secondary side to the primary side. The capacitor Cc is also discharged to provide the power.

Mode 2: When Q_a, Q_d are turned OFF, and Q_b, Q_c are ON (Fig. 5.3b); Ca, Cd, Cb, and Cc are charged using the energy, which was stored in the inductors La and Ld, while Q_a, Q_d were ON. During this time, Lb and Lc will release their energy.

Figure 5.3c,d shows the current direction when the load current flows in the opposite direction. The description for these two modes is omitted due to the similarity with Fig. 5.3a,b.

5.2.2 Analysis

Although the nonisolated inverter has already been proposed [10], detailed analysis and design of the isolated version have not appeared in any literature. The output of the inverter is the difference between two "sine-wave modulated PWM controlled" isolated Ĉuk inverters (Module 1 and Module 2), with their primary sides connected in parallel. The two diagonal switches of two modules are triggered by a same signal ($Q_a = Q_d$ and $Q_b = Q_c$), while the two switches in each module have complementary gate signals ($Q_a = /Q_b$ and $Q_c = /Q_d$). As we know, the output voltage of an isolated Ĉuk inverter can be expressed as follows:

$$V_o = V_i \cdot \frac{D}{N \cdot (1 - D)}, \tag{5.1}$$

where D is the duty ratio, N is the transformer turns ratio, and V_i is the input voltage. Because duty ratios for Modules 1 and 2 are complementary, the output difference between the two modules is

$$V_o = V_{c1} - V_{c2} = V_i \cdot \left(\frac{D}{N \cdot (1 - D)} - \frac{1 - D}{N \cdot D} \right). \tag{5.2}$$

The curves corresponding to the terms in (Eq. 5.2) with respect to the duty ratio D (assuming $N = 1$) are plotted in Fig. 5.4. The figure shows that although the gain-duty ratio curves of Modules 1 and 2 are not linear, their difference is almost linear. Therefore, if a sine-wave-modulated duty ratio D is used as a

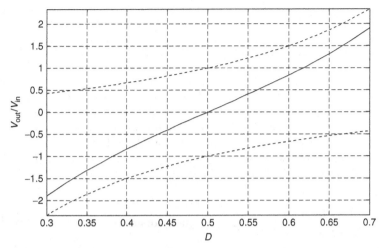

FIGURE 5.4 Voltage gain versus D [2] for Modules 1 and 2 (top and bottom), and their difference (middle).

control signal for the inverter, then its output voltage will be a sine wave with small distortion.

5.2.3 Design Issues

A 1 kW single-stage isolated dc–ac Ĉuk inverter prototype was designed and tested to verify its performance for fuel cell application, where stack voltage is 36 V. Some design issues are discussed later.

5.2.3.1 Choice of Transformer Turns-Ratio and Duty-Ratio Calculation

An inverter, normally, if operating at lower range of duty ratio (i.e., lower modulation index) with output power and input dc voltage fixed will produce lower output voltage, i.e., a higher current. This results in higher conduction losses and lower efficiency. Therefore, from the efficiency point of view, an inverter should usually operate at wide range of duty ratio. However, there is a duty-ratio limitation for proper operation on the dc–ac Ĉuk inverter. Unlike dc–dc, the duty ratios of the control PWM signals are not constant but sine-wave modulated. For a given input voltage (36 V dc, for instance) and output voltage (110 V ac, 60 Hz), the shape and magnitude of duty ratio (D) for dc–ac Ĉuk inverter will vary according to transformer turns ratio (N) [2]:

$$\frac{V_o}{V_i} = \frac{V_m \cdot \sin(wt)}{V_i} = \frac{2D - 1}{N \cdot D \cdot (1 - D)}. \tag{5.3}$$

Solving D, we obtain [2]

$$D = \frac{(\alpha \cdot \sin(\text{wt}) - 2) \pm \sqrt{4 + (\alpha \cdot \sin(\text{wt}))^2}}{2\alpha \cdot \sin(\text{wt})} = \frac{A \pm B}{C}, \qquad (5.4)$$

where $\alpha = V_m \cdot N/V_i$. It is a constant value. V_m represents the amplitude of the desired output sine wave. The numerical calculation shows that term B in (Eq. 5.4) is always larger than A. Thus, $A + B$ is always positive while $A - B$ is negative. Therefore, only $(A + B)/C$ is considered because D has to be a positive value. Note, when $\sin(\text{wt}) \to 0$, $C \to 0$, and $A + B \to 0$. In this case, D is calculated using L'Hospital's rule as [2]

$$D = \lim_{\sin(\text{wt}) \to 0} \frac{A + B}{C} = \frac{d(A + B)/d(\sin(\text{wt}))}{d(C)/d(\sin(\text{wt}))} = \frac{1}{2}. \qquad (5.5)$$

Figure 5.5 shows the plot of $D = (A + B)/C$ for three different transformer turns ratios. The results are summarized in Table 5.1. It is clearly shown that, there is a trade-off between output voltage distortion and duty-ratio range. An optimal transformer turns ratio, $N = 3$, is selected with corresponding D varying from 0.34 to 0.66.

5.2.3.2 Lossless Active-Clamp Circuit to Reduce Turn-Off Losses

There will be severe voltage spikes and ringing across the switches during turn-off. They are caused by the transformer and other parasitic leakage inductances combined with a very high current reverse going through the transformer primary side. The spike problem is more serious at the point where the output sine wave is at its peak because of the highest instantaneous current value at that point. The circuit inside the dotted block of Modules 1 and 2 in Fig. 5.6 shows a lossless active-clamp circuit, which can achieve zero-voltage turn-off, thereby reducing the turn-off losses and limiting the maximum voltage across the main switches. The circuit for each module contains two auxiliary diodes, one capacitor, one inductor, and one switch. The auxiliary switches S_{s1} and S_{s2} are triggered using the same gate signals as their corresponding main switches. The equations to calculate capacitance C_s and inductance L_s are listed as follows [2]:

$$C_s \geq L_{lk} \cdot \left(\frac{V_{in} \cdot D(\text{wt}) \cdot}{f \cdot (V_{c\,\max}(\text{wt}) - N \cdot V_0(\text{wt}) \cdot \text{Le})} \right)^2, \qquad (5.6)$$

$$\text{Le} = \frac{(1/N)^2 \cdot \text{La} \cdot \text{Lb}}{\text{La} + \text{Lb} \cdot (1/N)^2}, \qquad (5.7)$$

$$L_s \leq \frac{D^2(\text{wt})}{f^2 \cdot C_s \cdot [ar\cos(-\frac{1}{r(\text{wt})}) + \sqrt{r^2(\text{wt}) - 1}]}, \tag{5.8}$$

$$r(\text{wt}) = \frac{V_{c\,max}(\text{wt})}{V_{in}}, \tag{5.9}$$

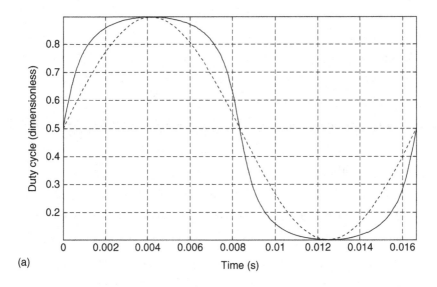

(a)

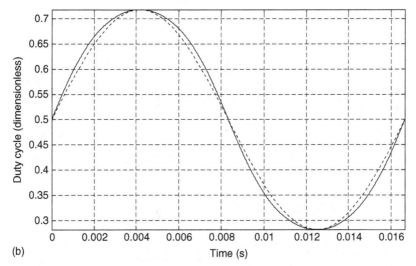

(b)

FIGURE 5.5 The plotted waveforms [2] of $D = (A + B)/C$ for variable transformer turns ratio (solid). (a) $N = 2{:}1$; (b) $N = 1{:}2$; and (c) $N = 1{:}5$, as compared with a standard sine wave (dash).

(Continued)

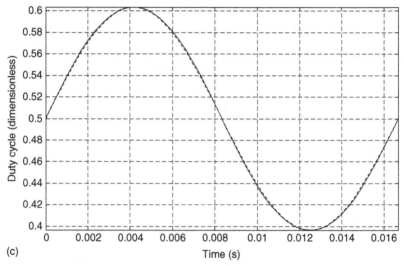

(c)

FIGURE 5.5, Cont'd

TABLE 5.1 Shape and range of D For Different Transformer Turns Ratio

Turns ratio	Shape of D	Magnitude of D	Figure number
2:1	Not a sine wave	0.1 ~ 0.9	5a
1:3	Close to a sine wave	0.34–0.66	5b
1:5	Very close to a sine wave	0.39 ~ 0.61	5c

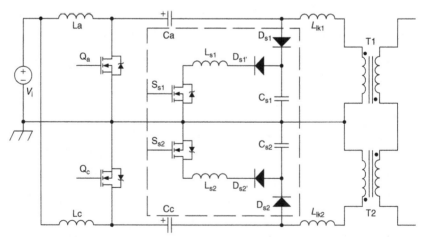

FIGURE 5.6 The single-stage Ĉuk inverter with lossless active-clamp circuit at the primary side [2].

where $D(\text{wt})$ is the sine wave modulated duty ratio, $V_o(\text{wt})$ is the sine wave output, $V_{c\,\text{max}}$ is the maximal clamped voltage, and Le is the effective inductance. With reference to Fig. 5.6, when the switch Q_a turns OFF, the clamp circuit will create an alternate path formed by diode D_{s1} and capacitor C_{s1} to divert the turn-off current from the primary switch Q_a. After switch Q_a and S_{s1} are turned ON, the energy stored in capacitor C_{s1} will eventually be fed back to the capacitor Ca as useful power. The performance of the active-clamp circuit along with inverter performance is provided in detail in [11].

5.3 RIPPLE-MITIGATING INVERTER [3, 4]

The inverter (see Fig. 5.7) described in this section comprises a dc–dc zero-ripple boost converter (ZRBC), which generates a high-voltage dc at its output followed by a soft-switched, transformer isolated dc–ac inverter, which generates a 110 V ac. The HF-inverter switches are arranged in a multilevel fashion and are modulated by a fully rectified sine wave to create a HF, three-level ac voltage as shown in Fig. 5.7. Multilevel arrangement of the switches is particularly useful when the intermediate dc voltage >500 V. The HF inverter is followed by the ac–ac converter, which converts the three-level ac to a voltage that carries the line-frequency sinusoidal information.

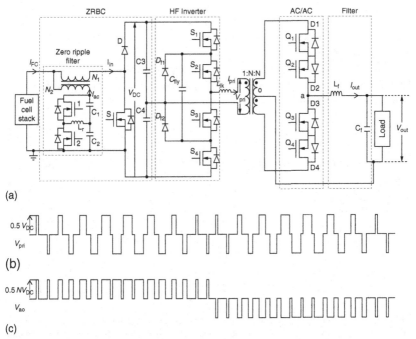

(a)

(b)

(c)

FIGURE 5.7 Schematic of the ripple-mitigating inverter [3]. The source can PV/battery/rectified wind as well.

5.3.1 Zero-Ripple Boost Converter (ZRBC)

The ZRBC is a standard nonisolated boost converter with the conventional inductor replaced by a hybrid zero-ripple filter (ZRF). The ZRF (shown in Fig. 5.8) is viewed as a combination of a coupled inductor (shown in Fig. 5.7) and a half-bridge active power filter (APF) (shown in Fig. 5.7). Such a hybrid structure serves the dual purpose of reducing the HF current ripples and the low-frequency current ripples. The coupled inductor minimizes the HF ripple from the source current ($I_{DC} + i_2 = i_1$) and the APF minimizes the low-frequency ripple from the source current ($I_{DC} + i_{ac} = i_{in}$). I_{DC} is the dc supplied by the source, i_2 is the HF ac supplied by the series combination of identical capacitors C_1 and C_2 (in Fig. 5.7), and i_{ac} is the low-frequency ac supplied by the APF storage reactor L_r. For effective reduction of the HF current from the source output, the value of the capacitors C_1 and C_2 should be as large as possible. However, the series combination should be small enough to provide a high-impedance path to the low-frequency current i_{ac}. Therefore, for a chosen value of capacitor, the values of the following expression hold true [3]:

$$C_1 = C_2 = 2C, \tag{5.10a}$$

$$f_{HF} = \frac{1}{\sqrt{L_2 C}}, \tag{5.10b}$$

$$f_{LF} = \frac{1}{\sqrt{4 L_r C}}, \tag{5.10c}$$

where f_{HF} is the switching frequency of the converter and f_{LF} is the lowest frequency component in i_{ac}.

Assuming the switching frequency is approximately 20 times the lowest frequency component, the value of ZRF passive components L_2 and L_r can be determined as follows [3]:

$$f_{HF} \geq 20 f_{LF}, \tag{5.10d}$$

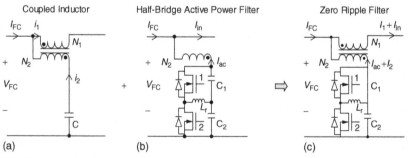

FIGURE 5.8 Schematic diagrams [3] and [4] of (a) coupled inductor structure for reducing the HF current ripple; (b) half-bridge active filter, which compensates for the low-frequency harmonic-current-ripple demand by the inverter; and (c) the proposed hybrid ZRF structure.

$$\frac{1}{\sqrt{L_2 C}} \geq \frac{10}{\sqrt{L_r C}}, \tag{5.10e}$$

$$L_r \geq 100 L_2. \tag{5.10f}$$

Therefore, the value of L_2 should be small in order to limit the value of L_r and also to minimize the phase shift in the injected low-frequency current i_{ac}.

In the following subsections, the HF and low-frequency ac-reduction mechanisms and the conditions to achieve the same are discussed. In addition to this, the effect of coupled inductor parameters on the bandwidth of the open-loop system will be discussed. For the purpose of analysis, the value of the capacitors C_1 and C_2 is assumed to be large. Hence, the dynamics of the APF is assumed to have minimal effect on the coupled-inductor analysis.

5.3.1.1 HF Current-Ripple Reduction

In this section, the inductance offered by the coupled inductor and the ripple reduction achievable is discussed. For that purpose, we need to derive an expression for the effective inductance of the coupled inductor. Because the value of the capacitors C_1 and C_2 is large and that of L_{22} is small, the dynamics of the APF is assumed to have minimal effect on the coupled-inductor analysis. The pi-model for the zero-ripple coupled inductor and the excitation voltage and the current for the primary and the secondary windings are shown in Fig. 5.9. The currents i_{1HF} and i_{2HF} are, respectively, the primary and the secondary ac shown in Fig. 5.9:

$$v_{FC} = (L_1 + L_M)\frac{di_{1HF}}{dt} + nL_M\frac{di_{2HF}}{dt}, \tag{5.11a}$$

$$v_C = (L_1 + L_M)\frac{di_{1HF}}{dt} + (L_2 + nL_M)\frac{di_{2HF}}{dt}, \tag{5.11b}$$

$$n = \frac{N_2}{N_1} \cong \sqrt{\frac{L_{22}}{L_{11}}}, \tag{5.11c}$$

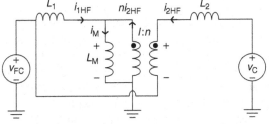

FIGURE 5.9 Ac model for the coupled inductor shown in Fig. 5.8a [3].

where L_{11} is the self inductance of the primary winding with N_1 turns. Solving (Eqs. 5.11a) and (5.11b), the expressions for $\frac{di_{1HF}}{dt}$ and $\frac{di_{2HF}}{dt}$ are obtained using

$$\frac{di_{1HF}}{dt} = \frac{(L_2 + nL_M)v_{FC} - nL_Mv_C}{(L_1 + L_M)L_2}$$

$$= \frac{v_{FC}}{(L_1 + L_M)} + \frac{nL_M(v_{FC} - v_C)}{(L_1 + L_M)L_2}, \qquad (5.12a)$$

$$\frac{di_{2HF}}{dt} = \frac{v_{FC} - v_C}{-L_2}. \qquad (5.12b)$$

By substituting Eq. (5.12a) in Eq. (5.12b), we obtain the following expression:

$$\frac{di_{1HF}}{dt} = \frac{(L_2 + nL_M)v_{FC} - nL_Mv_C}{(L_1 + L_M)L_2}$$

$$= \frac{v_{FC}}{(L_1 + L_M)} - \frac{nL_M}{(L_1 + L_M)} \frac{di_{2HF}}{dt}. \qquad (5.12c)$$

To reduce the ac component of the source current to zero, the following condition should hold:

$$\frac{di_{1HF}}{dt} = \frac{di_{2HF}}{dt}. \qquad (5.13)$$

Therefore, using the above condition and Eq. (5.12c), one obtains [3]

$$\frac{di_{1HF}}{dt} = \frac{v_{FC}}{L_{11}\left[1 + \frac{(1+n)}{n}k\right]} = \frac{V_{FC}}{L_{eff}}. \qquad (5.14)$$

The denominator of Eq. (5.14) is the effective inductance L_{eff} offered by the coupled-inductor structure of the hybrid filter. The effective inductance depends on the turns ratio n, the coupling coefficient k, and the self inductance L_{11} of the primary winding. For very small values of turns ratio ($n \ll 1$), significantly large values of effective inductances can be obtained. Figure 5.10 shows the effective inductance curves and the corresponding reduction in the ripple. Figure 5.10a shows the dependence of normalized L_{eff} on n as a function of k. For the values of effective inductance shown in Fig. 5.10a, the corresponding values of achievable ripple current in both the coupled-inductor windings are shown in Fig. 5.10b. Using Fig. 5.10b, a designer can decide on a value of HF current ripple, and using the corresponding values of n and k the normalized effective inductance can be chosen from Fig. 5.10a. While deciding the value of HF ripple, one should choose a small value for n (<0.25) to ensure that L_{22} is small enough to prevent significant variations in the voltage across capacitors C_1 and C_2. Also, the effective inductance should be chosen to meet the bandwidth requirements of the ZRBC. Increase in the effective input inductance has a two-pronged

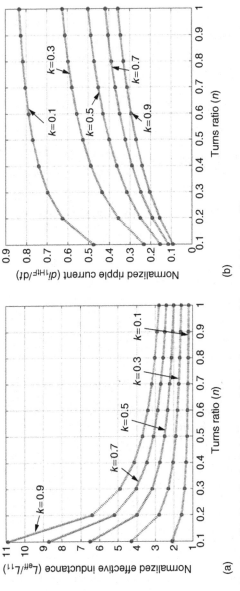

FIGURE 5.10 Normalized (a) effective inductance and (b) ripple current of the coupled inductor [3].

effect on the open-loop frequency response of the ZRBC. First, the bandwidth is reduced, and second, the RHP zero is drawn closer to the imaginary axis resulting in a reduction in the available phase margin and thereby the ZRBC stability.

5.3.1.2 Active Power Filter

The input current of the inverter comprises a dc component and a 120-Hz ac component and is expressed as [3]

$$I_{dc} + I_{ac} = \frac{V_{out}I_{out}}{V_{dc}} \cos(\theta) - \frac{V_{out}I_{out}}{V_{dc}} \cos(2\omega t - \theta), \qquad (5.15)$$

where, V_{out} are inverter output voltage

I_{out} are inverter output current
V_{dc} is the average value of voltage across the series capacitors C_1 and C_2
θ is the load power factor angle.

Here, we derive the condition for low-frequency current ripple elimination from the PCS input current. For the APF shown in Fig. 5.8, the voltage across the storage reactor L_r of the APF is expressed as

$$V_{ab} = V_a - \frac{V_{dc}}{2} = V_{dc}\left(S_a - \frac{1}{2}\right), \qquad (5.16)$$

where S_a is the modulating signal. The reactor current i_r is

$$i_r = \frac{V_{dc}\left(S_a - \frac{1}{2}\right)}{j\omega L_r}, \qquad (5.17)$$

where, $S_a = 0.5 + \sum\limits_{n=1}^{\infty} B_n \sin(n(\omega t + \phi))$ and $i_r = \frac{V_{dc}}{j\omega L_r}B_n \sin\left(n\left(\omega t + \phi - \frac{\pi}{2}\right)\right)$
(Considering only the fundamental component.)

The current injected by the APF is

$$i_{ac} = \left(S_a - \frac{1}{2}i_r\right), \qquad (5.18a)$$

$$i_{ac} = \frac{V_{dc}}{\omega L_r}B_n^2 \sin(\omega t + \phi)\sin\left(\omega t + \phi - \frac{\pi}{2}\right), \qquad (5.18b)$$

$$i_{ac} = \frac{V_{dc}}{2\omega L_r}B_n^2\left[\cos\left(\frac{\pi}{2}\right) - \cos\left(2\omega t + 2\phi - \frac{\pi}{2}\right)\right]. \qquad (5.18c)$$

In order to reduce the second harmonic in the input current to zero, $i_{ac} = I_{ac}$ [3]

$$i_{ac} = \frac{V_{dc}}{2\omega L_r}B_n^2 \cos\left(2\omega t + 2\phi - \frac{\pi}{2}\right) = \frac{V_o I_o}{V_{dc}}\cos(2\omega t - \theta). \qquad (5.19a)$$

This yields

$$B_n = \frac{\sqrt{2\omega L_r V_0 I_0}}{V_{dc}}, \tag{5.19b}$$

$$\phi = \frac{\pi}{4} - \frac{\theta}{2}. \tag{5.19c}$$

5.3.2 HF Two-Stage DC–AC Converter

The two-stage dc–ac converter (shown in Fig. 5.7) comprises a soft-switched, phase-shifted SPWM, multilevel HF converter (on the primary side of the transformer) and a line-frequency-switched ac–ac converter (on the secondary side of the transformer) followed by output low-pass filter. The multilevel arrangement of the HF converter switches reduces the voltage stress and the cost of the HF semiconductor switches. The ac–ac converter has two bidirectional switch pairs Q_1 and Q_2, and Q_3 and Q_4 for a single-phase output. To achieve a 60-Hz sine-wave ac at the output, a sine-wave modulation is performed either on the HF dc–ac converter or on the ac–ac converter. Therefore, two different modulation strategies are possible for the dc–ac converter. Both schemes result in the soft switching of the HF converter, while the ac–ac converter is hard-switched.

In the first modulation scheme, the ac–ac converter switches follow SPWM, while the HF converter switches are switched at fixed 50% duty ratio. The HF converter switches in this scheme undergo zero-voltage turn-on. In the second modulation scheme, the switches of the multilevel HF converter follow SPWM, and the ac–ac converter switches are switched based on the power-flow information. Unlike the first modulation scheme, which modulates the ac–ac converter switches at HF, in the second modulation scheme, ac–ac converter operates at line frequency. The switches are commutated at HF only when the polarities of output current and voltage are different. Usually this duration is very small, and therefore the switching loss of the ac–ac converter is considerably reduced compared with the conventional control method. Therefore, the heat-sinking requirement of the ac–ac converter switches is significantly reduced. The HF converter switches in this scheme undergo zero-current turn-off. Control signals for the second modulation scheme are shown in Fig. 5.11.

5.4 UNIVERSAL POWER CONDITIONER [1]

This approach achieves a direct power conversion and does not use any front-end dc–dc converter. As shown in Fig. 5.12, this approach has a HF dc–ac converter followed by a HF transformer and a forced ac–ac converter. Switches $(Q_1–Q_4)$ on the primary side of the HF transformer are sine-wave modulated to create a HF three-level bipolar ac voltage. The three-level ac at the output

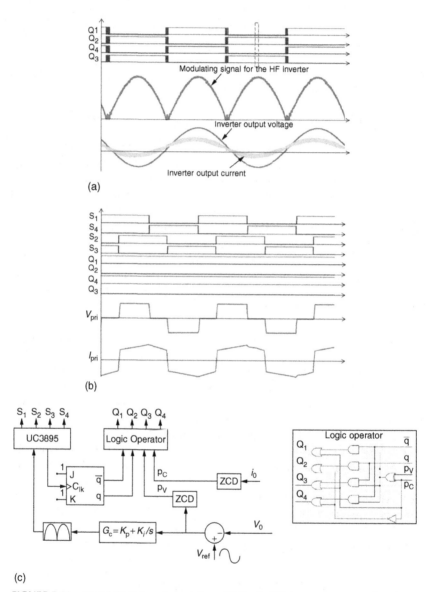

FIGURE 5.11 (a) and (b) Schematic waveforms [3] for the HF dc–ac converter on the primary side of the transformer and the ac–ac converter on the secondary side of the transformer. (c) Overall control scheme for the two-stage HF inverter.

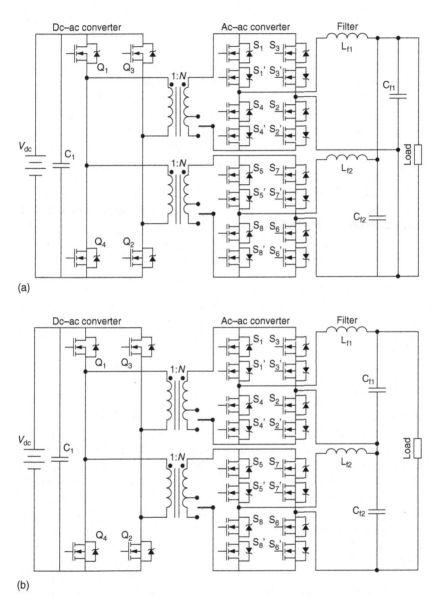

(a)

(b)

FIGURE 5.12   Circuit diagrams [1] of the proposed fuel-cell inverter for (a) 120 V/60 Hz ac outputs and (b) 240 V/50 Hz ac outputs. A single-pole-double-throw (SPDT) switch enables adaptive tapping of the transformer.

of the HF transformer is converted to 60/50-Hz line-frequency ac by the ac–ac converter and the output LC filter. For an input of 30 V, the transformation ratio of the HF transformer is calculated to be $N = 13$. Fabrication of a 1:13 transformer is relatively difficult. Furthermore, high turns-ratio yields enhanced secondary leakage inductance and secondary winding resistance, which result in measurable loss of duty cycle and secondary copper losses, respectively. Higher leakage also leads to higher voltage spike, which added to the high nominal voltage of the secondary necessitate the use of high-voltage power devices. Such devices have higher on-resistance and slower switching speeds. Therefore, a combination of two transformers and two ac–ac converters on the secondary side of the HF transformer is identified to be an optimum solution. For an input voltage in the range of 30–42 V, we use $N = 6.5$, while for an input voltage of more than 42 V, we use $N = 4.3$. To change the transformation ratio of the HF transformer, we use a single-pole double-throw (SPDT) relay, as shown in Fig. 5.12a,b. Such an arrangement not only improves the efficiency of the transformer but also significantly improves the utilization of the ac–ac converter switches for operation at 120/240 V ac and 60/50 Hz. For 120-V ac output, the two ac–ac-converter filter capacitors are paralleled (as shown in Fig. 5.12a), while for 240-V ac output, the voltage of the filter capacitor are connected in opposition (as shown in Fig. 5.12b). Finally, Fig. 5.13 shows the closed-loop control mechanism of the inverter for grid-parallel and grid-connected modes. It is described in detail in [1] and not repeated here.

5.4.1 Operating Modes

In this section, we discuss the modes of operation of the inverter in Fig. 5.12 for 120-V and 240-V ac output and for an input voltage in the range of 42–60 V (i.e., $N = 4.3$). The modes of operation less than 42 V (i.e., $N = 6.5$) remain the same. Figures 5.14 and 5.15 show the waveforms of the five operating modes of the phase-shifted HF inverter and a positive primary and a positive filter-inductor current. Modes 2 and 4 show the zero-voltage switching (ZVS) turn-on mechanism for switches Q_3 and Q_4, respectively. Unlike conventional control scheme for ac–ac converter [12], which modulates the switches at HF, the outlined ac–ac converter operates at line frequency. The switches are commutated at HF only when the polarities of the output current and voltage are different [12]. For unity-power-factor operation, this duration is negligibly small, and therefore, the switching loss of the ac–ac converter is considerably reduced compared with the conventional control method [13].

Five modes of the inverter operation are discussed for positive primary current. A set of five modes exists for a negative primary current as well. A similar set of five modes of operation for the 240 V ac exists for input voltage of more than 42 V ($N = 4.5$). Again, the mode of operation for input voltage of less than 42 V ($N = 6.5$) remains the same.

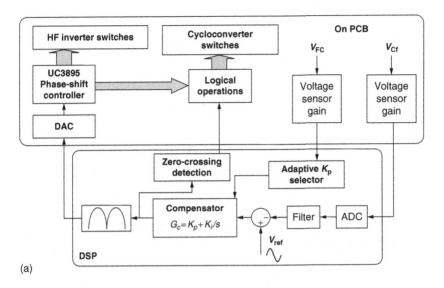

(a)

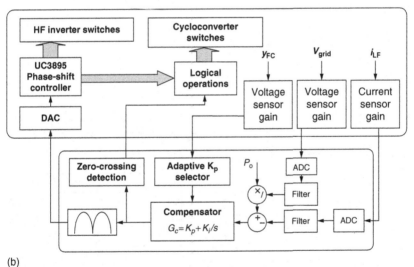

(b)

FIGURE 5.13 (a) and (b) Schematics for converter operation [1], respectively, at 120 V ac and 60 Hz and 240 V ac and 50 Hz. (c) and (d) Control schemes of the inverter in grid-parallel and grid-connected modes.

Mode 1 (Fig. 5.14a): During this mode, switches Q_1 and Q_2 of the HF inverter are ON, and the transformer primary current I_{p1} and I_{p2} is positive. The load current splits equally between the two cycloconverter modules. For the top cycloconverter module, the load current $I_{out}/2$ is positive and flows through the switches pair S_1 and S_1', the output filter L_{f1} and

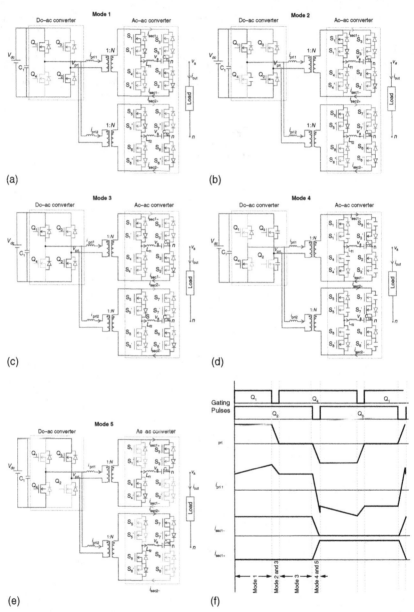

FIGURE 5.14 Modes of operation [1] for 120 V ac for input voltage in the range of 42–60 V ($N = 4.3$): (a–e) topologies corresponding to the five operating modes of the overall dc–ac converter for positive primary current and for power flow from the input to the load. (f) Schematic waveforms show the operating modes of the HF inverter when primary currents are positive. The modes of operation of less than 42 V (i.e., $N = 6.5$) remain the same.

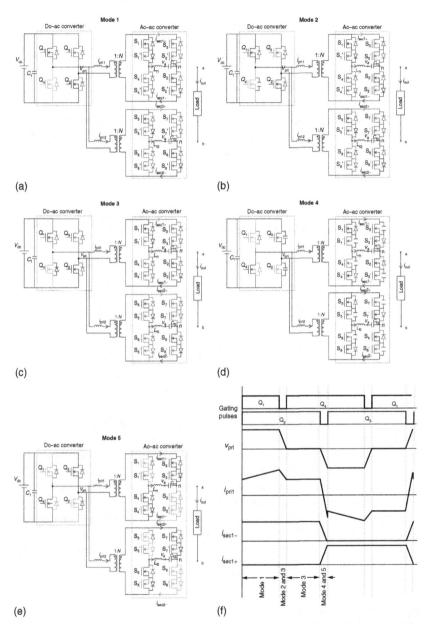

FIGURE 5.15 Modes of operation [1] for 240 V/50 Hz ac output for an input-voltage range of 42–60 V (corresponding to $N = 4.3$): (a–e) topologies corresponding to the five operating modes of the inverter for positive primary and positive filter-inductor currents. (f) Schematic waveforms show the operating modes of the dc–ac converter when primary currents are positive. The modes of operation of less than 42 V (corresponding to $N = 6.5$) remain the same.

C_{f1}, switches S_2 and S_2', and the transformer secondary. Similarly, for the bottom cycloconverter module, the load current $0.5 \times I_{out}$ is positive and flows through the switches pair S_5 and S_5', the output filter L_{f2} and C_{f2}, switches S_6 and S_6', and the transformer secondary.

Mode 2 (Fig. 5.14b): At the beginning of this interval, the gate voltage of the switch Q_1 undergoes a high-to-low transition. As a result, the output capacitance of Q_1 begins to accumulate charge and, at the same time, the output capacitance of switch Q_4 begins to discharge. Once the voltage across Q_4 goes to zero, it is can be turned on under ZVS. The transformer primary currents I_{p1} and I_{p2} and the load current I_{out} continue to flow in the same direction. This mode ends when the switch Q_1 is completely turned OFF and its output capacitance is charged to V_{DC}.

Mode 3 (Fig. 5.14c): This mode initiates when Q_1 turns OFF. The transformer primary currents I_{p1} and I_{p2} are still positive, and free wheels through Q_4 as shown in Fig. 5.14c. Also the load current continues to flow in the same direction as in Mode 2. Mode 3 ends at the commencement of turn off Q_2.

Mode 4 (Fig. 5.14d): At the beginning of this interval, the gate voltage of Q_2 undergoes a high-to-low transition. As a result, the output capacitance of Q_2 begins to accumulate charge and, at the same time, the output capacitance of switch Q_3 begins to discharge as shown in the Fig. 5.14d. The charging current of Q_2 and the discharging current of Q_3 together add up to the primary currents I_{p1} and I_{p2}. The transformer current makes a transition from positive to negative. Once the voltage across Q_3 goes to zero, it is turned ON under ZVS. The load current flows in the same direction as in Mode 3 but makes a rapid transition from the bidirectional switches S_1 and S_1' and S_2 and S_2' to S_3 and S_3' and S_4 and S_4', and during this process $I_{out}/2$ splits between the two legs of the cycloconverter modules as shown in Fig. 5.14d. Mode 4 ends when the switch Q_2 is completely turned OFF, and its output capacitance is charged to V_{DC}. At this point, it is necessary to note that because S_1 and S_2 are OFF simultaneously, each of them support a voltage of V_{DC}.

Mode 5 (Fig. 5.14e): This mode starts when Q_2 is completely turned OFF. The primary currents I_{p1} and I_{p2} are negative, while the load current is positive as shown in Fig. 5.14e.

5.4.2 Design Issues

5.4.2.1 Duty-Ratio Loss

As shown in Fig. 5.16, the finite slope of the rising and falling edges of the transformer primary current because of the leakage inductance (L_{lk}) will reduce the duty cycle (d). This duty-ratio loss is given by [1]

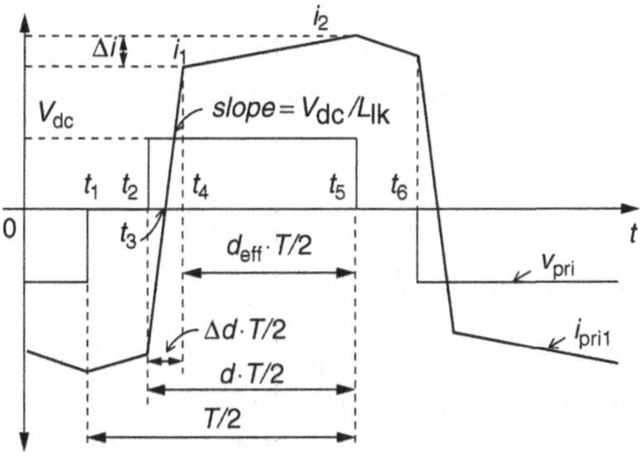

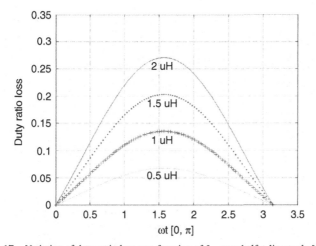

FIGURE 5.16 Transformer primary-side voltage and current waveforms [1].

$$\Delta d = \frac{N}{\frac{V_{dc}}{L_{lk}} \cdot \frac{T}{2}} \cdot \left(2i_{out} - \frac{v_{out}}{L_{f1}} \cdot (1 - d) \cdot \frac{T}{2} \right), \qquad (5.20)$$

where N is the transformer turns ratio, L_{f1} is the output filter inductance, i_{out} is the filter current, v_{out} is the output voltage, and T is the switching period. Assuming that L_{f1} is large enough, the second term in Eq. (5.20) can be omitted. Thus, the duty-ratio loss has a sinusoidal shape and is proportional to N and L_{lk}. One can deduce from Eq. (5.20) that, due to the high turns-ratio and low-input voltage, even a small leakage inductance will cause a big duty-ratio loss.

Figure 5.17 shows (for a 1-kW inverter) the calculated duty-ratio loss for an input voltage of 30 V and for $N = 6.5$. Four parametric curves correspond to four leakage inductances of 0.5, 1, 1.5, and 2 μH are shown. Figure 5.17 shows

FIGURE 5.17 Variation of duty-ratio loss as a function of L_{lk} over half-a-line cycle [1].

that, for a L_{lk} of 2 μH, the duty-ratio loss is more than 25%. Consequently, a transformer with even higher turns-ratio is required to compensate for this loss in the duty ratio, which increases the conduction loss and eventually decreases the efficiency.

5.4.2.2 Optimization of the Transformer Leakage Inductance

The leakage inductance of the HF transformer enhances the ZVS range of the dc–ac converter but reduces the duty ratio of the converter, which increases conduction loss. Thus, the leakage inductance of each transformer is designed to achieve the highest efficiency, as illustrated in Fig. 5.18. For the sinusoidally modulated dc–ac converter, the ZVS capability is lost twice in every line cycle. The extent of the loss of ZVS is a function of the output current. The available ZVS range (t_{ZVS}) as a percentage of the line cycle ($t_{LineCycle}$) is given by [1]

$$\frac{t_{ZVS}}{t_{LineCycle}} = \frac{2}{\pi} \sin^{-1} \left(\frac{1}{4} \frac{V_{dc}^2 \left(\frac{4}{3} C_{oss} + \frac{1}{2} C_T \right)}{i_{out}^2 L_{lk}} \right)^{1/2} , \qquad (5.21)$$

where C_{oss} is the device output capacitance and C_T is the interwinding capacitance of the transformer. When the dc–ac converter is not operating under ZVS condition, the devices are hard-switched. A numerical calculation of the total switching losses for the 1-kW inverter, as shown in Fig. 5.19, indicates that the optimal primary-side leakage inductance for the HF transformer should be between 0.2 and 0.7 μH. Clearly, as the leakage inductance of the HF

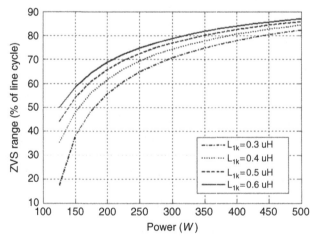

FIGURE 5.18   ZVS range of the dc–ac converter with variation in output power [1].

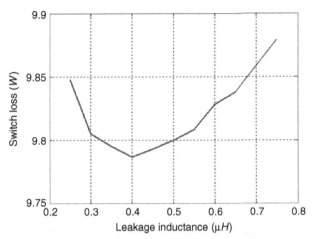

FIGURE 5.19 Variation of the total switch loss of the dc–ac converter with the leakage inductance of the HF transformer [1].

transformer increases, the total switching loss decreases due to an increase in the range of ZVS, while the total conduction loss increases with increasing leakage inductance.

5.4.2.3 Transformer Tapping

The voltage variation on the secondary side of the HF transformer necessitates high-breakdown-voltage rating for the ac–ac-converter switches and diminishes their utilization. For a step-up transformer with $N = 6.5$, the ac–ac-converter switches have to withstand at least 390 V nominal voltage when input ramps to the high end (60 V), while only 195 V is required when 30 V is the input. In addition to the nominal voltage, the switches of the ac–ac converter have to tolerate the overshoot voltage (as shown in Fig. 5.20) caused by the oscillation between the leakage inductance of the transformer and the junction capacitances of the power MOSFETs during turn-off [11]. The frequency of this oscillation is determined using $f_{\text{ring}} = \dfrac{1}{2\pi \sqrt{N^2 L_{\text{lk}} C_{\text{eq}}}}$, where C_{eq} is the equivalent capacitance of the switch output capacitance and the parasitic capacitance of the transformer winding. The conventional passive snubber circuit or active-clamp circuits can be used to limit the overshoot but they will induce losses, increase the system complexity, and component costs. One simple but effective solution is to adjust the transformer turns ratio according to the input voltage.

To change the turns ratio of the HF transformer, a bidirectional switch is required. Considering its simplicity and low conduction loss, a low-cost SPDT relay is chosen for the inverter, as shown in Fig. 5.12. For this prototype, for an input voltage in the range of 30–42 V, N equals 6.5. Hence, 500 V devices are used for the highest input voltage considering an 80% overshoot in the drain-to-

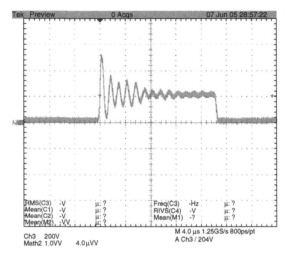

FIGURE 5.20   Drain-to-source voltage [1] across one of the ac–ac converter power MOSFETs.

source voltage that was observed in experiments. For an input voltage of more than 42 V, N equals 4.3, and hence the same 500 V devices can still be used to cover the highest voltage as the magnitude of the voltage oscillation is reduced. The relay is activated near the zero-crossing point (where power transfer is negligible) to reduce the inrush current. Such an arrangement improves the efficiency of the transformer and significantly increases the utilization of the ac–ac-converter switches for the full range of the input voltage. However, without adaptive transformer tapping, the minimal voltage rating for the devices is given by $V_{dc\,max} \cdot N \cdot (1 + 80\%) = 60 \cdot 6.5 \cdot 1.8 = 702$ V. In practice, power MOSFETs with 800 V or higher breakdown-voltage ratings are, therefore, required because of the lack of 700 V rating devices. The so-called rule of "silicon limit" (i.e., $R_{on} \propto BV^{2.5}$, where BV is the breakdown voltage) indicates that, in general, higher-voltage-rating power MOSFETs will have higher R_{on} and hence higher conduction losses. Furthermore, for the same current rating, the switching speed of a power MOSFET with higher breakdown-voltage rating is usually slower. As such, converter efficiency is expected to degrade further as a result of enhanced power loss.

5.4.2.4   Effects of Resonance between the Transformer Leakage Inductance and the Output Capacitance of the AC–AC-Converter Switches

Resonance between the transformer leakage inductance and the output capacitance of the ac–ac-converter devices causes the peak device voltage to exceed the nominal voltage (obtained in the absence of the transformer leakage inductance). This is demonstrated in Fig. 5.21. Considering $N = 6.5$, one can observe that the

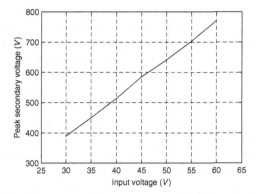

FIGURE 5.21 Peak voltage across the ac–ac converter [1] with varying input voltage for a transformer primary leakage inductance of 0.7 μH, output capacitance of 240 pF (for the devices of the ac–ac converter), and output filter inductance and capacitance of 1 mH and 2.2 μF, respectively.

peak secondary voltage can be around twice the nominal secondary voltage. Consequently, the breakdown-voltage rating of the ac–ac-converter switches needs to be higher than the nominal value. As power MOSFETs are used as switches, higher breakdown voltage entails increased on-resistance that yields higher conduction loss. So, the leakage affects the conduction loss and the selection of the devices for the ac–ac converter. The resonance begins only after the secondary current completes changing its direction and the ac–ac-converter switches initiate turn-on or turn-off. During this resonance period, energy is transferred back and forth between the leakage and filter inductances and the device and filter capacitances in an almost-lossless manner. The current through the switch that supports the oscillating voltage is almost zero. Thus, practically, no switching loss is incurred because of this resonance phenomenon although it may have an impact on the electromagnetic-emission profile.

5.5 HYBRID-MODULATION-BASED MULTIPHASE HFL HIGH-POWER INVERTER [5–8]

Recent high-voltage SiC DMOSFETs and SiC JBS diodes, with 100–400X lower on resistance and superior reverse recovery, respectively, along with projected >3X thermal sustenance and conductivity along with advanced high-permeability and high-efficiency nanocrystalline core (e.g., with >1 T flux density) transformers pave way for isolated high-power and HFL inverters. They have attained significant attention with regard to wide applications encompassing high-power renewable- and alternative-energy systems (e.g., photovoltaic, wind, and fuel-cell energy systems), DG/DER applications, active filters, energy storage, compact defense power-conversion modules for defense, as well as commercial electric/hybrid vehicles because of potential for significant reduction in materials and labor cost without much compromise in efficiency. Along that line, a new innovation in the form of a hybrid modulation (HM) [5,6] has been put forward by the author recently that significantly reduces the switching loss of HFL topologies (e.g., Fig. 5.22 [5–8]). The HM scheme is unlike all reported discontinuous-modulation (DM) schemes where the input to the final stage of the inverter is a dc and not a pulsating-dc; further, in the HM scheme, switches in two legs of the ac–ac converter do not change state in a 60° cycle, and switches in any one leg do not change state for an overall 240°. In contrast, for a conventional DM scheme, most switches of one leg of the ac–ac converter do not change stage in a 60 or 120° cycle. The present three-phase HM scheme is also different from earlier reported modulation schemes for single-phase, direct-power-conversion systems. The primary role of the modulation scheme for the single-phase ac–ac converter is to demodulate the rectifier output on a half-line-cycle basis to generate the output sine-wave-modulation pattern by switching all the ac–ac converter devices under low-frequency condition.

5.5.1 Principles of Operation [13]

5.5.1.1 Three-Phase DC–AC Inverter

Figure 5.23 illustrates the generation of switch-gate signals for the proposed converter. The bottom switches are controlled complimentarily to the upper ones, hence they are not described further. Three gate-drive signals UT, VT, and WT for primary side devices are obtained by phase shifting a square wave with respect to a 10-kHz square wave signal Q (shown in Fig. 5.24b). Q is synchronous with a 20-kHz saw-tooth carrier signal, shown in Fig. 5.24a. The phase differences are modulated sinusoidally using three 60 Hz references a, b, and c, respectively. Two gate signals for phase U and V are plotted in Fig. 5.24c and d. Because carrier frequency is much higher than the reference frequency,

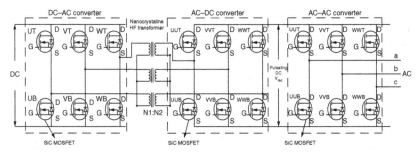

FIGURE 5.22 Schematic of the HM-based HFL topology. A conventional fixed-dc-link (FDCL) topology has the same architecture except that it has a filter capacitor after the ac–dc rectifier stage, and hence, the output ac–ac converter is fed with a dc voltage rather than a pulsating dc voltage (V_{rec}) for the HFL scheme. The HM scheme is implemented for the ac–ac converter stage. For the FDCL topology, the output stage is a voltage source inverter (VSI), which is operated using SVM scheme.

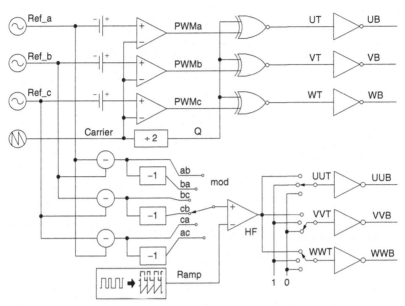

FIGURE 5.23 Diagram of gate-drive-signal generation for the HFL inverter [13].

UT, VT, and WT will be square wave with the frequency of 10 kHz, and their phases are modulated. The obtained output line–line voltages at the primary side of the transformers are bipolar waveforms. V_{uv} is plotted in Fig. 5.24e as an example. After passing through HF transformers, they are rectified by a three-leg diode bridge at the secondary side to obtain a unipolar PWM waveform, which has six-pulse as envelop. Its waveform is shown in Fig. 5.24g, and the mathematic expressions are:

$$V_{rec} = N \cdot V_{dc} \cdot \text{Max} \left(|UT - VT|, |VT - WT|, |WT - UT| \right), \quad (5.22)$$

$$UT = \overline{Q \otimes PWM_a} \quad VT = \overline{Q \otimes PWM_b} \quad WT = \overline{Q \otimes PWM_c}, \quad (5.23)$$

where PWM_x (x = a, b, or c) denotes the binary comparator output between reference and carrier for phase x. Symbol "$\otimes$" stands for XNOR operation. N is the transformer turns ratio. Divide the six-pulse rectified waveform into six segments named P1~P6 as shown in Fig. 5.25g. The rising and falling edges of V_{rec} are different for different segments. Figure 5.24a'–f' show a particular time interval within segment 2, where the rising and falling edges of V_{rec} (marked as $\Uparrow V_{rec}$ and $\Downarrow V_{rec}$) are determined by the edges of UT and VT, respectively. Other cases are summarized in Table 5.2.

5.5.1.2 Switching Strategy for the AC–AC Converter

Similar to the case of three-phase ac–dc rectifier, the rectified PWM output is contributed respectively by V_{wv}, V_{uv}, V_{uw}, V_{vw}, V_{vu}, and V_{wu} at each

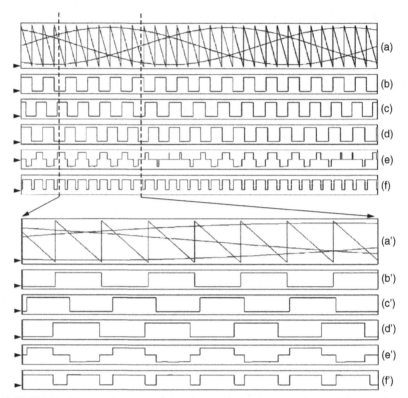

FIGURE 5.24 Key waveforms [13] of the primary-side dc–ac converter in one cycle and enlarged view of the interval between two dot lines; (a) three-phase sine-wave references and carrier signal; (b) Q: square ware with half frequency of the carrier; (c) UT: gate signal for the upper switch of phase U; (d) VT: gate signal for the upper switch of phase V; (e) V_{uv}: output of phase U and V; and (f) V_{rec}: output waveform of the rectifier.

segment from P1 to P6. The bottom part of the Fig. 5.23 shows the diagram of generating switching signals for three upper switches of secondary-side ac–ac inverter. During each segment, every switch will be either: permanently ON (1), permanently OFF (0), or toggling with 20 kHz. The switching pattern for the upper three switches in each segment for one cycle period is summarized in Table 5.3.

The switch positions illustrated in Fig. 5.23 are for the case of segment P2. Because the rectifier output has the same shape as V_{uv} within this interval, the line–line voltage V_{ab} at the output side of the ac–ac inverter can be directly obtained by keeping switches UUT and VVT at ON and OFF status, respectively. Another line–line voltage V_{cb}, however, needs to be achieved by operating switches on the third leg WWT and WWB under HF condition, where modulated signal (mod) is the difference between references c and b and the carrier signal (ramp) is a 20-kHz saw-tooth waveform synchronized with the PWM output

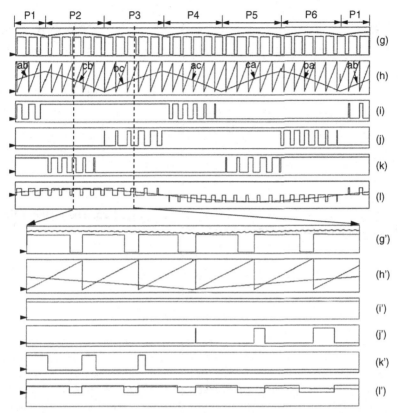

FIGURE 5.25 Key waveforms [13] of the secondary-side ac–ac inverter in one cycle and enlarged view of the interval between two dot lines; (g) V_{rec}: output PWM waveform of rectifier with six-pulse envelop; (h) mod: modulated signal and ramp: the carrier which is synchronous with (g); (i) UUT: gate signal for the top switch of phase a; (j) VVT: gate signal for the top switch of phase b; (k) WWT: gate signal for the top switch of phase c; and (l) PWM output of the line–line voltage V_{ab} and its envelop.

TABLE 5.2 The Edge Dependence of the Rectifier Output on Gate Signals

	P1	P2	P3	P4	P5	P6
$\uparrow V_{rec}$	wt	ut	ut	vt	vt	wt
$\downarrow V_{rec}$	vt	vt	wt	wt	ut	ut

TABLE 5.3 Switching Pattern for Upper Switches of the ac–ac Inverter

	P1	P2	P3	P4	P5	P6
V_{rec}	V_{wv}	V_{uv}	V_{uw}	V_{vw}	V_{vu}	V_{wu}
UUT	HF	ON	ON	HF	OFF	OFF
VVT	OFF	OFF	HF	ON	ON	HF
WWT	ON	HF	OFF	OFF	HF	ON
Mod	ab	cb	bc	ac	ca	ba

of the rectifier. The key waveforms are shown in Fig. 5.25. The mathematical expression for three line–line voltages is given as follows:

$$V_{ab} = V_{rec} \cdot (UUT - VVT)$$

$$V_{cb} = V_{rec} \cdot (WWT - VVT)$$

$$V_{ca} = V_{cb} - V_{ab} \tag{5.24}$$

An illustration of the ac–ac converter's HM scheme (as compared with several other SVM and SPWM schemes) is shown in Fig. 5.26. The outlined switching strategy is the best option for resistive load because the peaks of the currents follow the peaks of the fundamental voltages. Therefore, each phase leg does not switch just when the current is at its maximum value, thereby

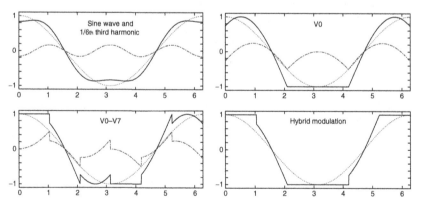

FIGURE 5.26 Modulation functions corresponding to sine wave (with 1/6th third harmonic), V0, V0–V7 SVMs, and HM schemes. Blue and red traces represent zero sequence and sinusoidal signals. Black trace is the modulating signal.

minimizing switching losses. For unity-power-factor load, the HM scheme for the ac–ac converter can be adjusted accordingly for minimizing the losses.

ACKNOWLEDGEMENT

The work described in this chapter is supported in parts by the National Science Foundation (NSF) under Award Nos. 0725887 and 0239131, by the Department of Energy (DOE) under Award No. DE-FC2602NT41574, and by the California Energy Commission (CEC) under Award No. 53422A/03-02. However, any opinions, findings, conclusions, or recommendations expressed herein are those of the authors and do not necessarily reflect the views of the NSF, DOE, and CEC.

COPYRIGHT DISCLOSURE

REFERENCES

[1] S.K. Mazumder, R. Burra, R. Huang, M. Tahir, K. Acharya, G. Garcia, S. Pro, O. Rodrigues, E. Duheric, A high-efficiency universal grid-connected fuel-cell inverter for residential application, IEEE Trans. Power Electron. 57(10) (2010) 3431-3447.

[2] S.K. Mazumder, R.K. Burra, R. Huang, V. Arguelles, A low-cost single-stage isolated differential Čuk inverter for fuel cell application, in: IEEE Power Electronics Specialists Conference, Rhodes, Greece, 2008, pp. 4426-4431.

[3] S.K. Mazumder, R.K. Burra, K. Acharya, A ripple-mitigating and energy-efficient fuel cell power-conditioning system, IEEE Trans. Power Electron. 22(4) (2007) 1437-1452.

[4] S.K. Mazumder, R.K. Burra, K. Acharya, Power conditioning system for energy sources, USPTO Patent# 7,372,709 B2, Awarded in May 13, 2008.

[5] S.K. Mazumder, A novel hybrid modulation scheme for an isolated high-frequency-link fuel cell inverter, Invited NSF Panel Paper, IEEE Power and Energy Society General Meeting, 2008, pp. 1-7, doi: 10.1109/PES.2008.4596911.

[6] S.K. Mazumder, R. Huang, Multiphase converter apparatus and method, USPTO Patent Application US2009/0196082 A1, Filed by University of Illinois at Chicago, 2009.

[7] R. Huang, S.K. Mazumder, A soft switching scheme for multiphase dc/pulsating-dc converter for three-phase high-frequency-link PWM inverter, IEEE Trans. Power Electron. 25(7) (2010) 1761-1774.

[8] R. Huang, S.K. Mazumder, A soft-switching scheme for an isolated dc/dc converter with pulsating dc output for a three-phase high-frequency-link PWM converter, IEEE Trans. Power Electron. 24(10) (2009) 276-2288.

[9] S.K. Mazumder, R. Huang, A high-power high-frequency and scalable multi-megawatt fuel-cell inverter for power quality and distributed generation, in: IEEE Power Electronics, Drives, and Energy Systems Conference, New Delhi, India, 2006, pp. 1-5.

[10] R.D. Middlebrook, S. Ĉuk, Advances in Switched-Mode Power Conversion, vols. I-III, TESLACO, Irvine, CA, 1983.

[11] J.A. Pomilio, G. Spiazzi, Soft-commutated Ĉuk and SEPIC converters as power factor pre-regulators, in: 20th International Conference on Industrial Electronics, Control and Instrumentation, Bologna, Italy, 1994, pp. 256-261.

[12] T. Kawabata, H. Komji, K. Sashida, High frequency link dc/ac converter with PWM cyclo-converter, in: IEEE Industrial Application Society Conference, Seattle, WA, USA, 1990, pp. 1119-1124.

[13] S. Deng, H. Mao, A new control scheme for high-frequency link inverter design, in: IEEE Applied Power Electronics Conference and Exposition, Miami, FL, USA, 2003, pp. 512-517.

Chapter 6

Energy Storage

Sheldon S. Williamson and Pablo A. Cassani
Power Electronics and Energy Research (PEER) Group, P. D. Ziogas Power Electronics Laboratory, Department of Electrical and Computer Engineering, Concordia University, Montreal, QC, Canada

Srdjan Lukic
Department of Electrical and Computer Engineering, North Carolina State University, Raleigh, NC, USA

Benjamin Blunier
Universite de Technologie de Belfort-Montbeliard, Belfort Cedex, France

Chapter Outline

Alternative Energy in Power Electronics

267

6.1 INTRODUCTION

It is well known that energy storage devices provide additional advantages to improve stability, power quality, and reliability of the power-supply source. The major types of storage devices being considered today include batteries, ultracapacitors, and flywheel energy systems, which will be discussed in detail in this chapter. It is empirical that precise storage device models are created and simulated for several applications, such as hybrid electric vehicles (HEV) and various power system applications.

The performances of batteries, ultracapacitors, and flywheels have, over the years, been predicted through many different mathematical models. Some of the important factors that need to be considered while modeling various energy storage devices include storage capacity, rate of charge/discharge, temperature, and shelf life.

The electrical models of two of the most promising renewable energy sources, namely fuel cells and PV cells, are also described in this chapter. They are now being studied widely as they do not produce much emission and are considered to be environment friendly. However, these renewable energy sources are large, complex, and expensive at the same time. Hence, designing and building new prototypes is a difficult and expensive affair. A suitable solution to overcome this problem is to carry out detailed simulations on accurately modeled devices. This chapter discusses the various equivalent electrical models for fuel cells and PV arrays and analyzes their suitability for operation at system levels.

6.2 ENERGY STORAGE ELEMENTS

6.2.1 Battery Storage

6.2.1.1 Lead Acid Batteries

The use of lead acid batteries for energy storage dates back to mid-1800s for lighting application in railroad cars. Battery technology is still prevalent in cost-sensitive applications where low-energy density and limited cycle life are not an issue but ruggedness and abuse tolerance are required. Such applications include automotive starting lighting and ignition (SLI) and battery-powered uninterruptable power supplies (UPS).

Lead acid battery cell consists of spongy lead as the negative active material, lead dioxide as the positive active material, immersed in diluted sulfuric acid electrolyte, with lead as the current collector:

$$Pb + SO_4^{-2} \underset{charge}{\overset{discharge}{\rightleftharpoons}} PbSO_4 + 2e^-$$

$$PbO_2 + SO_4^{-2} + 4H^+ + 2e^- \underset{charge}{\overset{discharge}{\rightleftharpoons}} PbSO_4 + 2H_2O$$

During discharge, $PbSO_4$ is produced on both negative and positive electrodes. If the batteries are overdischarged or are kept at a discharged state, the sulfate crystals become larger and are more difficult to break up during charge. In addition, the large size of lead sulfate crystals leads to active material disjoining from the plates.

Due to the production of hydrogen at the positive electrode, lead acid batteries suffer from water loss during overcharge. To deal with this problem, distilled water may be added to the battery as is typically done for flooded lead acid batteries. Also, maintenance-free versions are available to deal with this problem whereby inserting a valve keeps the gasses within the battery and minimizes water loss by recombination.

Current collectors in lead acid batteries are made of lead, leading to the low-energy density. In addition, lead is prone to corrosion when exposed to the sulfuric acid electrolyte. SLI applications make use of flat-plate grid designs as the current collectors, whereas more advanced batteries use tubular designs.

Recent advances aim to replace lead with lighter materials, such as carbon, to reduce the system weight.

6.2.1.2 Nickel-Cadmium (Ni-Cd) and Nickel-Metal Hydride (Ni-MH) Batteries

Nickel-cadmium (Ni-Cd) batteries were the chemistry of choice for a wide range of high-performance applications between 1970 and 1990. Recently, they were replaced by lithium-ion (Li-ion) and nickel-metal hydride (Ni-MH) chemistries in most applications. The Ni-Cd battery uses nickel oxyhydroxide for the positive electrode and metallic cadmium for the negative electrode. The chemical reaction is as follows:

$$Cd + OH^- \underset{charge}{\overset{discharge}{\rightleftharpoons}} Cd\,(OH)_2 + 2e^-$$

$$2NiO\,(OH) + 2H_2O + 2e^- \underset{charge}{\overset{discharge}{\rightleftharpoons}} 2Ni\,(OH)_2 + 2OH^-$$

As can be seen from this chemical reaction, there is a balance of reactions that implies that the electrolyte is always of the same concentration. This leads to relatively constant performance during discharge. In addition, the balance of reactions leads to very good overcharge characteristics where the additional power is used up as heat rather than water loss.

Ni-Cd batteries have a higher energy density and longer cycle life than lead acid batteries, but are inferior to chemistries such as Li ion and Ni-MH, that are also becoming cheaper than Ni-Cd batteries. Other disadvantages of using Ni-Cd batteries compared to Ni-MH include shorter cycle life, more pronounced "memory effect," toxicity of cadmium requiring complex recycling procedure, and lower energy density. Moreover, a flat discharge curve and negative temperature coefficient may cause thermal runaway in voltage-controlled charging.

For the aforementioned reasons, in the recent past, Ni-MH batteries have gained prominence over Ni-Cd batteries. The Ni-MH batteries use nickel oxyhydroxide for the positive electrode and metallic cadmium for the negative electrode. The chemical reaction is as follows:

$$MH + OH^- \underset{charge}{\overset{discharge}{\rightleftharpoons}} H_2O + M + e^-$$

$$NiO\,(OH) + H_2O + e^- \underset{charge}{\overset{discharge}{\rightleftharpoons}} Ni\,(OH)_2 + OH^-$$

Here, M stands for the metallic group. At the negative electrode, hydrogen is released from the metal to which it was temporarily attached, and reacts, producing water and electrons. Note that the reaction at the negative electrode is similar to that of a fuel cell, which will be discussed later in this chapter.

Ni-MH batteries have been the chemistry of choice for electric and hybrid electric vehicle (EV and HEV) applications in the 1990s and 2000s, respectively, due to their relatively high power density, proven safety, good abuse tolerance, and very long life at a partial state of charge. One of the disadvantages is the relatively high self-discharge rate, though the introduction of novel separators has mitigated this problem.

When overcharged, Ni-MH batteries use excess energy used to split and recombine water. Therefore, the batteries are maintenance free. However, if the batteries are charged at an excessively high charge rate, hydrogen gas buildup can cause the cell to rupture. If the battery is overdischarged, the cell can be reverse-polarized, thus reducing the battery capacity.

6.2.1.3 Lithium-Ion (Li-Ion) Batteries

Due to their high specific energy and the potential to be produced at low cost, Li-ion batteries are expected to widely replace Ni-MH batteries for future electric propulsion applications. The Li-ion battery consists of oxidized cobalt material on the positive electrode, carbon on the negative electrode, and lithium salt in an organic solvent as the electrolyte. Even though the widespread use of this chemistry is fairly novel, it is interesting to note that the processes in the battery are fairly simple to model compared to other chemistries.

Li-ion batteries have been instrumental in increasing the performance standards of batteries, and more recently, even in predicting battery deterioration. The promising aspects of lithium-based battery chemistries include low memory effect, high specific energy of nearly 100 Wh/kg, high specific power of 300 W/kg, and battery life of at least 1000 cycles. The key barriers include calendar life, cost, operation at temperature extremes, and abuse tolerance. A breakthrough in the development of advanced electrodes is needed to further increase the specific energy [1]. The characteristics of some of the major battery chemistries being considered for propulsion, storage, and renewable energy systems are enlisted in Table 6.1.

This chapter will later discuss the major advantages and disadvantages of the various battery chemistries, especially those of Li-ion batteries, in detail. Major power-electronics-based solutions to overcome electrochemical-related barriers for batteries will also be discussed later in this chapter.

6.2.2 Ultracapacitor (UC)

Electrochemical double-layer capacitors (EDLCs) or ultracapacitors (UCs) work in much the same way as regular capacitors, in that there is no ionic or electronic transfer, resulting in a chemical reaction (there is no *Faradic* process). In other words, energy is stored in the electrochemical capacitor by simple charge separation. Therefore, the energy stored in the electrochemical capacitor can be calculated using the same well-known equation that is used for conventional capacitors:

$$Q = CV = \frac{A\varepsilon}{d}V$$

TABLE 6.1 Summary of Major Battery Chemistries and their Characteristics [2–4]

Type	Electrolyte	Energy efficiency (%)	Energy density (Wh/kg)	Sustained power density (W/kg)	Cycle life (cycles)	Operating temperature (°C)	Self-discharge
Pb-acid	H_2SO_4	70–80	20–35	25	200–2000	−20 to 60	Low
Ni-Cd	KOH	60–90	40–60	140–180	500–2000	−40 to 60	Low
Ni-MH	KOH	50–80	60–80	220	<3000	10–50	High
Li-ion	$LiPG_6$	70–85	100–200	360	500–2000	−20 to 60	Medium
Li-polymer	Li-β-Alu	70	200	250–1000	>1200	−20 to 60	Medium
NaS	β-Al_2Cl_2	70	120	120	2000	300–400	Medium
VRB		80	25		>16,000		Negligible
Electrochem. capacitor		95+	<50	4000+	>50,000		Very high
Pumped hydro		65–80	0.3				Negligible
Compressed air		40–50					
Flywheels (steel)		95	5–30		>20,000		Very high
Flywheels (composite)		95	>50		>20,000		Very high

As for the conventional capacitor, the capacitance, C, is proportional to the area, A, of the plates, the permittivity of the dielectric, ε, and inversely proportional to the distance, d, between the plates. In general, UCs are designed to have a very high electrode surface area and use high-permittivity dielectric while keeping the current collectors very close. Therefore, UCs attain very high capacitance ratings (kilo-Farads versus milli- and micro-Farads for conventional capacitors). This is achieved by using porous carbon as the current collector rather than metallic strips. The porous carbon collector exhibits a very large surface area, allowing a relatively large amount of energy to be stored at the collector surface. The two electrodes are separated by a very thin porous separator and immersed in an electrolyte, such as propylene carbonate. Due to the high permeability and close proximity of the electrodes, UCs have a low voltage withstand capability (typically 2.5 V).

Currently, there exist five UC technologies in development: carbon/metal fiber composites, foamed (aerogel) carbon, carbon particulate with a binder, doped conducting polymer films on carbon cloth, and mixed metal oxide coatings on metal foil. Current trends indicate that higher energy densities are achievable with a carbon composite electrode, using an organic electrolyte, rather than carbon/metal fiber composite electrode devices, with an aqueous electrolyte.

As described earlier, a UC stores energy by physically separating unlike charges. This has profound implications on cycle life, efficiency, energy, and power density. It must be noted that, typically, a UC depicts long cycle life due to the fact that (ideally) there exist no chemical changes on the electrodes, under normal operation. Furthermore, overall efficiency is superior; it is only a function of the *ohmic* resistance of the conducting path. In addition, power density is exceptional because the charges are physically stored on the electrodes. Conversely, energy density is low because the electrons are not bound by chemical reactions. This lack of chemical bonding also implies that the UC can be completely discharged, leading to larger voltage swings as a function of the state of charge (SOC).

6.2.3   Flow Batteries and Regenerative Fuel Cells (RFC)

Flow batteries, also called redox (reduction-oxidation) batteries, comprise two electrolytes, separated by ion or proton exchange membrane. Energy can be stored in the electrolytes by increasing the potential difference between the two liquids – in other words, by oxidising one and reducing the other. Alternatively, electricity can be produced by reversing the process. The oxidation/reduction process is performed by the proton exchange membrane.

Flow batteries have a number of advantages, such as easy scalability (capacity proportional to tank size, whereas power output is proportional to PEM surface area), no detrimental effects of a deep discharge, very low self–discharge, low cost for a large system compared to batteries, and long cycle life. In contrast, the energy and power densities are quite low and the system is complex, requiring pumps and plumbing to circulate the electrolyte through the membrane. Therefore, this technology has found commercial application only for large-scale storage such as utility applications [3,4].

Three types of flow batteries have shown commercial promise: vanadium redox, polysulfide bromide, and zinc bromide. Vanadium redox has the added advantage of using the same material for both electrolytes, removing the threat of cross-contamination through the PEM. Polysulfide bromide and zinc bromide use bromide, presenting the threat of releasing the highly toxic bromine gas [4].

Regenerative fuel cells (or unitized regenerative fuel cells) are sometimes grouped in with flow batteries, as the power-producing process is quite similar. Fuel cells consume hydrogen and oxygen to produce water and electricity. Unitized regenerative fuel cells are also able to function to separate water back into hydrogen and oxygen. The hydrogen is then stored as hydrogen gas or as methanol for future use, to generate electricity. Current research aims to use PEM-type fuel cells, with hydrogen or methanol as the main storage. The issue is to design a system that is efficient in both directions – in fact, current unitized fuel cells are less efficient in hydrogen production than other methods, such as conventional electrolysis.

6.2.4   Fuel Cells (FC)

A fuel cell is typically similar in operation to a conventional battery, although it has some distinct physical differences. Primarily, a fuel cell is an electrochemical device wherein the chemical energy of a fuel is converted directly into electric power [5]. The main difference between a conventional battery and a fuel cell is that, unlike a battery, a fuel cell is supplied with reactants externally. As a result, whereas a battery is discharged, a fuel cell never faces such a problem as long as the supply of fuel is provided. As depicted in Fig. 6.1, electrodes and electrolyte are the main parts of a fuel cell. The most popular type of fuel cell is the hydrogen-oxygen fuel cell.

As shown in Fig. 6.1, hydrogen is used as the fuel to be fed to the anode. The cathode, in contrast, is fed with oxygen, which may be acquired from air. The hydrogen atom is split up into protons and electrons, which follow different paths, but ultimately meet at the cathode. For the splitting-up process, we need to use a suitable catalyst. The protons take up the path through the electrolyte, whereas the electrons follow a different external path of their own. This, in turn, facilitates a flow of current, which can be used to supply an external electric load. The electrode reactions are given as follows:

$$\text{Anode: } 2H_2 \rightarrow 4H^+ + 4e^-$$
$$\text{Cathode: } O_2 + 4e^- \rightarrow 2O^{2-}$$
$$\text{Overall reaction: } 2H_2 + O_2 \rightarrow 2H_2O$$

From these simple and basic expressions describing the operation of a typical fuel cell, it is clear that there exists no combustion and hence no production of emissions. This makes the fuel cell environmentally suitable.

A typical $i - v$ curve of fuel cells is shown in Fig. 6.2. The output voltage decreases as the current increases. Moreover, the efficiency of a fuel cell is defined as the ratio of electrical energy generated to the input hydrogen energy. Generally, cell efficiency increases with higher operating temperature and pressure [5].

Fuel cells have many favorable characteristics for energy conversion. As explained earlier, they are environmentally acceptable due to a reduced value of carbon dioxide (CO_2) emission for a given power output. Moreover, the usage of fuel cells reduces transmission losses, resulting in higher efficiency. Typical values of efficiency range between 40 and 85%. Another advantage of fuel cells is their modularity. They are inherently modular, which means they can be configured to operate with a wide range of outputs, from 0 to 50 MW, for natural gas fuel cells, to 100 MW or more, for coal gas fuel cells. Another unique advantage of fuel cells is that hydrogen, which is the basic fuel used, is easily acquirable from natural gas, coal gas, methanol, and other similar

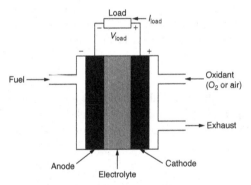

FIGURE 6.1   Typical diagram of a fuel cell.

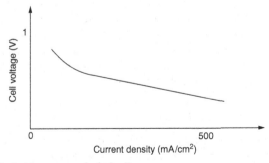

FIGURE 6.2   Typical $i - v$ curve of a fuel cell.

fuels containing hydrocarbons. Finally, the waste heat/exhaust can be utilized for cogeneration and for heating and cooling purposes. This exhaust is useful in residential, commercial, and industrial cogeneration applications. Basically, for cogeneration, the fuel cell exhaust is used to feed a mini- or a microturbine generator unit. These turbines are generally gas turbines. Because the waste thermal energy is recovered and converted into additional electrical energy, the overall system efficiency is improved. The gas turbine fulfills this role suitably. The typical sizes of such systems range from 1 to 15 MW.

6.3 MODELING OF ENERGY STORAGE DEVICES

6.3.1 Battery Modeling

As mentioned earlier, precise battery models are required for several applications such as for the simulation of energy consumption of electric vehicles, portable devices, or for power system applications. The major challenge in modeling a battery source is the nonlinear characteristic of the equivalent circuit parameters, which require lengthy experimental and numerical procedures. The battery itself has some internal parameters, which need to be taken care of [6]. In this section, three basic types of battery models will be presented: ideal, linear, and Thevenin. Finally, a simple lead acid battery model for traction applications that can be simulated in a CAD software will also be presented.

6.3.1.1 Ideal Model

The ideal model of a battery basically ignores the internal parameters and, hence, is very simple. Figure 6.3a depicts an ideal model of a battery, wherein it is clear that the model is primarily made up of only a voltage source [6].

6.3.1.2 Linear Model

This is by far the most commonly used battery model. As is clear from Fig. 6.3b, this model consists of an ideal battery with open-circuit voltage, E_0, and an equivalent series resistance, R_s [6]. "V_{batt}" represents the terminal voltage of the battery. This terminal voltage can be obtained from the open-circuit tests as well as from load tests conducted on a fully charged battery.

Although this model is quite widely used, it still does not consider the varying characteristics of the internal impedance of the battery with the varying state of charge (SOC) and electrolyte concentration.

6.3.1.3 Thevenin Model

This model consists of electrical values of the open-circuit voltage (E_o), internal resistance (R), capacitance (C), and the overvoltage resistance (R_o) [6]. As observed in Fig. 6.3c, capacitor C depicts the capacitance of the parallel plates and resistor R_o depicts the nonlinear resistance offered by the plate to the electrolyte.

In this model, all the elements are assumed to be constants. However, in reality, they depend on the battery conditions. Thus, this model is also not the most accurate, but is the most widely used. In this view, a new approach to evaluate batteries is introduced. The modified model is based on operation over a range of load combinations. The electrical equivalent of the proposed model is depicted in Fig. 6.4. As is clear from Fig. 6.4, the main circuit model consists of the following five subcircuits:

1. E_{batt}: This is a simple dc voltage source designating the voltage in the battery cells.
2. E_{pol}: It represents the polarization effects due to the availability of active material in the battery.

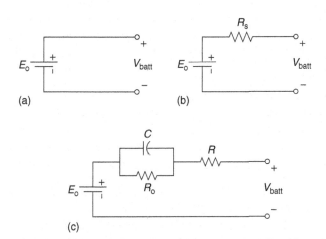

FIGURE 6.3   Battery models: (a) ideal model, (b) linear model, and (c) Thevenin model.

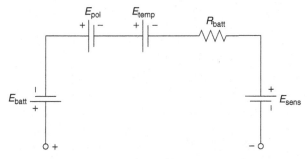

FIGURE 6.4   Main circuit representation of a modified battery model.

3. E_{temp}: It represents the effect of temperature on the battery terminal voltage.
4. R_{batt}: This is the battery's internal impedance, the value of which depends primarily on the relationship between cell voltage and state of charge (SOC) of the battery.
5. V_{sens}: This is a voltage source, with a value of 0 V. It is used to record the value of battery current.

Thus, this model is capable of dealing with various modes of charge/discharge. It is comparatively more precise and can be extended for use with Ni-Cd and Li-ion batteries, which could be applied to hybrid electric vehicles and other traction applications. Only a few modifications need to be carried out in order to vary the parameters such as load state, current density, and temperature [6].

6.3.2 Electrical Modeling of Fuel Cell Power Sources

Over the past few years, there have been great environmental concerns shown with respect to emissions from vehicles. These concerns, along with the recent developments in fuel cell technology, have made room for the hugely anticipated fuel cell market [6]. Fuel cells are today being considered for applications in hybrid electric vehicles, portable applications, renewable power sources for distributed generation applications, and other similar areas where the emission levels need to be kept to a minimum.

The proton exchange membrane (PEM) fuel cell has the potential of becoming the primary power source for HEVs utilizing fuel cells. However, such fuel cell systems are large and complex and, hence, need accurate models to estimate the auxiliary power systems required for use in the HEV. In this section, the fuel cell modeling techniques will be highlighted, thus avoiding the need to build huge and expensive prototypes. To have a clearer picture, refer to Fig. 6.5, which shows the schematic of a fuel cell/battery hybrid power system.

The battery pack in Fig. 6.5 is used to compensate for the slow start-up and transient response of the fuel processor [6]. Furthermore, the battery can also be used for the purpose of regenerative braking in the HEV.

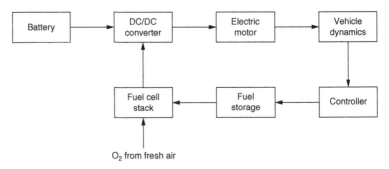

FIGURE 6.5 Schematic of a fuel cell/battery hybrid power system.

As mentioned previously, because fuel cell systems are large, complex, and expensive, designing and building new prototypes is difficult. Hence, the feasible alternative is to model the system and examine it through simulations. The fuel cell power system consists of a reformer, a fuel cell stack, and a dc/dc (buck/boost) or dc/ac power converter. The final output from the power electronic converter is in the required dc or ac form acquired from the low-voltage dc output from the fuel cell stack. An electrical equivalent model of a fuel cell power system is discussed here, which can be easily simulated using a computer simulation software.

In the electrical equivalent model, a first-order time-delay circuit with a relatively long time-constant can represent the fuel reformer. Similarly, the fuel cell stack can also be represented by a first-order time-delay circuit, but with a shorter time constant. Thus, the mathematical model of the reformer and stack is represented as:

$$\frac{V_{cr}}{V_{in}} = \frac{\frac{1}{C_r \cdot S}}{R_r + \frac{1}{C_r \cdot S}} = \frac{1}{1 + R_r C_r \cdot S}$$

$$\frac{V_{cs}}{V_{cr}} = \frac{\frac{1}{C_s \cdot S}}{R_s + \frac{1}{C_s \cdot S}} = \frac{1}{1 + R_s C_s \cdot S}$$

Here, $R_r C_r = \tau_r$ is the time constant of the reformer and $R_s C_s = \tau_s$ is the time constant of the fuel cell stack. The equivalent circuit is shown in Fig. 6.6.

By simulating the equivalent circuit of Fig. 6.6, the system operation characteristics can be investigated. In order to achieve a fast system response, the dc/dc or dc/ac converter can utilize its short time constant for control purposes. However, eventually the fuel has to be controlled despite its long time-delay [6]. The inputs to the chemical model of a fuel cell include mass flows of air (O_2) and hydrogen (H_2), cooling water, relative humidity of oxygen and hydrogen, and load resistance. The outputs from the chemical model include temperature of the cell, power loss, internal resistance, heat output, efficiency, voltage, and total power output. Generally, in case of excess of hydrogen supply, it is recirculated in order to avoid any wastage.

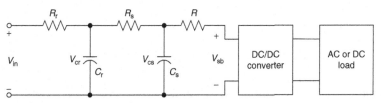

FIGURE 6.6   Equivalent circuit model of a fuel cell power system.

6.3.3 Electrical Modeling of Photovoltaic (PV) Cells

As mentioned earlier, photovoltaic systems have been studied widely as a renewable energy source, because not only are they environment friendly, but they also have infinite energy available from the sun. Although the PV systems have the above-mentioned advantages, their study involves precise management of factors such as solar irradiation and surface temperature of the PV cell [6]. The PV cells typically show varying $v - i$ characteristics depending on the above-mentioned factors. Figure 6.7 shows the output characteristics of a PV cell with changing levels of illumination.

As is clear from Fig. 6.7, the current level increases with increase in the irradiation level. Figure 6.8 shows the $v - i$ curves with varying cell temperatures.

As depicted in Fig. 6.8, the output curves for varying cell temperatures show higher voltage level as the cell temperature increases. Therefore, while modeling the PV cell, adequate consideration must be given to these two characteristics in particular.

Keeping the above-mentioned factors in mind, the electrical equivalent circuit modeling approach is proposed here. This model is basically a nonlinear distributed circuit, in which the circuit elements consist of familiar semiconductor device parameters. Eventually, running a suitable computer simulation can easily simulate this model.

The PV cell can basically be considered a current source with the output voltage primarily dependent on the load connected to its terminals [6]. The equivalent circuit model of a typical PV cell is shown in Fig. 6.9.

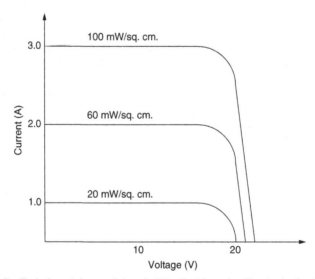

FIGURE 6.7   Typical $v - i$ characteristics of a PV cell with varying illumination levels.

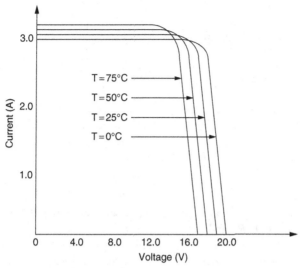

FIGURE 6.8 Typical $v - i$ characteristics of a PV cell with varying cell temperatures.

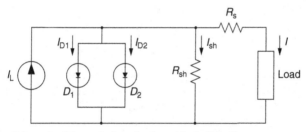

FIGURE 6.9 Schematic of the equivalent circuit model of a PV cell.

Suitable forward-bias (illuminated) and reverse-bias (dark) tests can be performed on the PV cell, in order to generate the $v - i$ curves similar to those in Figs. 6.7 and 6.8. As is clear from Fig. 6.9, there are various parameters involved in the modeling of a typical PV cell. These parameters are

I_L, light-generated current (A);
I_{D1}, diode saturation current (A);
I_{D2}, additional current due to diode quality constant (A);
I_{sh}, shunt current (A);
R_s, cell series resistance (Ω);
R_{sh}, cell shunt resistance (Ω); and
I, cell-generated current (A).

The model depicted in Fig. 6.9 examines all the characteristic measurements of the $p–n$ junction cell type. From the above circuit, the following equation for cell current can be obtained:

$$J = J_L - J_{o1} \left\{ \exp \left[\frac{q\,(V + J \cdot R_s)}{kT} \right] - 1 \right\}$$

$$- J_{on} \left\{ \exp \left[\frac{q\,(V + J \cdot R_s)}{A \cdot kT} \right] - 1 \right\} - G_{sh}\,(V + J \cdot R_s)$$

Here, q is the electron charge and k is the Boltzmann's constant. The voltage at the terminals of the diodes in Fig. 6.9 can be expressed as follows:

$$V = V_{oc} - IR_s + \frac{1}{\Delta} \log_n \left\{ \frac{\beta(I_{sc} - I) - V/R_{sh}}{\beta \cdot I_{sc} - V_{oc}/R_{sh}} + \exp\left[\Delta\,(I_{sc} \cdot R_s - V_{oc})\right] \right\}$$

Here, β is the voltage change temperature coefficient (V/°C). For the PV cell model of Fig. 6.9, R_s and R_{sh} are usually estimated when the cell is not illuminated. The series resistance, R_s, represents the ohmic losses in the front surface of the PV cell, whereas the shunt resistance, R_{sh}, represents the loss due to diode leakage currents. Thus, these values can be approximated from the dark characteristic curve of the cell. The generated light current (I_L) is calculated by the collective probability of free electrons and holes. It can be expressed as follows:

$$I_L = q \cdot N \left[\sum f_c\,(x_N) + \sum f_c\,(x_P) + 2l \right]$$

Here, $f(x)$ is the probability distribution function and N is the rhythm of generated electrons and holes. Once the equations of the cell model are formulated, the efficiency of the PV cell can be obtained as

$$\text{Efficiency} = \frac{P_{out}}{P_{in}} = \frac{f \cdot I_{sc} \cdot V_{oc}}{P_{in}}$$

A distinct advantage of such a computer model is the fact that, with a very few number of changes, it can receive data from different kinds of PV cells maintaining satisfactory results [6]. An example of such a PV model used in conjunction with a power electronic intensive system is shown in Fig. 6.10.

The vital part of the system, as depicted in Fig. 6.10, is the dc/dc converter. The dc/dc converter could be a boost, buck, or buck–boost converter. Although, in the system shown in Fig. 6.10, it is valid to assume a buck or buck-boost dc/dc converter, it becomes erroneous to assume usage of a boost converter. This is because the dc/dc boost converter does not make use of the full voltage range.

6.3.4   Electrical Modeling of Ultracapacitors (UCs)

Ultracapacitors (also known as double-layer capacitors) work on the electrochemical phenomenon of very high capacitance/unit area using an interface between electrode and electrolyte. Typical values of such capacitors range from

400 to 800°F and have low values of resistivity (approximately $10^{-3}\Omega$-cm) [6]. These UCs operate at high-energy densities, which are commonly required for applications such as space communications, digital cellular phones, electric vehicles, and hybrid electric vehicles. In some cases, usage of a hybridized system employing a battery alongside the UC provides an attractive energy storage system, which offers numerous advantages. This is particularly due to the fact that the UC provides the necessary high-power density, whereas the battery provides the desired high-energy density. Such a hybridized model will be discussed here.

6.3.4.1 Double Layer UC Model

A simple electrical equivalent circuit of a double-layer UC is shown in Fig. 6.11. Its parameters include equivalent series resistance (ESR), equivalent parallel resistance (EPR), and the overall capacitance. The ESR in Fig. 6.11 is important during charging/discharging as it is a lossy parameter, which in turn causes the capacitor to heat up. In contrast, the EPR has a leakage effect and, hence, it only affects the long-term storage performance. For the purpose of simplification in calculations, the EPR parameter is dropped.

Furthermore, the dropping of the EPR parameter does not have any significant impact on the results. The circuit for analysis is thus simply an ideal capacitor in series with a resistance and the corresponding load [6]. Hence, the value of resistance can be written as

$$R = n_{\text{s}}\frac{\text{ESR}}{n_{\text{p}}}$$

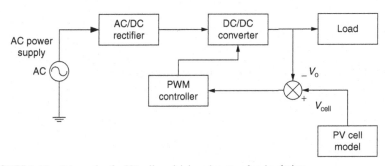

FIGURE 6.10   Schematic of a PV cell model–based system for simulation.

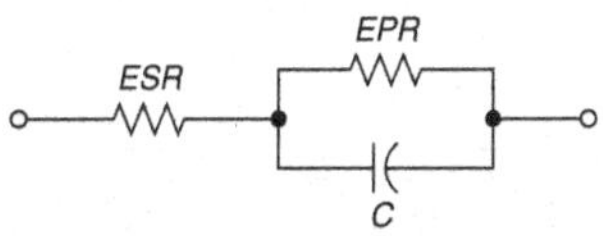

FIGURE 6.11   Electrical equivalent circuit of an ultracapacitor.

Here, R is the overall resistance (Ω), n_s is the number of series capacitors in each string, and n_p is the number of parallel strings of capacitors. Furthermore, the value for the total capacitance can be expressed as

$$C = n_p \frac{C_{rated}}{n_s}$$

Here, C is the overall value of capacitance and C_{rated} is the capacitance of individual capacitor. This model can be used in conjunction with a dc/dc converter, which in turn acts as a constant power load, as shown in Fig. 6.12.

The capacitor bank can be used in stand-alone mode or can be operated in parallel with a battery of suitable size for the applications mentioned earlier. A brief description of such a hybrid model is described in the following section.

6.3.4.2 Battery/UC Hybrid Model

Combining a battery and an UC to operate in parallel makes an attractive energy storage system with many advantages. Such a hybrid system uses both the high-power density of the UC and the high-energy density of the battery. In this section, an electrical equivalent model of such a system will be presented, which can be used to evaluate its voltage behavior. This model is depicted in Fig. 6.13.

The equivalent circuit of Fig. 6.13 shows an equivalent series resistance (R_c) and a capacitor (C) as a model of the UC, whereas the Li-ion battery can be modeled simply by using a series resistance (R_b) and a battery. The values of R_c and C depend on the frequency due to the porous nature of the electrodes of the

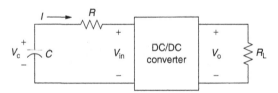

FIGURE 6.12 Circuit showing UC connected to a constant power load.

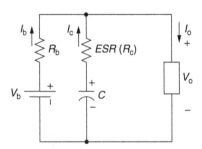

FIGURE 6.13 Equivalent circuit model of a battery/UC hybrid system.

UC [6]. When the pulse width (T) is varied, the discharge rate of the UC can be varied and can be shown to be equal to a frequency of $f = 1/T$. The following equations can be written for I_0 and V_0 from the equivalent circuit model of Fig. 6.13:

$$I_0 = I_c + I_b$$

$$V_0 = V_b - I_b R_b = \left[V_b - \frac{1}{C} \int_0^T I_c \cdot dt \right] - I_c R_c$$

Here, I_0 and V_0 are the output current and voltage delivered to the load, respectively. From the above two equations, it is possible to achieve a voltage drop $\Delta V = V_b - V_0$ due to a pulse current of I_0. This voltage drop can be finally expressed as

$$\Delta V = \frac{I_0 R_b R_c}{R_b + R_c + \frac{T}{C}} + \frac{I_0 R_b \frac{T}{C}}{R_b + R_c + \frac{T}{C}}$$

The currents delivered by the battery (I_b) and capacitor (I_c) can also be derived, and their ratio can be expressed as

$$\frac{I_c}{I_b} = \frac{R_b}{R_c + \frac{T}{C}}$$

It can be seen that for a long pulse, I_c can be limited by the value of C. Furthermore, it can be concluded that during the pulsed discharge, about 40%–50% of the total current is delivered by C. Upon computer simulation of the equivalent circuit model, it is possible to study the fact that, during peak power demand, UC delivers energy to assist the battery, whereas, during low power demand, UC receives energy from the battery.

Due to the advanced energy storage capabilities of the UC, it can be used for applications requiring repeated short bursts of power such as in vehicular propulsion systems. In a typical scenario, both the battery and the UC provide power to the motor and power electronic dc/ac inverter during acceleration and overtaking, whereas they receive power via regenerative braking during slow down/deceleration. Two of the most popular topologies for inserting batteries and UCs into drivetrains are shown in Figs. 6.14a, b.

As is clear in the topology of Fig. 6.14a, the UC bank is placed on the dc bus, whereas in Fig. 6.14b, the dc bus houses the battery. Among the two topologies, the one shown in Fig. 6.14a has a much more degraded energy efficiency because the whole of the battery energy has to go through the dc/dc converter. Another drawback worth highlighting is that a very high-voltage UC bank is required, which is extremely expensive. Hence, more often than not, the topology of Fig. 6.14b is generally considered for HEV applications [6]. In a typical brushless dc (BLDC) motor-driven electric vehicle propulsion

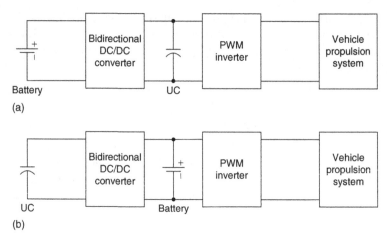

FIGURE 6.14   Typical topologies of batteries and UCs in drivetrains.

system, an UC bank could be used to achieve a wider drive range, good acceleration/deceleration performance, and low cost.

Future projections regarding performance of UCs show that energy densities as high as 10–20 Wh/kg are easily achievable using carbon electrode materials with specific capacitance values of nearly 150–200 F/g [6]. Currently, extensive R&D on UCs is being carried out in the United States, Canada, Europe, and Japan. As mentioned earlier, most of the research on UCs focuses on EV and HEV applications and on medical and power system applications.

6.3.5   Electrical Modeling of Flywheel Energy Storage Systems (FESS)

Flywheels are most definitely finding numerous applications as energy storage devices in various power system configurations. Furthermore, the constant improvement in digital signal processing (DSP) and microprocessor technologies in conjunction with the recent development in magnetic material technology makes this fact a distinct possibility. A flywheel energy storage system (FESS) is advantageous in a system comprising other secondary storage devices such as batteries as it is capable of generating optimum charge/discharge profiles for specific battery characteristics [6]. This fact facilitates the exploration of the benefits for optimizing battery management.

A rotating flywheel can store mechanical energy in the form of kinetic energy based on its inertial properties. Essentially, a FESS consists of a rotor, a motor/generator system, and a suitable enclosure [6]. An example of a flywheel energy storage system used as a voltage regulator and a UPS system is shown in Fig. 6.15.

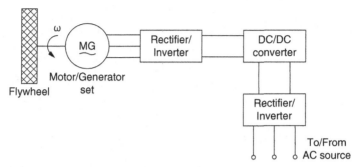

FIGURE 6.15 Typical FESS employed as voltage regulator and UPS.

The system of Fig. 6.15 essentially operates in three modes: charging, voltage regulation, and UPS. The motor/generator (M/G) set is required for energy storage purposes in the form of the inertia of the rotor. At some suitable point in the operation of the system, it retrieves this stored energy as demanded by the load. The M/G set is a high-speed device that basically operates in the motoring mode when charging the flywheel and in the generating mode when discharging it. The motor used for the M/G set could be a brushless dc motor of appropriate rating. The FESS can be easily simulated using the following equation:

$$V_x = R \times i_x + (L - M)\frac{di_x}{dt} + E_x$$

Here, V_x, i_x, and E_x are the voltages, stator currents, and back EMFs for the three phases of the BLDC motor, respectively. In addition, R, L, and M are the resistance, self-inductance, and mutual inductance of the stator winding. The back EMF is directly proportional to the mechanical speed, ω_m, and the rotor angle, θ_r.

In order to electrically simulate the same flywheel energy storage system operating in conjunction with power electronic intensive systems, it is essential to derive an equivalent electrical model of the same. For this purpose, it is critical to note the important mathematical equations that describe the above system. These are as follows:

$$v = R \cdot i + L\frac{di}{dt} + a\omega$$

$$T_{em} = a \cdot i = J_r\frac{d\omega}{dt} + b\omega + T_L$$

$$T_L = J\frac{d\omega}{dt}$$

Here, v is the voltage across the motor terminals; i is the electric current through the motor; ω is the rotor speed; T_{em} is the electromagnetic torque imposed on the rotor; T_L is the mechanical torque imposed on the rotor by the flywheel;

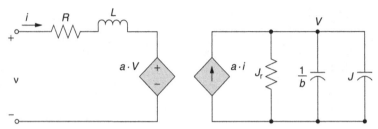

FIGURE 6.16 Electrical equivalent circuit of a flywheel energy storage system.

J_r is the equivalent moment of inertia of the rotor; J is the moment of inertia of the flywheel; and R and L are the armature resistance and self-inductance. Furthermore, a indicates the ratio of the rated voltage of the motor to its rated speed, whereas b indicates the mechanical drag coefficient. The electrical equivalent circuit generated by combining the above three equations is depicted in Fig. 6.16.

It is essential to note that the circuit parameters used are basically the parameters employed for the definition of the mathematical model of the FESS. Thus, Fig. 6.16 describes the FESS system in its entirety via an electrical equivalent. Again, as mentioned earlier, the task of simulating a FESS with any electrical system becomes immensely simplified because such an electrical model can be constructed in any popular electrical CAD simulation software and can be appropriately analyzed.

6.4 HYBRIDIZATION OF ENERGY STORAGE SYSTEMS

Certain applications require a combination of energy and power density, cost, and life cycle specifications that cannot be met by any single energy storage device. To satisfy such applications, hybrid energy storage devices (HESD) have been proposed. HESD electronically combine the power output of two or more devices with complementary characteristics. HESD all share a common trait of combining high-power devices (devices with quick response) and high-energy devices (devices with slow response).

The proposed HESDs are listed next, with the energy-supplying device listed first, followed by the power-supplying device:

- Battery/UC
- Fuel cell/battery or UC
- Battery and flywheel
- Battery and superconducting magnetic energy storage (SMES)

Note that batteries can be considered either as the energy-supplying device or as the power-supplying device depending on the application. Also, note that references [7–9] consider fuel cells rather than regenerative fuel cells. However,

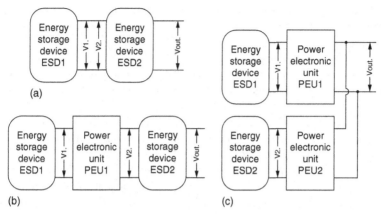

FIGURE 6.17 Proposed topologies for hybrid energy storage devices.

the system operation principle would be identical for the regenerative fuel cell, with the difference that the fuel cell would be bidirectional. However, HESDs have been proposed for utilization as an energy source for propulsion applications or grid support [10–12].

In order to have two or more energy storage devices act as a single power source, conditioning circuitry is required to combine their outputs. Numerous topologies have been proposed to achieve this task, ranging from simple to very flexible. In general, the proposed topologies can be grouped into three categories, as shown in Fig. 6.17. A discussion of merits of each topology and typical uses follows.

Direct parallel connection of two energy storage devices is shown in Fig. 6.17a. Utilization of this topology requires that the voltage outputs of the two power sources match ($V_1 = V_2$). Direct parallel connection of batteries and EDLC has been proposed [13–15] for low-voltage, cost-sensitive applications, such as the automotive 42 V PowerNet system [14]. The automotive 42V PowerNet application power profile consists of a high-power pulse (engine cranking) followed by a constant low-power demand over a longer time period (while the vehicle is in operation). A direct parallel connection of batteries and EDLC makes use of the source impedance mismatch causing the low-impedance UC to provide power during high-power pulses, while the high-energy battery supplies the long low-power demand. Note that the output V_{out} varies as the system charges and discharges. In addition, the range of power that is used from either energy source is limited by the voltage swing of the other. In other words, individual maximum power point tracking is not possible for each source.

A more complex but flexible solution is to place an additional converter between the two power sources, as shown in Fig. 6.17b. PEU1 controls the current output of ESD1, allowing its voltage to vary, whereas ESD2 supplies the remaining power requirement to the load. Therefore, this system allows for the decoupling of the two power sources. Typically, the energy storage device with a stronger dependency is utilized as ESD1. Another criterion may be to put the more sensitive device in place of ESD1 to prolong the life of the system by conditioning the current output of ESD1. Systems that make use of this topology include battery and EDLC; fuel cell and battery or EDLC; SMES and battery [16]. The commonly used topology for combining the two systems is the single leg (two switches in series) converter, which can act as a boost in the forward operation mode and as a buck in the reverse operation mode. In reference [17], the authors propose to use a variation of this system, where the battery and EDLC are connected to the load one at a time, allowing the system controller to pick which source should power the load. Source switching results in step changes of the bus voltage that requires an appropriate flexible modulation strategy. Recently, researchers have proposed the use of an isolated topology to allow for a larger voltage gain between the input and output. A superconducting magnetic energy storage (SMES) device could be connected to the midpoint of two converter legs, allowing it to charge or discharge. The battery is connected to the bus to make use of the relatively invariant battery voltage.

Finally, researchers [18,19] have looked at using the topology shown in Fig. 6.17c, where each power source is connected to a dedicated power converter with the converters connected to the common output bus. Such a system provides the highest level of flexibility as each power source is allowed to operate at its optimal voltage – in essence, maximum power point tracking can be implemented for each source. Having dedicated converters for each power source allows for a wide range of topologies and control strategies to be implemented. The simplest topology that allows an acceptable degree of flexibility is to use the single leg (two switches in series) converter, which can act as a boost in the forward operation mode and as a buck in the reverse operation mode [18,19]. Other topologies have been proposed that introduce a transformer either for isolation or to allow efficient voltage boosting.

6.5 ENERGY MANAGEMENT AND CONTROL STRATEGIES

In addition to properly sizing the battery pack to meet the power and energy demand of the vehicle, the system designer needs to ensure that the batteries perform their function as expected by the designer. To ensure this, the designer must incorporate the following: undervoltage protection, overvoltage protection, short-circuit protection (maximum current limit), thermal protection, state of charge (SOC), state of health (SOH), and state of function (SOF) monitoring, and cell equalization (balancing) on cell, as well as module level.

The state of the battery pack is monitored by measuring the SOC of the pack. Thermal management and strict manufacturing tolerances ensure that all parts of the pack are at the same state of charge. In addition, the designer may introduce some active methods of balancing the cells/modules.

6.5.1   Battery State Monitoring

A good knowledge of the state of the battery is essential for meaningful energy management [20–22]. The difficulty in measuring the condition of a battery in an operating system stems from the fact that the rate and the efficiency of the chemical reaction that produces the current depend on a number of factors, including the temperature, age, and manufacturing conditions [20]. Therefore, various figures of merit have been used to define the state of the battery. Battery SOC is defined as

$$SOC = \frac{\text{Actual amount of charge}}{\text{Total amount of usable charge at a given C-rate}}$$

The issue with this metric is that the actual amount of charge is very difficult to measure. For instance, the total amount of charge that is available for utilization changes as the battery ages. Also, the capacity scatter due to manufacturing variations makes the total amount of charge hard to determine, even for a new cell.

State of health (SOH) measures the ability of a battery to store energy, source and sink high currents, and retains charge over extended periods, relative to its initial or nominal capability. This quantity is closely related to battery age and SOC.

State of function (SOF) is the capability of the battery to perform a specific duty which is relevant to the functionality of a system powered by the battery. The SOF is a function of the battery SOC, SOH, and operating temperature. For example, a new battery (high SOH) at a lower SOC, and higher operating temperature, may perform better (higher SOF) than an older battery (low SOH) at a higher SOC, and lower operating temperature.

There are a number of methods that allow for the determination of the SOC, SOH, and SOF as outlined in Table 6.2. The simplest and most commonly used method is measuring of the open-circuit voltage of the battery and relating it to the SOC (lookup table method). In a dynamic application, this method will be very imprecise when the battery is under a load.

A simple equivalent circuit model that uses the cell voltage and current to estimate the open-circuit voltage can be implemented (model-based estimation). This method requires a current sensor and processing power. Depending on the complexity of the algorithm, the processor requirements range from minute to substantial. Alternatively, a comparator can be used to identify the points where the current is zero and measure only these points (voltage at zero current method). This method introduces imprecision because the battery does not reach its equilibrium voltage for a longer period after the zero current condition. Therefore, in dynamic situations, the error becomes substantial. More complex, processor-intensive procedures can be used and give much higher precision. For example, a more complex equivalent circuit model can be used, which also uses information from impedance and resistance measurements (model, dc

TABLE 6.2 Summary of Techniques to Measure Battery Condition

Technique	SOC	SOH	SOF	Advantages	Disadvantages
Discharge test	+	+	+	Easy, accurate, and independent of SOH	Offline; time intensive; modifies battery state
Current integration	+	−	−	Online; accurate; recalibrated often	Needs a model for the losses; sensitive to parasitic reactions, and their changes; processing power required; needs recalibration
Electrolyte measurements	+	+	+	Online	Sensitive to acid stratification; slow dynamics; temperature sensitive
Model	+	+	+	Online and flexible	Processor intensive
Impedance spectroscopy	+	+	+	Online; little processing required	Temperature sensitive; expensive
dc resistance	+	+	+	Cheaper than impedance measurement; online	Requires resistance changes that are substantial
Kalman filter	+	+	+	Online; precise in dynamic situations	Needs computing capacity; needs a suitable battery model
Voltage at zero current	+	−	−	No current sensor required	Limited precision especially in dynamic situations; needs many zero current situations
Artificial neural network	+	+	+	Online; has the potential to be very precise	Needs training on similar battery; complex and expensive to implement
Fuzzy logic	+	+	+	Online	Complex and expensive to implement

(+/− indicates whether the method is able to estimate a particular figure of merit) [23–25]

resistance, and impedance spectroscopy methods). Other options include the use of fuzzy logic, neural networks, or Kalman filters. Discharge test is the only certain way of determining the values of all three figures of merit. However, to administer this test, the battery needs to be removed from the application and discharged. Therefore, this method is only used for precision validation of other monitoring methods.

Ni-MH batteries present a bigger challenge for the determination of figures of merit because the voltage versus SOC plot is not linear. In fact, the voltage is almost flat throughout the 20–80% SOC range. Also, the batteries exhibit a memory effect. The most common way of determining the SOC is by current integration. However, this method does not consider charge inefficiencies or the effect of temperature. Fuzzy logic method has been used with success for monitoring Ni-MH state of charge.

6.5.2 Cell Balancing

Cell balancing is critical for systems that consist of long strings of cells in series. Because the cells are exposed to different conditions within the pack, without equalization, the individual states of charge and, therefore, cell voltages, will gradually drift apart. In the worst-case scenario, this leads to a catastrophic event such as ignition in the case of Li-ion batteries, and in the best-case scenario, this leads to the degradation of pack life. The sources of cell imbalance stem from manufacturing variance, leading to variations in internal impedance and differences in self-discharge rate. Another source of variation is the thermal differential across the pack, resulting in differing thermodynamics in the cells. Variations in the SOC can be minimized by designing a good thermal management system and with tight manufacturing controls.

The equalization methods can be considered to be active or passive. Passive methods are effective for lead acid and Ni-MH batteries which can be overcharged safely. However, overcharge equalization is only effective on a small number of cells in series, because equalization problems grow exponentially with the number of cells in series, and extensive overcharge does lead to cell degradation.

Many active methods are available ranging from minimally effective to exorbitantly expensive. Table 6.3 gives an extensive list of available cellbalancing methods. Generally, four methods of cell equalization exist that are defined by the method of current distribution: resistive (dissipative), switch (bypass method), capacitor, and transformer-based equalization. Typically, the cost and bulkiness and efficiency increase in the same order (resistive, switchbased, capacitive, and transformer-based).

An example of a dissipative system is the use of dissipative resistors. Dissipative resistor in continuous mode is good for low-power application.

TABLE 6.3 Summary of Cell-Balancing Strategies

Name	Description	Advantages	Disadvantages
Dissipative resistor	Dissipate power in accordance with voltage	Cheap, simple to incorporate	Not very effective; inefficient
c	Current shunted around the cell in proportion to cell voltage	Cheap; can be operated in both charging and discharging	Dissipative; not very effective; only works during charging
PWM + inductor shunting	By applying a PWM square wave on the gating of a pair of MOSFETS, the circuit controls the current difference of the two neighboring cells	Soft switching makes balancing highly efficient	Needs accurate voltage sensing; could be operated in charging mode only
Buck-boost shunting	By using a buck-boost converter the circuit shunts the current from single cell to the rest of cells	Control strategy relatively easy; relatively low cost; easy for modular design; also need intelligent control unit	Voltage sensing needed
Complete shunting	Complete shunting when cell reaches max voltage	Simple and effective	Can be only used in charging; special mass charger is needed when string is long
Switched capacitors	Balance adjacent cells by equalizing their voltages via adjacent capacitor	Simple control; operates in all modes; only two switches and one capacitor needed for each cell	Needs large capacitor bank large switched because of capacitor inrush current

Single capacitor	Balance adjacent cells by equalizing their voltages via single capacitor	Simple control; operates in all modes; many switches, but only one capacitor	Long time to balance cells
Step-up converter	Each cell is equipped with a step-up converter for cell balancing	Easy for modular design	Intelligent control needed; high cost
Multi-winding transformer	A shared transformer has a single magnetic core with secondary taps for each cell. The secondary with the least reactance will have the most induced current	Possible integration of trickle charging and equalization	High cost; inability to modularize the system; requires transformers
Multiple transformer	Several transformers can be used with the same core	Can be modularized	High cost; requires transformers
Multilevel converters	Each cell/module has a dedicated converter. The resulting topology can act as the motor driver	Ideal for transportation applications	High cost

Because the resistors are operating in continuous mode, the resistors can be small and do not need much thermal management. Another advantage of this method is the low price.

Analog current shunting is another inexpensive method of cell balancing. In this case, current is shunted around the cell in proportion to cell voltage. This system requires a comparator and a switch per cell or module. In contrast, the system is more efficient as the current is shunted rather than dissipated. This circuit has another advantage that the current could potentially be shunted both during charge and discharge, protecting the batteries from over-discharge as well as overcharge. An alternative to analog shunting is complete shunting. This system is simpler than analog shunting, as the cell is completely bypassed when the voltage reaches a certain voltage, rather than administering a PWM signal in proportion to the cell/module voltage. Dedicated buck-boost converter at each cell/module is another alternative. This system requires the use of a switch, inductor, and a diode per cell, and is only effective during charging.

Switched capacitor method uses capacitors to balance the voltage at each cell. This is achieved by having $(n - 1)$ capacitors connected in parallel to n batteries and using capacitors to equalize the current over the two adjacent cells. Further improvements of this method consider the use of two levels of capacitors for faster equalization or only one capacitor to improve system reliability. This equalization method is advantageous for hybrid applications where the batteries are never or seldom fully charged, as the equalization takes place at any voltage and operating condition.

Another approach is to use transformers to administer the charge to the batteries, with the charger on the primary, and each cell/module as multiple secondary. Each cell will then absorb a varying current that is inversely proportional to its voltage, thus ensuring voltage equalization. The issues with this system are the bulkiness and expense associated with the transformer, as well as the complex system control, and the inability to modularize the system once it is designed.

In summary, dissipative resistors in continuous mode, buck-boost shunting, and switched capacitors are the three most effective methods for different applications. Buck-boost shunting is appropriate for either high- or low-power applications, has relatively low cost, and is simple to control. The switched-capacitor method is suitable for HEV applications because it is effective in both charging and discharging regimes.

6.6 POWER ELECTRONICS FOR ENERGY STORAGE SYSTEMS

In order to appreciate the role of power electronics in battery energy management, it is essential to first identify the various issues with typical battery packs, especially those of recently touted Li-ion batteries for electric, hybrid electric, and plug-in hybrid electric vehicles.

6.6.1   Advantages and Disadvantages of Li-Ion Battery Packs for HEV/PHEV Applications

Lithium rechargeable batteries are today at the top of their wave. For instance, a 20-kWh Li-ion battery pack weighs about 160 kg (at the rate of 100–140 kWh/kg), which is acceptable for HEV applications. In contrast, current HEV nickel-metal hydride (Ni-MH) batteries weigh between 275 and 300 kg for the same application. Moreover, Li-ion batteries also depict excellent power densities (400–800 W/kg), allowing more than 2C discharge rate (at the rate of 40–80 kW peak power in a 20-kWh pack) [26]. However, they also have many drawbacks. One of the drawbacks is the cost (projected at about $250–$300/kWh), which is the most expensive of all chemistries [26]. Another drawback is that lithium is a very flammable element, whereby its flame cannot be put off with a normal ABC extinguisher. Finally, Li-ion batteries have a cycle life between 400 and 700 cycles, which does not satisfy HEV expectations. Therefore, finding a solution to these issues is crucial.

In order to resolve the safety issue, some manufacturers have modified the chemistry of the battery to the detriment of capacity and cost. Others have developed chemistries that improve the cycle life and the calendar life to the detriment of capacity and power availability. In summary, an overall success in all the aspects has not been achieved for the moment [26].

With reference to cycle life, the battery can suffer significant degradation in its capacity, depending on its usage. Furthermore, the internal resistance also increases with each charge cycle. Also, according to the chemistry and the quality of the cells, a battery typically loses about 20% of its initial capacity after about 200–1000 full cycles, also known as the 100% DOD (depth of discharge) cycles. The cycle life can be greatly increased by reducing DOD and by avoiding complete discharges of the pack between recharging or full charging. Consequently, a significant increase is obtained in the total energy delivered, whereby the battery lasts longer. In addition, overcharging or over-discharging the pack also drastically reduces the battery lifetime.

An alternative way to solve the above-mentioned problems, which are essentially common to the all lithium rechargeable batteries, is using electronic control in the form of cell voltage equalizers. Few of the control rationales are briefly listed next.

- *Overvoltage protection*: This functionality cuts the charging current when the total voltage is more than 4.3 V/cell. This is because, at higher voltages, metallic lithium is formed inside the cell, which is highly flammable, as explained earlier. For the sake of simplicity, this protection is sometimes applied to the whole pack of cells, instead of measuring the voltage of each cell.
- *Undervoltage protection*: This functionality cuts the discharging current when the voltage is under 2.5 V/cell. Under this voltage, some capacity fades, and a specific quantity of unwanted copper plating is formed inside the cell.

This unwanted copper may generate internal short circuits. Also in this case, for the sake of simplicity, the total voltage might be measured, instead of verifying the voltage of each cell.

- *Short-circuit or overcurrent protection*: This protection scheme disconnects the charging/discharging current if it is over a certain limit (2C–50C, depending on the cell technology).
- *Overheating protection*: In this case, the current is disconnected if the pack temperature rises over a certain value (about 60°C).

Although these protection functionalities are useful, they prove to be highly insufficient. In fact, the differences in capacity and internal resistance from cell to cell, within the same pack, may result in unwanted voltage peaks, especially during the final stages of charge and discharge. Consider a battery pack of 14.4 V (4 cells, of 3.6 V each) normally charging at 16.8 V (4×4.2 V). Due to the differences among the cells, the smaller capacity cell ends up with a voltage higher than the average. Consequently, if the total voltage is 16.8 V at the end of the charging cycle, the cell voltage distribution might be 4.3 V + 4.2 V + 4.2 V + 4.1 V = 16.8 V, where 4.3 V corresponds to the smaller capacity cell and 4.1 V corresponds to the larger capacity cell. Depending on the protection circuitry, this situation may or may not be detected, and the resistive equalizer gradually downgrades the voltage from 4.3 to 4.2 V. The net result is a considerable downgrade in overall capacity, considering that the overvoltage occurs in the smaller capacity cell, and that this cell is later discharged by a specific amount, in order to draw level with the other cells.

In contrast, during discharge, if the lower cut-off voltage is 3 V per cell, the pack will discharge up to 12 V. As the lower capacity cell discharges faster than the rest of the cells, the voltage distribution might be 2.7 V + 3 V + 3 V + 3.3 V = 12 V. Again, in this case, the lower capacity cell suffers from an over discharge, which is not detected by the protection circuit. Thus, the cell incurs an additional capacity reduction due to the resultant over-voltage and under-voltage, which downgrade the capacity of the entire pack.

6.6.2 Operational Characteristics of Classic and Advanced Power Electronic Cell Voltage Equalizers

A battery cell voltage equalizer is an electronic controller that takes active measures to equalize the voltage in each cell. In addition, by some more complex methods, such as measuring the actual capacity and internal resistance of each cell, it equalizes the DOD of each cell. As a result, each of the cells will have the same DOD during charging and discharging, even in conditions of high dispersion in capacity and internal resistance. This causes each cell of the pack to act as an average cell. In the example presented in the previous section, instead of 282 cycles, the pack would last 602 cycles, an increment of more than 100% in the cycle life. For the same application, the requirement of current though the equalizer at a discharge of 50 A, over a 100-Ah pack, would be 6-A drain/source, over the extreme capacity cells (the ones with bigger and smaller capacity), and a total transferred power of 350 W (in the entire chain), over the complete charging/discharging time. In addition to the increment in cycle time, depending on the internal resistance of each cell, the equalizer also has the potential to improve the output power. This requires a detailed analysis of the internal resistance versus DOD, in order to have precise improvement. In the following analyses of various equalizers, it will be realized that only a few of them are capable of accomplishing this requirement.

In principle, there exist three basic families of equalizers: resistive, capacitive, and inductive. Resistive equalizers simply burn the excess power in higher voltage cells, as depicted in Fig. 6.18. Consequently, it is the cheapest option and is widely utilized for laptop batteries. As is obvious, due to inherent heating problems, resistive equalizers tend to have low equalizing currents in the range 300–500 mA and work only in the final stages of charging and flotation.

In contrast, capacitive equalizers use switched capacitors, as shown in Fig. 6.19, in order to transfer the energy from the higher voltage cell to the lower voltage cell. It switches the capacitor from cell to cell, allowing each cell to physically have the same voltage. Besides, it also depicts higher current capabilities than a resistive equalizer. At the same time, the main drawback of capacitive equalizers is that it cannot control inrush currents, when big differences in cell voltages exist. In addition, it does not allow any desired voltage difference, for example, when equalizing DOD.

Finally, inductive or transformer-based equalizers use an inductor to transfer energy from the higher voltage cell to the lower voltage cell. In fact, this is the most popular family of high-end equalizers, and because of its capability to fulfill most of the needs expressed above, it is explored in more detail in the following sections.

6.6.2.1 Basic Inductive Equalizer

A basic inductive equalizer is shown in Fig. 6.20. These equalizers are relatively straightforward and can transport a large amount of energy. At the same time,

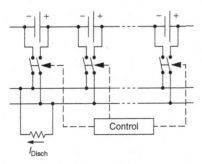

FIGURE 6.18   Schematic of a typical resistive equalizer.

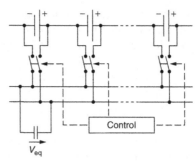

FIGURE 6.19 Schematic of a typical capacitive equalizer.

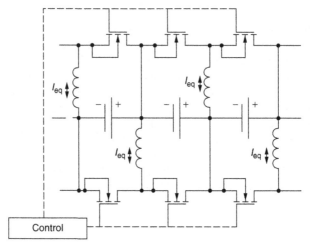

FIGURE 6.20 Schematic of a typical inductive equalizer.

they are also capable of handling more complex control schemes, such as current limitation and voltage difference control [26].

In contrast, it takes some additional components to avoid current ripples from getting into the cell. Typically, this requires two switches (and control) per cell. Also, due to switching losses, the distribution of current tends to be highly concentrated in adjacent cells. Hence, a high-voltage cell will distribute the current largely among the adjacent cells, instead of equally in all cells. In this case, the switching scheme could be replaced by a more complex, global scheme, with the additional cost of more processing power.

6.6.2.2 Cuk Equalizer

As the name indicates, this is an inductive-capacitive type of equalizer, primarily based on the *Cuk Converter* topology. It shares almost all the characteristics of inductive equalizers, but it does not suffer from high-current ripples, or cost of additional power capacitors, or double rated switches. A schematic of a typical *Cuk* equalizer is shown in Fig. 6.21.

The *Cuk* equalizer does incur minor losses due to the series capacitor, but this is far from a major shortcoming. This equalizer also possesses high current and complex control capability, at the expense of additional processing power.

6.6.2.3 Transformer-Based Equalizer

The solutions provided by transformer-based equalizers permit the right current distribution along all cells. One such popular arrangement is depicted in Fig. 6.22.

This topology poses an additional issue of a very complex multi-winding transformer. This transformer is very difficult to mass produce, because all windings must have the compatible voltages and resistances. Hence, it is not a practical solution for high-count HEV cell packs, and moreover, it also lacks the capability of handling complex control algorithms such as current and voltage control. An alternative solution is presented in Fig. 6.23, using separate transformers for each cell.

This solution is modified here, in order to use 1:1 transformers. Although, through this topology, a substantial amount of redundancy is obtained, only a very small dispersion can be accepted, with the added risk of experiencing current imbalance.

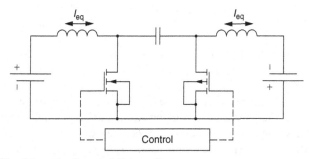

FIGURE 6.21 Schematic of a typical *Cuk* equalizer.

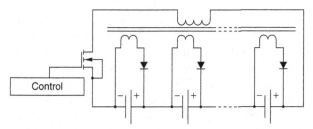

FIGURE 6.22 Schematic of a multi-winding transformer equalizer.

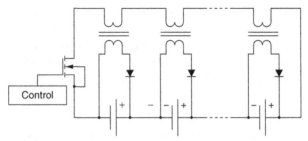

FIGURE 6.23   Schematic of a multiple transformer equalizer.

6.7   PRACTICAL CASE STUDIES

6.7.1   Hybrid Electric and Plug-in Hybrid Electric Vehicles (HEV/PHEV)

It is an obvious fact that future vehicles are moving toward more electrification. It is widely agreed that more electric vehicles will depict major benefits over conventional vehicles in terms of improved performance, less harmful emissions, and higher drive train efficiency. Since the legislation of strict ultralow-emission standards by California Air Resource Board (CARB) in 1990, auto industries and research laboratories around the world are pursuing electric and hybrid electric vehicle research seriously. Motivation of auto manufacturers toward more electric and hybrid electric vehicles (HEVs) increased in 1993, by declaring the historic partnership program for next generation vehicles (PNGV) between the big three automakers (GM, Ford, and Chrysler) and the U.S. government.

Primarily, a hybrid electric vehicle improves total overall drive train efficiency over a standard drive system by supplying electric energy from an electric energy storage system to assist the main power source and reusing braking energy that would otherwise be lost. High-quality energy storage system (ESS) is one of the most crucial components that affects vehicles' performance characteristics. The energy storage device charges, during low-power demands, and discharges, during high-power demands. Thus, it basically acts as a catalyst for providing an energy boost. Therefore, the engine ideally operates at its most efficient speed. Because of this intended operation, the energy storage device is sometimes referred to as the load-leveling device.

The most important traits that customers look for in the vehicles loaded by LLDs are acceleration rate, fuel economy, level of maintenance, safety, and cost [27]. All these requirements need to be supported by an efficient, fast responding, and high-capacity ESS. Therefore, improved energy storage devices are a key technology for next generation HEVs.

The introduction of large battery packs in advanced vehicles presents a major shift in how batteries are used; the profile is truly unique and varies greatly depending on the vehicle in question. In this section, typical battery load profiles are investigated in advanced vehicles. Depending on the battery usage, various battery designs are better suited for the application. In some cases, high-power batteries are required, whereas in other cases, high-energy density batteries are more appropriate. A number of simulations were performed to demonstrate the requirements from the batteries as a function of the application. The vehicles considered are the following:

- Electric vehicle with a 50- and 100-mile range
- Series HEV
- Parallel HEV
- Series PHEV with a 10-, 20-, and 30-mile all-electric range

The parameters of the simulated vehicles are given in Table 6.4, power-train specifications in Table 6.5, and the battery pack parameters in Table 6.6.

Note that the batteries are sized in accordance to the type of vehicle and the targeted all-electric range. The battery voltage was assumed to be 314 V, and the battery capacity was varied to achieve the total battery capacity required. This is equivalent to connecting strings of batteries in parallel. Battery pack specifications were chosen to be 60 Wh/kg (energy density) and 375 W/kg (power density). These specifications are well within what is achievable practically. The

TABLE 6.4 Chassis Specifications

Parameter	Value
Glider mass (kg)	900
Frontal area (m^2)	1.8
Coefficient of drag	0.3
Wheel radius (m)	0.3
Rolling resistance coefficient	0.009
Battery capacity (Wh/kg)	60
Battery power density (W/kg)	375

TABLE 6.5 Power-train Specifications

Parameter	Series HEV and PHEV	Parallel HEV and PHEV	EV
E-motor (kW)	75	30	75
Generator (kW)	40	0	0
Engine (kW)	40	45	0

energy density is defined at the 1C rate, and the power is defined by the HPPC test at 20% DOD. The driving cycle is the urban dynamometer driving schedule (UDDS), representing typical urban driving. The vehicle is driven over 30 miles (four UDDS cycles), representing a typical daily commute.

Based on the parameters enlisted in Tables 6.4–6.6, the vehicles were simulated using the Advanced Vehicle Simulator (ADVISOR) software. Figure 6.24 shows simulation results for two electric vehicles with a 50-mile (EV50) and

TABLE 6.6 Battery Specifications

Vehicle	Battery energy (kWh)	Battery peak power at 20% DOD (kW)	Battery energy rating (Ah)
HEV	2.5	15	8
PHEV10	4.8	30	15
PHEV20	7.8	52	25
PHEV30	12	75	40
EV50	13.2	82.5	42
EV100	26.4	172.5	85

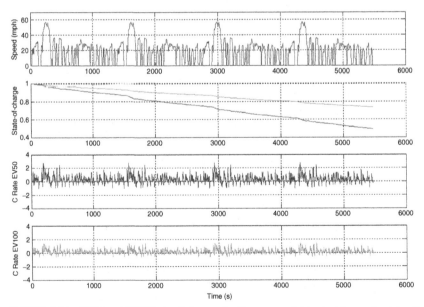

FIGURE 6.24 EV50 and EV100 simulation results (from top to bottom): drive cycle, state of charge, EV50 current, normalized to the battery C-rate, EV100 current, normalized to the battery C-rate.

100-mile (EV100) electric range. For each vehicle, the battery pack size was increased until the driving range reached 50 and 100 miles respectively. Because we are concerned with the battery operation, we will compare the current outputs out of the batteries, and the currents are normalized to the battery 1C rate to more easily compare the relative stresses on the battery. In the case of EV50, the currents reach a 2C rate quite often, whereas in the case of EV100, the peak current is about 1C. This stresses the need for different battery designs for the two vehicles. For EV50, the designer must chose a battery capable of supplying and absorbing a 2C rate without affecting battery life. For EV100, the designer can opt for a very high energy density battery as only 1C rate will be required. A higher density battery would in turn extend vehicle range due to the smaller vehicle mass.

Another issue that has to be considered for EVs and PHEVs is the recharge time of the battery. If the battery pack is designed for a maximum current of 1C, it would take over an hour to charge the battery. This may not be acceptable to the user in some cases.

In the next test, HEVs are considered, as shown in Fig. 6.25. Here, three cases are investigated: Parallel HEV, Thermostat Series HEV, and Load-Following Series HEV. The Parallel HEV uses the electrical system to assist the engine with

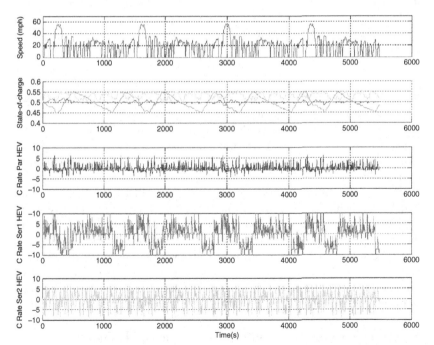

FIGURE 6.25 Parallel HEV, thermostat series HEV, and load following series HEV (from top to bottom): drive cycle, state of charge, parallel HEV current, normalized to the battery C-rate, thermostat series HEV current, normalized to the battery C-rate, load following series HEV current, normalized to the battery C-rate.

vehicle acceleration and to propel the vehicle at low speeds where the engine operation is inefficient. The state of charge of the vehicle is kept close to the target value of 50%. As the electrical system is small, the batteries provide a relatively small portion of the total vehicle power. The batteries supply currents up to 5C, and the total energy demand from the battery is also very small. Therefore, a high specific power battery would be ideal for this application, and even ultracapacitors may be considered, as the energy throughput is minimal.

The second case is the thermostat series hybrid. Here, the internal combustion engine is only operated at its most efficient point. The internal combustion engine is turned on when the battery state of charge reaches the minimum allowed (in this case, 45%) and is turned off when the battery state of charge reaches the maximum value (in this case, 55%). Such a control system is very demanding on the battery; in this simulation, the currents are shuffled at a 10C rate. Therefore, a practical implementation of this system would require a larger battery pack to reduce peak currents.

Finally, in the load-following series hybrid, the internal combustion engine tries to follow the road load as closely as possible, while still optimizing the use of the internal combustion engine to minimize fuel consumption. In this vehicle, the current usage drops as compared to the series thermostat. Note that the battery use in a fuel cell vehicle would be the same as in a series hybrid. If the load-following control strategy is implemented, the fuel cell is stressed more as its operating point changes, while for the thermostat control the battery pack is stressed more.

Recently, plug-in hybrids have received a great deal of attention. These vehicles act as a hybrid vehicle, where the battery pack can be recharged via the grid to supply some all-electric range. In these vehicles, the battery pack is discharged from a full charge to some target state of charge, where the vehicle is then operated as a regular hybrid. In simulations, as shown in Fig. 6.26, the target SOC is chosen to be 40%, as this value makes the most of the all-electric range, while ensuring that the life of the battery pack is not compromised by over-discharge.

Three vehicles are considered in this study, PHEV10, PHEV20, and PHEV30, which give a 10-, 20-, and 30-mile all-electric range, respectively. The vehicle of choice is the series hybrid, as it is capable of all-electric operation, wherein all of the propulsion power comes from the electric motor. As is clear, the C-rates are substantial on the PHEV10 and decrease for PHEV20 and PHEV30. PHEV10 proposes an interesting power versus energy density problem, where both the power and the energy density must be high. In contrast, as the all-electric range of the vehicle increases, high specific energy is critical.

6.7.2 Fuel Cells for Automotive and Renewable Energy Applications

For automotive applications, proton exchange membrane fuel cells (PEMFCs) appear to be most suitable compared to other fuel cell technologies, such as

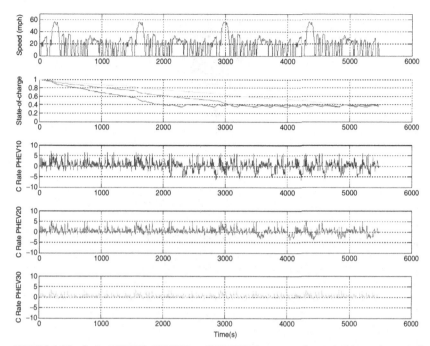

FIGURE 6.26 Series PHEV10, PHEV20, and PHEV30 (from top to bottom): drive cycle, state of charge, series PHEV10 current, normalized to the battery C-rate, series PHEV20 current, normalized to the battery C-rate, series PHEV30 current, normalized to the battery C-rate.

alkaline fuel cells (AFC) or solid oxide fuel cells (SOFC). This is due to the fact that the PEMFC working conditions at low temperature allow the system to start up faster than with those technologies using high-temperature fuel cells. Moreover, the solid state of their electrolyte (no leakages and low corrosion) and their high power density make PEMFCs extremely convenient for transport applications [28]. PEMFCs also provide a very good tank-to-wheel efficiency compared to heat engines. At the same time, PEMFCs need a great deal of auxiliary equipment as shown in Fig. 6.27.

Automotive fuel cell technology continues to make substantial progress. However, fuel cell vehicles (FCVs) have not yet proven to be commercially viable nor have they been proven to be efficient. More recently, technological and engineering advancements have improved, simplified, and even eliminated components of the fuel cell system. The technical challenges and objectives for fuel cell systems in transportation applications have been well highlighted by the U.S. Department of Energy (DOE). Cost and durability are the major challenges for fuel cell systems. Air, thermal, and water management for fuel cells are also key issues. Power density and specific power are now approaching set targets. However, further improvements are needed to meet packaging requirements of commercial systems. The objective, by 2010, is to develop a 60% peak-efficient,

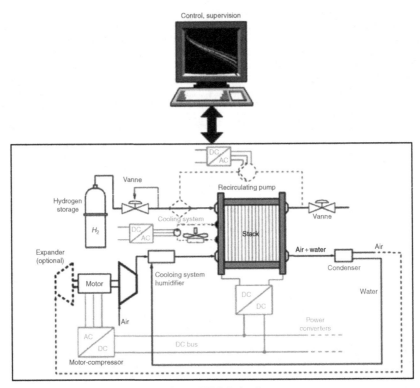

FIGURE 6.27 Proton exchange membrane (PEM) fuel cell system.

durable, direct hydrogen fuel cell power system (including all auxiliaries) for transportation at a cost of $45/kW and, by 2015, at a cost of $30/kW, to become competitive with conventional internal combustion engine vehicles.

Another critical issue with fuel cells is their start-up times as well as operating temperatures. More recently developed fuel cell systems have been able to start and operate in temperatures ranging between −40°C and +40°C. In these temperature conditions, start-up times of up to 50% of the rated power have been depicted: 30 s at −20°C and 5 s at 20°C. However, the size and the weight of current fuel cell systems have to be reduced drastically to meet the automotive compactness requirements, which apply both for the fuel cell stack and for auxiliary components such as the compressor, expander, humidifiers, pumps, sensors, and hydrogen storage [28]. The power mass density and the power volume density requirements for the fuel cell system are 650 W/kg, 650 W/L, and 2000 W/kg or 2000 W/L, for the stack itself. The transient response of the stack is also a key issue and depends mainly on the air supply system inertia. Ideally, the transient response from 10 to 90% of the maximum power should be less than 1 s.

The hydrogen fuel is stored on-board and is supplied by a hydrogen production and fuelling infrastructure. For other applications (such as for distributed stationary power generation), hydrogen can be fuelled with reformate, produced from natural gas, liquefied petroleum gas, or renewable liquid fuels. For portable electronic devices in small equipment, methanol, and sometimes hydrogen, is the fuel of choice, using microfuel cell systems.

The objectives of hydrogen storage are the volume, weight, cost, durability, cycle life, and transient performance. On-board hydrogen storage solutions are summarized in Fig. 6.28. Some of the popular storage systems include high-capacity metal hydrides, high-surface area sorbents, chemical hydrogen storage carriers, low-cost and conformable tanks, compressed/cryogenic hydrogen tanks, and new materials or processes, such as conducting polymers, spillover materials, metal organic frameworks (MOFs), and other nanostructured materials. In general, there are two principal types of on-board storage systems. (1) *On-board reversible systems*, which can be refueled on-board the vehicle from a hydrogen supply at the fueling station. Compressed/cryogenic tanks, some of the metal hydrides, and high-surface area sorbents represent these solutions. (2) *Regenerative off-board systems*, which involve materials that are not easily refilled with hydrogen, while on-board the vehicle. These solutions include chemical hydrogen storage materials and some metal hydrides, where temperature, pressure, kinetics, and/or energy requirements are such that the processes must be conducted off-board the vehicle [28].

A majority of FCVs that are being proposed more recently use 350-bar hydrogen tanks, which have system gravimetric and volumetric capacities of 2.8–3.8% weight and 17–18 g/L, respectively. Cryo-compressed hydrogen and liquid hydrogen systems are also fast approaching U.S. DOE targets, as shown in

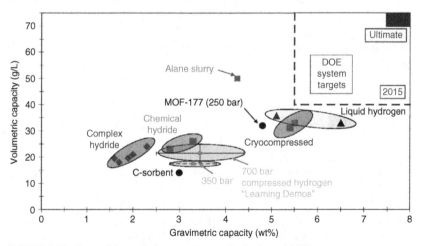

FIGURE 6.28   State-of-the-art hydrogen storage technologies and DOE system targets.

Fig. 6.28. However, these solutions are still not affordable for mass production, as other requirements such as cost, hydrogen charging/discharging rates, and durability are also key issues for hydrogen storage. Liquid hydrogen storage is being demonstrated as workable, but with limitations. It provides both higher gravimetric and volumetric density advantages over compressed gas storage, but depicts issues with boil off and dealing with cryogenic liquids. Hence, it is not likely to be widely accepted by automobile manufacturers.

A more practical application for fuel cells is distributed generation (DG) power plants. The popular choices for DG are Phosphoric Acid FC, Molten Carbonate FC, Solid Oxide FC, and the Proton Exchange Membrane FC. Their applications in DG are briefly discussed here.

6.7.2.1   Phosphoric Acid Fuel Cell (PAFC)

Typically, PAFC units of the order 250–300 kW are available for commercial cogeneration, and more than 150 of such units are in operation worldwide. The typical efficiencies of such units are 40–50%, and the cost of production of power is approximately $4200/kW, which is beyond the economic margin unless a financial benefit can be demonstrated from the PAFC's emissions, power quality, and reliability merits [29].

6.7.2.2   Molten Carbonate Fuel Cell (MCFC)

MCFC operates at higher temperatures, of the order of 600°C, and was initially marketed for 1- to 5-MW plant applications. This system has a much higher efficiency compared to PAFCs with values reaching as high as 50%. The MCFC-based systems are typically rated at around 200 MW and exhibit an efficiency of about 75%. Upon further research, the MCFC's cost of power production is estimated to be as low as $1000/kW [29]. A hybrid MCFC/turbine cycle for maximizing the efficiency is shown in Fig. 6.29.

In such a system, about 70% of the power is produced by fuel cell and about 15% comes from the gas turbine generator.

6.7.2.3   Solid Oxide Fuel Cell (SOFC)

Comparatively speaking, a 100-kW MCFC system uses about 100–110 cells in stacks, whereas a similarly rated SOFC system would use about 1150 cells [29]. An example of a SOFC-based cogeneration plant is shown in Fig. 6.30.

Here, pressurized air and fuel are the inputs to the SOFC because pressurized SOFC is being preferred for cogeneration purposes. The exhaust quality obtained by these units is comparatively higher. The hybrid fuel cell system, using a pressurized fuel cell, combined with the use of gas turbines, provides high efficiency and low emissions. A higher efficiency is gained by using the thermal exhaust from the fuel cell to power a noncombusting gas turbine.

6.7.2.4   Proton Exchange Membrane Fuel Cell (PEMFC)

This technology has attracted the attention of most utility companies as it has produced extremely low-cost power compared to other fuel cells. The cost of power production in PEMFC-based automobiles is as low as $100–$150/kW, which provides a competitive potential for stationary power production. It is expected that a successful market for PEMFC will have a significant impact on power generation because it could shift the primary role of the present-day grid to back up and peaking power [29]. In addition, PEMFCs find applications in cogeneration, in providing premium power, and in households.

It is apparent that several fuel cell types have strong potentials for entering the DG market. They can basically provide cost-effective cogeneration, grid support, and asset management. Also, long-term plans for fuel cell are underway with advanced designs for combined-cycle plants, which could eventually compete for a share of the DG market.

6.7.3   Fuel-Cell-Based Hybrid DG Systems

Individual fuel cell units ranging from 3 to 250 kW can be used in conjunction with high-speed microturbines, for high-power DG applications. The other popular DG technology is the PV power generation system, which is suited to provide up to 250 kW of power. These topologies are discussed in detail here.

6.7.3.1   Fuel Cell/Microturbine Hybrid DG System

Emission specification for the pressurized SOFC design is less than 1.0 ppm (parts per million) NOx and almost zero level of SOx. Another unique advantage of this unit is that a small percentage (about 15%) of fuel is wasted. The exhaust thermal energy is used to drive the microturbine. The hot exhaust from the plant supplants the microturbine combustor during the normal steady operation. It must be noted here that microturbine forms no additional pollutants [29]. The SOFC type of fuel cell is chosen for this application as it operates at the highest known temperature among fuel cells, at about 1000°C, and can be operated at

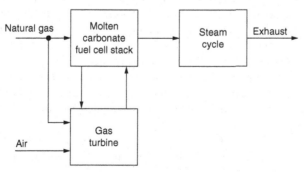

FIGURE 6.29   An MCFC-hybrid power cycle.

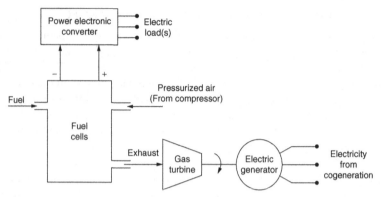

FIGURE 6.30   Typical layout of a SOFC-based cogeneration plant.

high pressure. All these features are added up to provide additional thermal and electrical efficiency for the hybrid unit.

A diagrammatic representation of such a unit is shown in Fig. 6.5. Typical rating of such a hybrid system is about 250 kW and efficiencies of greater than 60% are targeted for the future. An important point to be noted here is that the fuel cell supplies about 80% of the output power, whereas the microturbine supplies the remaining 20%. Hence, microturbine functions primarily as a turbocharger for SOFC, with additional shaft power coming from microturbine for electrical power generation.

The National Critical Technologies (NCT) panel has identified fuel cells and microturbines as 2 of 27 key technologies in the United States for maintaining economic prosperity and national security. Several utilities facing with the dilemma of increasing transmission capacity are opting for DG technology. Such efficient packages of about 250-kW size could avoid the need to increase transmission capacity for years to come.

6.7.3.2   Fuel Cell/Photovoltaic (PV) Hybrid DG System

Fuel cells are attractive options for intermittent sources of generation such as PV, because of their high efficiency, fast response to loads, modularity, and fuel flexibility. Such a system is able to smoothen the PV problem of intermittent power generation by utilizing the fast ramping capabilities of fuel cells. Unlike batteries, which get discharged after a short time of operation, fuel cells can continuously provide the required amounts of power as long as the reactants (fuel + air) are supplied. Thus, the quality of power fed to the utility system by the hybrid system is of improved nature. A simple schematic of a PV-fuel cell hybrid system is depicted in Fig. 6.31.

Figure 6.31 illustrates a grid-connected PV-fuel cell power plant including two feedback controllers, which basically can control the power conditioner switches. These power electronic switches, in turn, control the maximum

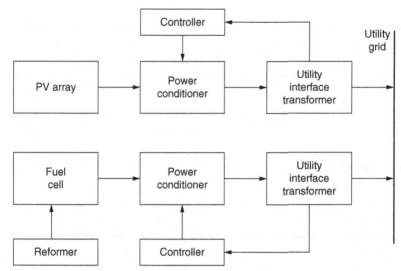

FIGURE 6.31   Block diagram of a fuel cell-PV hybrid system.

power point and active and reactive power flows. In Fig. 6.31, the hybrid system consists of a PV array, a fuel cell stack, a reformer for purifying the hydrocarbon-based fuel, and power conditioners, which consist of dc–dc and dc–ac power electronic converters. The PV generator operates independently and is controlled to produce maximum available solar power. The fuel cell generating system is used as a supplement to this PV system to meet the system's power demand [29].

The PV array in Fig. 6.31 is made up of 80–100 series-connected and 450–500 parallel-connected solar modules, which produces 1.5–2 MW of power at 1400 V. In contrast, the fuel cell system made up of a PAFC unit generates about 2 MW and satisfies the system's demand for active and reactive power. Such a combination of fuel cells with PV arrays proves to be feasible for solving the inherent problems of standalone PV systems. Furthermore, because conventional fossil fuel energy sources are diminishing at a fast rate, such energy sources are attracting even more attention from utility companies. Much research is being devoted to such hybrid systems to bring down their O&M costs and render them favorable over conventional gas turbines and diesel engines.

SUMMARY

This chapter dealt with the electrical characteristics and modeling techniques of major types of renewable energy systems and competent energy storage devices, namely batteries, fuel cells, PV cells, ultracapacitors, and flywheels. The introduced models can easily be simulated and validated for their performance by running a simple computer simulation. In addition, this chapter summarized the various up-and-coming applications for renewable energy systems and storage devices, thus making their modeling and simulation studies worthwhile. Future research and development work includes hardware-in-the-loop (HIL) implementation of the various topologies with novel control strategies for hybrid energy storage systems.

REFERENCES

[1] S. G. Chalk, J. F. Miller, Key challenges and recent progress in batteries, fuel cells, and hydrogen storage for clean energy systems, J. Power Sources 159(1) (2006) 73–80.
[2] B. Sorensen, Renewable Energy, third ed., Academic Press, London, UK, 2004.
[3] B.R. Williams, T. Hennessy, Energy oasis, Power Eng. 19(1) (2005) 28–31.
[4] F. Kreith, D.Y. Goswami, Energy Conversion. CRC Press, Boca Raton, FL, 2007.
[5] A. Emadi, S.S. Williamson, Status review of power electronic converters for fuel cell applications, J. Power Electron. 1(2) (2001) 133–144.
[6] S.S. Williamson, S.C. Rimmalapudi, A. Emadi, Electrical modeling of renewable energy sources and energy storage devices, J. Power Electron. 4(2) (2004) 117–126.
[7] Z. Jiang, R.A. Dougal, A compact digitally controlled fuel cell/battery hybrid power source, IEEE Trans. Ind. Electron. 53(4) (2006) 1094–1104.
[8] M.H. Todorovic, L. Palma, P.N. Enjeti, Design of a wide input range DC–DC converter with a robust power control scheme suitable for fuel cell power conversion, IEEE Trans. Ind. Electron. 55(3) (2008) 1247–1255.
[9] M. Ortuzar, J. Moreno, J. Dixon, Ultracapacitor-based auxiliary energy system for an electric vehicle: implementation and evaluation, IEEE Trans. Ind. Electron. 54(4) (2007) 2147–2156.
[10] S. Lemofouet, A. Rufer, A hybrid energy storage system based on compressed air and supercapacitors with maximum efficiency point tracking (MEPT), IEEE Trans. Ind. Electron. 53(4) (2006) 1105–1115.
[11] T. Ise, M. Kita, A. Taguchi, A hybrid energy storage with a SMES and secondary battery, IEEE Trans. Appl. Supercond. 15(2) (2005) 1915–1918.
[12] H. Tao, J.L. Duarte, M.A.M. Hendrix, Multiport converters for hybrid power sources, in: Proceeding of the IEEE Power Electronics Specialists Conference, Rhodes, Greece, June 2008, pp. 3412–3418.
[13] W. Henson, Optimal battery/ultracapacitor storage combination, J. Power Sources 179(1) (2008) 417–423.
[14] J.P. Zheng, T.R. Jow, M.S. Ding, Hybrid power sources for pulsed current applications, IEEE Trans. Aerosp. Electron. Syst. 37(1) (2001) 288–292.
[15] K. Jin, X. Ruan, M. Yang, M. Xu, A hybrid fuel cell power system, IEEE Trans. Ind. Electron. 56(4) (2009) 1212–1222.
[16] O. Briat, J.M. Vinassa, W. Lajnef, S. Azzopardi, E. Woirgard, Principle, design and experimental validation of a flywheel-battery hybrid source for heavy-duty electric vehicles, IET Electr. Power Appl. 1(5) (2007) 665–674.
[17] L. Shuai, K.A. Corzine, M. Ferdowsi, A new battery/ultracapacitor energy storage system design and its motor drive integration for hybrid electric vehicles, IEEE Trans. Veh. Technol. 56(4) (2007) 1516–1523.

[18] H. Matsuo, L. Wenzhong, F. Kurokawa, T. Shigemizu, N. Watanabe, Characteristics of the multiple-input DC–DC converter, IEEE Trans. Ind. Electron. 51(3) (2004) 625–631.

[19] L.L. Solero, A. Pomilio, Design of multiple-input power converter for hybrid vehicles, IEEE Trans. Power Electro. 20(5) (2005) 1007–1016.

[20] K. Kutluay, Y. Cadirci, Y.S. Ozkazanc, I. Cadirci, A new online state-of-charge estimation and monitoring system for sealed lead-acid batteries in telecommunication power supplies, IEEE Trans. Industrial Electronics, 52(5) (2005) 1315–1327.

[21] A. Affanni, A. Bellini, G. Franceschini, P. Guglielmi, C. Tassoni, Battery choice and management for new-generation electric vehicles, IEEE Trans. Ind. Electron. 52(5) (2005) 1343–1349.

[22] L.R. Chen, A design of an optimal battery pulse charge system by frequency-varied technique, IEEE Trans. Ind. Electron. 54(1) (2007) 398–405.

[23] N.H. Kutkut, H.L.N. Wiegman, D.M. Divan, D.W. Novotny, Design considerations for charge equalization of an electric vehicle battery system, IEEE Trans. Ind. Appl. 35(1) (1999) 28–35.

[24] L.A. Tolbert, F.Z. Peng, T. Cunnyngham, J.N. Chiasson, Charge balance control schemes for cascade multilevel converter in hybrid electric vehicles, IEEE Trans. Ind. Electron. 49(5) (2002) 1058–1064.

[25] A.C. Baughman, M. Ferdowsi, Double-tiered switched-capacitor battery charge equalization technique, IEEE Trans. Ind. Electron. 55(6) (2008) 2277–2285.

[26] P.A. Cassani, S.S. Williamson, Status review and suitability analysis of cell-equalization techniques for hybrid electric vehicle energy storage systems, IEEE Power Electron. Soc. Newslett. 20(2) (2008) 8–12 (Second Quarter).

[27] A. Emadi, K. Rajashekara, S.S. Williamson, S.M. Lukic, Topological overview of hybrid electric and fuel cell vehicular power system architectures and configurations, IEEE Trans. Veh. Technol. 54(3) (2005) 763–770.

[28] B. Blunier, M. Pucci, G. Cirrincione, A. Miraoui, A scroll compressor with a high-performance induction motor drive for the air management of a PEMFC system for automotive applications, IEEE Trans. Ind. Appl. 44(6) (2008) 1966–1976.

[29] S.S. Williamson, A. Emadi, M. Shahidehpour, Distributed fuel cell generation in restructured power systems, in: Proceedings of the IEEE Power Engineering Society General Meeting, Denver, CO, June 2004, pp. 2079–2084.

Chapter 7

Electric Power Transmission

Ir. Zahrul Faizi bin Hussien, Azlan Abdul Rahim,
and Noradlina Abdullah
Transmission and Distribution, TNB Research, Malaysia

Chapter Outline

7.1   ELEMENTS OF POWER SYSTEM

The growth of electricity production and usage in the world is at an all time high. It is the fastest growing form of end-use energy worldwide. The world net electricity generation in the year 2006 is estimated at 18 trillion kWh. It

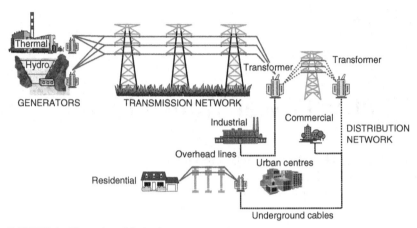

FIGURE 7.1   Illustration of the basic components of a power system.

is forecasted to be 23.2 trillion kWh in the year 2015, and 31.8 trillion in the year 2030 [1]. Electricity production from fossil fuels raises concern on climate change and hence alternative energy sources such as solar, geothermal, wind, wave, tidal, and biomass are being developed rapidly around the world.

Modern electric power system consists of complex interconnected network of components generally divided into generation, transmission, distribution, and load. The invention of the transformer at the end of the nineteenth century enables electrical energy to be transmitted at higher voltages and hence higher capacity. In this chapter, the components that make up the modern power system are briefly reviewed, namely generators, transformers, and transmission line. The chapter will focus on the transmission of electric power and how to optimize its power transfer capability. The chapter will also discuss the phenomena of overvoltages and the insulation requirement of transmission lines.

7.2   GENERATORS AND TRANSFORMERS

Generators are the starting point in a power system where electricity is generated. The most commonly used type of generator is the synchronous generator, driven by a prime mover. In a thermal power station, the prime mover is powered by steam turbines using coal, gas, nuclear, or oil as the fuel. Steam turbines usually operate at high speed to optimize its power output. For a 60 Hz system, the speed of rotation is 3600 rpm (two-pole machine) or 1800 rpm (four-pole machine). For a 50 Hz system, the speed of rotation is 3000 (two-pole machine) or 1500 rpm (four-pole machine). Hydro power station uses the potential energy of water to power the hydraulic turbines. The hydraulic turbines usually operate at relatively low speed, coupled to generators with salient type rotor with many

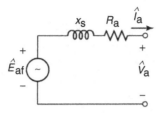

FIGURE 7.2 Equivalent circuit of a synchronous generator.

poles. In power generating stations, the electrical power is generated at high voltages typically between 10 and 30 kV while the size of the generators varies from 50 to 1500 MW.

A synchronous generator produces alternating current in the armature winding (usually a three-phase winding on the stator) with dc excitation supplied to the field winding (usually on the rotor). The per-phase equivalent circuit of a synchronous generator is shown below where the armature voltage, V_a, can be written as

$$V_a = R_a I_a + j X_s I_a + E_{af}$$

Where

V_a is the armature voltage;
R_a is the armature resistance;
I_a is the armature current;
X_s is the synchronous reactance;
E_{af} is the internal generated voltage.

The open-circuit characteristic (occ) of a synchronous generator as shown below represents the relationship between the field current I_f and the internal generated voltage E_{af}. Note that with the armature winding open-circuited, the armature voltage V_a is equal to the internal generated voltage E_{af}. As the field current I_f is increased from zero, the armature voltage V_a increases linearly. Near the rated voltage, the saturation effect is clearly seen with the departure of the occ from the straight line (air-gap line).

Another characteristic of the synchronous generator is illustrated by the V curve shown in Fig. 7.4. For a constant real-power loading, the field current and hence excitation can be adjusted to vary the reactive power VAr generated, hence power factor (p.f).

The voltage output of the generator is transformed to a higher voltage before transmitting the power over a long distance to reduce the transmission losses, $I^2 R$ losses. Transformers are used to step up or step down the voltage.

Transmission system voltages vary from country to country. For example, in Britain they are 400 and 275 kV, in the U.S.A. they are 765, 500, and 345 kV, while in Malaysia they are 500 and 275 kV. The subtransmission network is at 115 KV in the U.S.A., while it is 132 kV in Britain and Malaysia. This network in

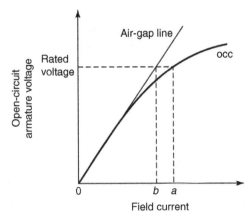

FIGURE 7.3   Open-circuit characteristic (occ) of a synchronous generator.

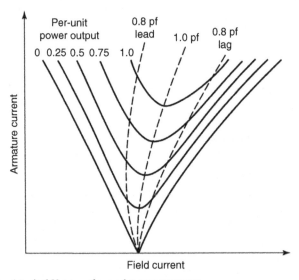

FIGURE 7.4   A typical V curve of a synchronous generator.

turn supplies the distribution network to the customers in a given area. In Britain and Malaysia, the distribution network operates at 33 and 11 kV supplying the customer's feeders at 400 V three-phase, 230 V per phase. Transformer used in the power system ranges from small 500 kVA distribution transformers to 1000 MVA supergrid transformers.

Detailed theoretical analysis of transformers has been covered in numerous textbooks. It is suffice to show here the equivalent circuit of a transformer to appreciate its electrical circuit representation. The component R_c represents the core losses that consist of eddy current losses and hysteresis losses, while the

FIGURE 7.5 Transmission lines out of Connaught Bridge Power Station, Malaysia.

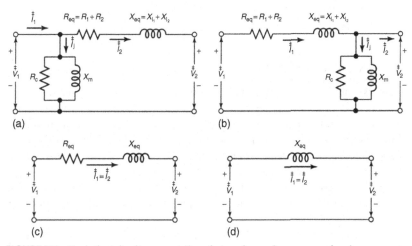

FIGURE 7.6 Equivalent circuit representation of a transformer from a comprehensive representation to most simplified.

component X_m represents the flow of the magnetization current. Core losses also known as no-load losses are essentially the power required to energize the core. It depends primarily on the voltage and frequency and does not vary much with system operation variations. On the other hand, R_{eq} and X_{eq} represent

FIGURE 7.7   Steel lattice structure of a transmission line tower.

the primary and secondary winding of the transformer where the load losses originate. Load losses consist of the I^2R losses mainly in the windings and stray losses that include winding eddy losses and the effect of leakage flux entering internal metallic structures.

7.3 TRANSMISSION LINE

The bulk transfer of electrical energy from generating power plants to substations located near population centers is achieved through high-voltage transmission lines. Interconnected transmission lines are referred to as high-voltage transmission networks. Electricity is transmitted at high voltage to reduce the energy lost in long distance transmission. It is important to understand the characteristic of the important component, which is the conductor, to appreciate the limiting factors to the power transfer capability of the transmission line. The power transfer can then be optimized using latest sensor and communication technology.

The conductor is one of the major components of a transmission line design. It is essential that the most appropriate conductor type and size be selected for optimum operating efficiency. There are four common types of conductors used for many years in electric utilities: (i) all-aluminum conductor, AAC, (ii) aluminum conductor steel-reinforced, ACSR, (iii) aluminum conductor alloy-reinforced, ACAR, and (iv) all-aluminum alloy conductor, AAAC. Regardless of the type of metal used in the make-up of the conductor, the strands are always round and have a concentric lay.

7.3.1 Aluminum Conductor Steel-Reinforced, ACSR

Electric utility companies have utilized ACSR as a common choice of conductor in transmission lines for many years. ACSR is used extensively on long spans as both ground and phase conductors because of its high mechanical strength-to-weight ratio and good current-carrying capacity. ACSR consists of solid or stranded coated steel core surrounded by one or more layers of 1350-H19 aluminum. Because of the presence of the 1350 aluminum in the construction, ACSR has equivalent or higher thermal ratings to equivalent sizes of AAC [2].

The cross-section area of ACSR is specified according to the cross-sectional area of aluminum to be contained in the construction. For example, a 428 mm^2 − 54/7 ACSR has 428 mm^2 of aluminum, the equivalent aluminum area content of 428 mm^2 AAC. The steel content of ACSR typically ranges from 11 to 18% by weight for the conductor stranding of 18/1, 45/7, 72/7, or 84/19. However, it can vary up to 40% depending on the desired tensile strength. It is desirable for ground wires in extra long spans crossing rivers, for example, to have a stranding of 8/1, 12/7, or 16/19, giving them higher tensile strength. Figure 7.8 shows the standard stranding of ACSR [3].

The high tensile strength combined with the good conductivity gives ACSR several advantages such as:

- The high tensile strength of ACSR allows it to be installed in areas subject to extreme wind loading.
- Because of the presence of the steel core, lines designed with ACSR elongate less than other standard conductors, yielding less sag at a given tension. Therefore, the maximum allowable conductor temperature can be increased to allow a higher thermal rating when replacing other standard conductors with ACSR.
- ACSR is less likely to be broken by falling tree limbs.

7.4 FACTORS THAT LIMIT POWER TRANSFER IN TRANSMISSION LINE

Factors that limit the power transfer are voltage drop, voltage stability, and thermal rating. For short transmission line, which is approximately less than 80 km, the power transfer is limited by the thermal rating. The thermal rating is limited by the maximum operating temperature of the conductor. This maximum operating temperature is limited by the maximum line sag limit and maximum allowable operating temperature of the conductor material.

7.4.1 Static and Dynamic Thermal Rating

Static thermal rating of the overhead transmission line is a fix rating in terms of current carrying capacity. It is a conservative rating where worse case weather for contributing for maximum heat onto the conductor is assumed.

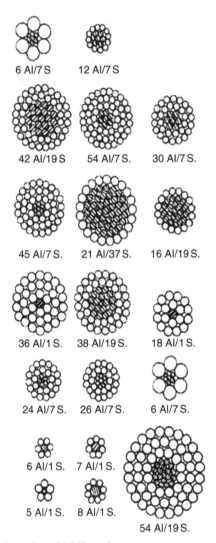

FIGURE 7.8 Standard stranding of ACSR conductor.

Dynamic thermal rating of the overhead transmission line is a time dependence rating where the actual weather and loading of the conductor are measured and the thermal rating is calculated in real time.

7.4.2 Thermal Rating

Bare overhead conductor is normally exposed to environment when it is in service. Due to this reason, the conductor temperature is normally influenced by weather condition as well as the electrical current loading of the conductor.

In thermal rating calculation, meteorological parameters such as wind speed and direction, solar radiation and ambient temperature are required for the calculation where the calculation is based on heat balance energy equation; in steady-state condition, the conductor heat gain is equal to the heat loss (no heat energy is stored in the conductor). The mathematical representation of the heat balance energy equation as in Eq. (7.1) and by reorganizing the equation as in Eq. (7.2), the thermal rating can be calculated [4]. The heating elements are solar radiation and internal heating by the electrical current flow through the conductor resistance or ohmic loss (I^2R). The heat loss is due to convection and radiation.

$$q_c + q_r = q_s + I^2 \cdot R(T_c) \tag{7.1}$$

$$I_{\text{rating}} = \sqrt{\frac{q_c + q_r - q_s}{R}} \tag{7.2}$$

where

q_s Heat input due to solar, W/m

$I^2R(T_c)$ Heat input due to line current (R is a function of conductor temperature)

q_c Heat output due to convection (a function of wind, air temperature, conductor temperature), W/m

q_r Heat output due to radiation (a function of wind, air temperature, conductor temperature), W/m

7.4.3 Convection Heat Loss

Convection heat loss calculation can be grouped into natural convection and forced convection. The natural convection heat loss, q_{cn}, is calculated in the condition where there is no wind. This happens in all thermal systems where there is a thermal gradient. Heat will migrate from the hotter region to the cooler region until the temperature is uniform across the entire system. The natural convection heat loss (watt per unit length, W/m) for the bare stranded overhead conductor is a function of air density, ρ_f, conductor diameter, D, and temperature different between conductor and ambient, $(T_c - T_a)$ as given by Eq. (7.3) [4]

$$q_{cn} = 0.0205\rho_f^{0.5}D^{0.75}(T_c - T_a)^{1.25} \tag{7.3}$$

In conditions where there is wind, wind blowing against the bare overhead conductor introduces cooling effect where the heat from the conductor will be transferred over the temperature gradient, from high to low, to the moving air, this is called forced convection heat loss. There are two equations used in the calculation for the forced convection heat loss. One is used for low wind speed, q_{c1}, and another for high wind speed, q_{c2} [4]. The reason that there are two similar, yet independent, equations is because the corrections factors incorporated into the equations are proprietary to a certain band of conditions.

Equation (7.4) is for low wind speed and Eq. (7.5) is for high wind speed.

$$q_{c1} = \left[1.01 + 0.0372\left(\frac{D\rho_f V_w}{\mu_f}\right)^{0.52} \cdot k_f \cdot k_{angle} \cdot (T_c - T_a)\right] \text{W/m} \qquad (7.4)$$

$$q_{c2} = \left[0.0119\left(\frac{D\rho_f V_w}{\mu_f}\right)^{0.6} \cdot k_f \cdot k_{angle} \cdot (T_c - T_a)\right] \text{W/m} \qquad (7.5)$$

Where air density, ρ_f, air viscosity, μ_f, and coefficient of thermal conductivity of air, k_f, are calculated using Eqs. (7.8), (7.9), and (7.10), respectively at T_{film} where

$$T_{film} = \frac{T_c + T_a}{2} \qquad (7.6)$$

At any wind speed, the highest of the two calculated forced convection heat loss rates is used. The angle at which the wind is blowing relative to the conductor's axis is also considered. The convective heat loss rate is multiplied by the wind direction factor, K_{angle}, where ϕ is the angle between the wind direction and the conductor axis. This modifying factor is used to create a modified forced convective heat loss value. The maximum heat loss will occur when the wind is perfectly perpendicular. The wind direction factor is calculated using Eq. (7.7)

$$K_{angle} = 1.194 - \cos(\phi) + 0.194\cos(2\phi) + 0.368\sin(2\phi) \qquad (7.7)$$

$$\rho_f = \frac{1.293 - 1.525 \times 10^{-4}H_e + 6.379 \times 10^{-9}H_e^2}{1 + 0.00367T_{film}} \text{ kg/m}^3 \qquad (7.8)$$

$$\mu_f = \frac{1.458 \times 10^{-6}(T_{film} + 273)^{1.5}}{T_{film} + 383.4} \text{ Pa.s} \qquad (7.9)$$

$$k_f = 2.424 \times 10^{-2} + 7.477 \times 10^{-5}T_{film}$$
$$- 4.407 \times 10^{-9}T_{film}^2 \text{ W/m}^\circ\text{C} \qquad (7.10)$$

7.4.4 Radiative Heat Loss

Radiative heat loss is usually a small fraction of the total heat loss. It is depending on conductor diameter, D, emisivity, ε, conductor temperature, T_c, and ambient temperature, T_a, as given by Eq. (7.11) [4]. If the ambient temperature is kept constant, e.g., at 28°C, the radiative heat loss increases exponentially with conductor temperature as shown in Fig. 7.9. Emissivity, ε, is dependent on the conductor surface and varies from 0.27 for new stranded conductor to 0.95 for industrial weathered conductors [5]. It is equal to the ratio of the heat radiated by the conductor and to the heat radiated by a perfect black body of the same shape

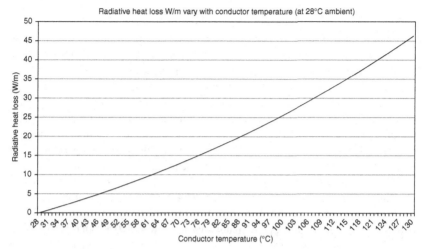

FIGURE 7.9   Radiative heat loss (W/m) as function of conductor temperature.

and orientation. Typical values for new bare overhead conductor are between 0.2 and 0.3. The solar emissivity of a new bare overhead conductor increases with age. A typical value for a conductor that has been in use for more than 5 (in industrial environments) to 20 years (in rural clean environments) is 0.7. A value of 0.5 is used if nothing is known about the conductor emissivity [4].

$$q_r = 0.0178 D\varepsilon \left[\left(\frac{T_c + 273}{100} \right)^4 - \left(\frac{T_a + 273}{100} \right)^4 \right] \text{W/m} \qquad (7.11)$$

7.4.5 Solar Heat Gain

The solar heat gain, q_s, depends on the conductor diameter, D, absorptivity, α, and global solar radiation, S, as given by Eq. (7.12)

$$q_s = \alpha S D \qquad (7.12)$$

The conductor solar absorptivity, α, is a number that varies between 0.27 for bright stranded aluminum conductor and 0.95 for weathered conductor in industrial environment. It is equal to the ratio of the solar heat absorbed by the conductor to the solar heat absorbed by a perfect black body of the same shape orientation. Typical values for new conductor are between 0.2 and 0.3. The solar absorptivity of an overhead conductor increases with age. A typical value for a conductor that has been in use for more than 5 (in industrial environments) to 20 years (in rural clean environment) is 0.9. A value of 0.5 is often used if nothing is known about the conductor absorptivity [4].

7.4.6 Ohmic Losses ($I^2 R(T_c)$) Heat Gain

The ohmic loss of the bare overhead conductor is depending on the amount of current flows and conductor electrical resistance whereas the conductor electrical resistance is depending on conductor temperature. For all bare overhead conductors, controlled lab tests have been carried out; high and low temperature and electrical resistance values have been established. This is then published as a guide for electric utilities to use as a reference in their estimation of their system's performance. As upper and lower limits have been defined, it is merely a matter of interpolation to obtain values between the given temperatures. For values outside the envelope, extrapolation can be readily used to obtain the result with reasonable accuracy. For example, lab tests conducted for 1350 H19 aluminum shows that for entry resistance values at temperatures of 25 and 75°C the errors are negligible [6]. For entry resistance values at temperatures of 25 and 175°C, the errors are approximately 3% too low at 500°C but 0.5% too high at 75°C. It is concluded that the use of resistance data for temperature of 25 and 75°C is adequate for rough calculation of steady-state and transient thermal rating for conductor temperature up to 175°C. The formula that is used to obtain the resistance of the conductor, per unit length, based on the established lab test results are given in Eq. (7.13)

$$R(T_c) = \left[\frac{R(T_{High}) - R(T_{Low})}{T_{High} - T_{Low}} \right] \cdot (T_c - T_{Low}) + R(T_{Low}) \qquad (7.13)$$

7.5 EFFECT OF TEMPERATURE ON CONDUCTOR SAG OR TENSION

7.5.1 Conductor Temperature and Sag Relationship

The relationship of the sag to the temperature of a homogenous conductor is almost linear. A third-degree equation gives a very close approximation of the temperature as a function of the sag, Eq. (7.14) [8]

$$\text{Temperature} = T_0 + A(S - S_0) + B(S - S_0)^2 + C(S - S_0)^3 \qquad (7.14)$$

The line sag can be determined from the measured tension or direct measured ground clearance. For practical purposes, the sag can be considered to be inversely proportional to the tension in most spans. For a higher degree of accuracy, the sag/tension relationship can be determined to any desired degree of accuracy based on catenary equations. For level conductor spans as in Fig. 7.10, the low point is in the center of the total sag, D given by Eq. (7.15).

$$D = \frac{H}{w} \left[\cosh\left(\frac{ws}{2H}\right) - 1 \right] \qquad (7.15)$$

The horizontal component of tension, H, is located at the point in the span where the conductor slope is horizontal, or at the midpoint for level spans. The

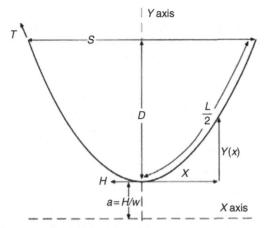

FIGURE 7.10   Catenary sag model – level span [7,8].

conductor tension, T, is found at the ends of the spans at the point of attachment and is calculated using Eq. (7.16) and it corresponds to a conductor length, L, given by Eq. (7.17).

$$T = H + wD \qquad (7.16)$$

$$L = \left(\frac{2H}{w}\right) \sin h\left(\frac{Sw}{2H}\right) \qquad (7.17)$$

Equation (7.17) describes the behavior of ideal (with perfectly elastic stress and strain characteristics) concentric-lay stranded conductors [6]. The actual conductors such as ACSR conductors exhibit nonlinear behavior when loaded from initial tension to some final value due to wind or temperature loading. Permanent elongation from creep and heavy loading also affects the resulting sag. Also, high-temperature operations result in thermal elongation of the steel and aluminum strands, thus affecting sag. Therefore, in order to calculate the correct sag, it is necessary to separate the effects of conductor elongation due to tensile loading as well as thermal loading. This process requires an iterative procedure in which the mathematical formulas describing the conductor elongation caused by the temperature change are solved simultaneously with the tension and conductor length relationship. To calculate the change in length due to temperature loading Eq. (7.18) is used.

$$L_T = L_{T_{REF}} [1 + \alpha_{AS}(T - T_{rRef})] \qquad (7.18)$$

Where

L_T	= Final length of conductor
$L_{T_{REF}}$	= Reference length of the conductor
α_{AS}	= Coefficient of linear thermal elongation for AL/SW strands, given in Table 7.1
$(T - T_{rRef})$	= Change in temperature

TABLE 7.1 Coefficients of thermal expansion for bare stranded conductors [7]

Conductor	α (per degree C)
AAC	23.0×10^{-6}
36/1 ACSR	22.0×10^{-6}
18/1 ACSR	21.1×10^{-6}
45/7 ACSR	20.7×10^{-6}
54/7 ACSR	19.5×10^{-6}
26/1 ACSR	18.9×10^{-6}
30/7 or 30/19 ACSR	17.5×10^{-6}

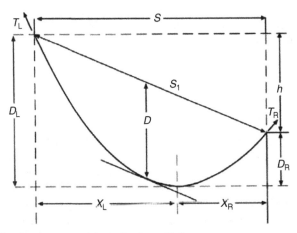

FIGURE 7.11 Catenary sag model – inclined span [7,8].

For inclined spans, the length of the conductor between the supports is divided into two separate sections as in Fig. 7.11. The same equation in (7.15) can be used for the calculation of the height of the conductor. In each part of the span, the sag is dependent upon the vertical distance between support points and can be described by Eqs. (7.19) and (7.20) [8].

$$D_R = D\left(1 - \frac{h}{4D}\right)^2 \tag{7.19}$$

$$D_L = D\left(1 + \frac{h}{4D}\right)^2 \tag{7.20}$$

Most of the sag and tension calculations are typically performed with line rating software, e.g., PLS-CADD software from Power Line Systems Inc.,

USA. To demonstrate the process on how sags and tensions occur for overhead line, simplified sag calculations and conductor length can be done by using Eqs. (7.21) and (7.22), respectively [6].

$$D = \frac{H}{w}\left[\cos h\left(\frac{ws}{2H}\right) - 1\right] \approx \frac{wS}{8H} \tag{7.21}$$

$$L = \left(\frac{2H}{w}\right) \sin h\left(\frac{Sw}{2H}\right) \approx S + \frac{8D^2}{3S} \tag{7.22}$$

The example of the relationship between conductor temperature and conductor sag can be describe by the following example [7]:

A transmission line with ACSR conductor 795 kcmil (405 mm^2) and a 1000 ft level span. The conductor weight per unit length is 1.094 lbs/ft (1.6 kg/m), ambient temperature is 77°F (25°C) and horizontal tension component, H, is 25% of the ultimate tension strength (31,500 lbs)

$$H = 0.25 \times 31,500\,\text{lbs} = 7875\,\text{lbs}(35,196\,\text{N})$$

Conductor sag, D, is calculated by using Eq. (7.21).

$$D = \frac{wS^2}{8H} = \frac{1.094(1000)^2}{(8)\,7875} = 17.37\,\text{ft}(5.29\,\text{m})$$

Conductor length, L is calculated by using Eq. (7.22).

$$L = S + \frac{8D^2}{3S} = 1000 + \frac{8(17.37)^2}{3(1000)} = 1000.8\,\text{ft}(305.05\,\text{m})$$

Equation (7.18) is used to calculate the conductor length when the conductor temperature increase from ambient to 50°C (122°F) and then to 150°C (302°F). The coefficient of linear thermal expansion is $10.7 \times 10^{-6}/°\text{F}$.

$$L_T = L_{T_{REF}}\left[1 + \alpha_{AS}(T - T_{rRef})\right]$$

$$L_{(122)} = 1000.80\left[1 + 10.7 \times 10^{-6}(122 - 77)\right]$$

$$= 1001.28\,\text{ft}(305.2\,\text{m})$$

$$L_{(302)} = 1000.80\left[1 + 10.7 \times 10^{-6}(302 - 77)\right]$$

$$= 1003.21\,\text{ft}(305.8\,\text{m})$$

The conductor sag is calculated by rearranging Eq. (7.22)

$$D = \sqrt{\frac{3S(L - S)}{8}}$$

$$D_{(122)} = \sqrt{\frac{3(1000)(1.28)}{8}} = 21.9\,\text{ft}(6.68\,\text{m})$$

$$D_{(302)} = \sqrt{\frac{3(1000)(3.21)}{8}} = 34.7\,\text{ft}(10.58\,\text{m})$$

TABLE 7.2 Relationship Between Conductor Temperature and Conductor Sag for the ACSR Conductor 405 mm^2 for a 1000 ft Level Span [6]

Conductor temperature (°C)	Conductor sag, D (m)
25	5.29
50	6.68
150	10.58

The relationship of the conductor temperature and the conductor sag for the above example can be summarized in Table 7.2.

The conductor temperature corresponds to the measured sag given by Eq. (7.14), since the current conductor temperature is known, the dynamic thermal rating for the maximum conductor temperature for example at 75°C can be calculated by using Eq. (7.2).

From the known conductor temperature, the value of q_r can be calculated by the thermal rating Eq. (7.11). The value of q_s can be calculated by Eq. (7.12), the conductor resistance at the known temperature can be calculated by Eq. (7.13). By rearranging the thermal balance equation and with the known value of conductor current the convective heat loss q_c can be calculated using Eq. (7.23).

$$q_c = I^2 R(t) + q_s - q_r \qquad (7.23)$$

From the known value of q_c the equivalent perpendicular wind speed can be calculated by Eq. (7.4) or (7.5). Assuming the equivalent perpendicular wind speed is constant then the thermal rating at the maximum conductor temperature can be calculated.

7.6 STANDARD AND GUIDELINES ON THERMAL RATING CALCULATION

Reference standard for thermal rating calculation is published by IEEE, the first standard is IEEE Std 738-1993, IEEE Standard for Calculation of Bare Overhead Conductor Temperature and Ampacity Under Steady-State Conditions. Later the new revision of this standard is published; IEEE Std 738-2006. The revised standard has additional information that addresses fault current and transient rating calculation, SI units were used throughout, the solar heating calculation was extensively revised to cater for most part of the world instead of only 20° North and above in the first standard IEEE Std 738-1993. The IEEE standard presents a method of calculating the current–temperature relationship of overhead lines given the weather conditions. The heat balance energy equation used for thermal rating calculation is described in details in this standard. The equation relating electrical current to conductor temperature described in this standard can be used in two ways:

- To calculate the conductor temperature when the electrical current is known.
- To calculate the current that yields a given maximum allowable conductor temperature.

Another guidelines on thermal rating calculation were developed by CIGRE working group WG 22.12, published in Electra No. 144 in October 1992, The Thermal Behavior of Overhead Conductors, Sections 1 and 2: Mathematical model for evaluation of conductor temperature in the steady state and the application thereof. And another one published in Electra No. 174 in October 1997, The Thermal Behavior of Overhead Conductors, Section 3: Mathematical model for evaluation of conductor temperature in the unsteady state. In these guidelines, the heat balance energy equation is discussed in more details compared to in IEEE standard where other addition elements are added in the heat balance energy equation such as magnetic heating, corona heating, and evaporating cooling.

7.7   OPTIMIZING POWER TRANSMISSION CAPACITY

7.7.1   Overview of Dynamic Thermal Current Rating of Transmission Line

Real-time thermal rating methods can be classified into two basic types: weather model (WM) and conductor temperature model (CTM) [4]. The difference between the two models is the method by which the conductor convective heat transfer coefficient is calculated. The WM approach calculates the convective heat transfer coefficient using ambient temperature, wind speed, and wind direction data together with empirical expressions for convective heat transfer for air flow past cylindrical objects. The CTM computes the convective heat transfer coefficient directly using the conductor heat balance energy equation, conductor temperature, air temperature, and conductor current. There are five main dynamic thermal rating techniques in WM and CTM that is weather monitoring, line tension monitoring, sag monitoring, line temperature monitoring, and conductor replica technique. The most appropriate technique depends upon a particular application and is based on various issues including accuracy, cost, and ease of installation [7].

In general, the overview of DTCR system is as shown in Fig. 7.12. There are three main components that form the Dynamic Thermal Current Rating (DTCR) system; remote monitoring station, communication system, and the system computer. Remote monitoring station gathers information on the environment in which the transmission lines are located and also information on the line sag/clearance or tension or conductor temperature. The remote monitoring

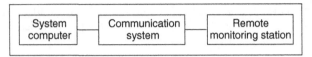

FIGURE 7.12   Dynamic thermal current rating (DTCR) system block diagram.

station is normally placed at a critical location along the transmission line where the wind cooling is minimal or where there is an increased probability of contact between the conductor and objects on the ground.

The communication system used in this real-time thermal rating system can be wireless or wired system. Normally, wireless communication system is more suitable for the thermal rating system compared to wired system due to connectivity issues and cost. The wireless system is easy to be implemented; it is robust and suitable for use in the outdoor environment. Data transfer from the remote monitoring station to the system computer can be based on dial-up connection at certain time intervals.

The system computer normally consists of two software modules; thermal rating calculation module and thermal rating display module. It has two primary functions, namely to process the weather and line clearance data then calculate the thermal rating of the line and to display the thermal rating in real-time.

As electrical current in an overhead conductor increases, the line temperature increases and therefore the line sags. Each line has a minimum clearance to ground, which must never be violated for safety reasons. The thermal rating is the maximum current which results in the line sagging down exactly to the minimum clearance. Any additional current would result in too much sag and therefore breaches the safety requirement [9].

There are a few methods of transmission line sag monitoring that has been developed. One of the methods proposed is based on Global Positioning System (GPS). The method is known as differential GPS [10]. This mode consists of the use of two GPS receivers, the base and the rover as shown in Fig. 7.13. The actual position of the base is known (e.g., by precise surveying) and compared to the readings received at the same base point. With the estimated error, the readings obtained at the rover can be compensated by simple subtraction.

Another method to monitor the line sag is by using smart camera. This method is developed by Electric Power Research Institute (EPRI), USA, [4]. The system uses an imaging system to monitor the location of the conductor or a target attached to it as shown in Fig. 7.14. The imaging system is installed on either a transmission line structure or any other structure with a view of the span being monitored. The field of view of the imaging system remains the same at all time and the location of the conductor or the target in the view changes as the conductor moves up, down, or horizontally. Image processing technique is used to determine the location of the conductor or the target attached to the conductor in the cameras field of view. The search algorithm uses a pre-selected sub-image of the target and searches the image to determine the most likely location of the

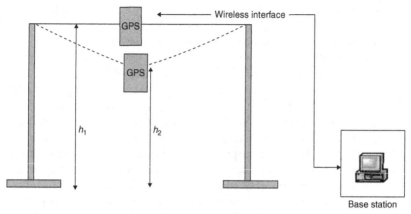

FIGURE 7.13   Line sag measurement using differential GPS technique [6].

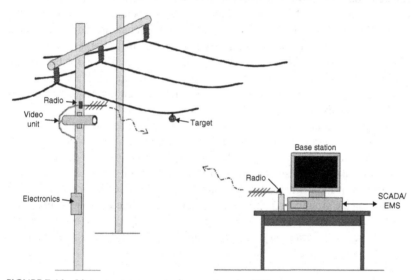

FIGURE 7.14   Line sag measurement using smart camera [11].

target in the camera image. The key for the success of this approach is to have a target with unique features that cannot be easily found in the background. The change in vertical position of the conductor in the image is directly related to the change in ground clearance or sag. At the time of installation, location of the conductor or the target attached to the conductor in the imaging system's field of view is calibrated to the measured ground clearance/sag. Ground clearance or sag at any later time is obtained from the measured change in location of the conductor using calibration constants obtained at the time of system calibration. The resulting clearance information may be made available on a real-time basis using telemetry and/or logged in a data logger for historical study. An effective conductor temperature is calculated using sag-tension routines of line rating software such as PLS-CADD software from Power Line Systems Inc., USA and the real-time ratings is calculated using IEEE standard 738 [4].

Tension monitors consist of solar-powered line tension monitoring system located along the transmission line that communicate the measured data via spread spectrum radios to receiving units at substations which are further linked to the utility's EMS/SCADA systems. The conductor tension is measured using load cell installed at the dead-end insulators. There are also sensors that measure the ambient temperature and the net radiation at the conductor. All data are transmitted to the center processor computer for the line thermal rating calculation. The calibration of the system allows determination of the conductor temperature based on the measured tension [2].

Direct measurement of conductor temperature method uses temperature sensor that is directly attached to the conductor. It uses solid-state thermoelectric sensors and it is power-up using line current transformer. The system was developed by Niagra Mohawk in 1980s and the temperature sensor was called Power Donut. This method can provide conductor temperature data in real-time. Power-Donut has ambient and conductor temperature measurement accuracy of ±2°C over a range of −40–125°C [12]. The advantage of this method is that the user has a direct measurement of conductor temperature. If the line rating is intended to limit the loss of conductor strength at high temperature, then this method is mostly appropriate.

Ground clearance has direct relationship with line sag as they depend on conductor temperature. Therefore, thermal rating can also be determined by monitoring the ground clearance. Laser distance measurement sensor can be used to measure the distance between the transmission line and ground. The installation of the laser distance measurement sensor must be at the middle of the span on the ground directly under the transmission line as shown in Fig. 7.15.

To enable continuous monitoring of the line clearance, the sensor can be set to operate in automatic mode which will take measurement

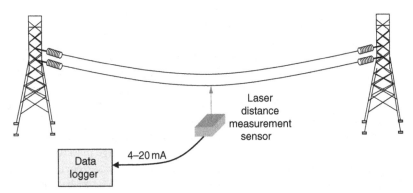

FIGURE 7.15 Laser distance measurement sensor installation.

FIGURE 7.16 Weather monitoring using a weather station for transmission line in Malaysia.

continuously for a certain time interval. The line thermal rating can be calculated based on the conductor temperature determined by the clearance measurement.

A number of studies have shown that it is possible to calculate the thermal rating of an overhead conductor in a single span if the weather conditions in the span are known. This can be done using historical weather data records in order to select appropriately conservative static ratings or it can be done in real-time as a basis to dynamically rate the line. By accounting for variations in span orientation along a transmission line, it was found that the weather data from a single weather station as far as 20 miles away could be used to predict the statistical distribution of thermal ratings for that line [5]. For a real-time monitoring, the dynamic thermal rating of the line is calculated using real-time value of the line load current and the weather data. Example of a weather station installation at the transmission line is shown in Fig. 7.16.

7.7.2 Example of Dynamic Thermal Current Rating of Transmission Line

The dynamic ratings profile as shown in Fig. 7.17 is based on DTCR calculations using weather data recorded by a weather station for one of the transmission line in Malaysia [13]. The figure shows three curves:

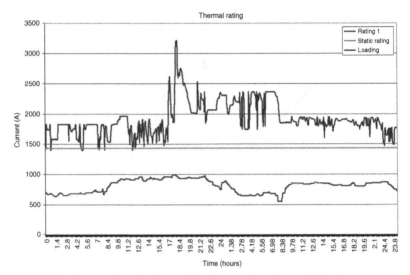

FIGURE 7.17   Dynamic thermal rating profile of a transmission line in Malaysia.

- Measured load on the line
- Static rating
- Calculated dynamic rating from DTCR software

It is important to note that:

- The dynamic rating is almost always greater than the static rating
- The measured load is always significantly lower than the static rating

For statistical representation of the thermal rating, the thermal rating data for certain duration, e.g., 6 months can be represented using probability density plot. The probability density plot shows the frequency of the thermal rating in a certain period of time. It can be used as a guide for operating the transmission line or revising the static rating. Example of the probability density plot is shown in Fig. 7.18.

The graphs show that the load is always below the static rating. Only during emergencies the load would exceed the normal static rating. The dynamic thermal rating is mostly 20% higher than the static rating which is shown at the tip of the dynamic thermal rating probability density graph in Fig. 7.18. This is the available potential extra capacity ready to be utilized when DTCR system is used to monitor the line rating.

7.8   OVERVOLTAGES AND INSULATION REQUIREMENTS OF TRANSMISSION LINES

The consideration for insulation requirements of transmission lines are critical because overvoltages can occur as a result of lightning strokes, switching, faults,

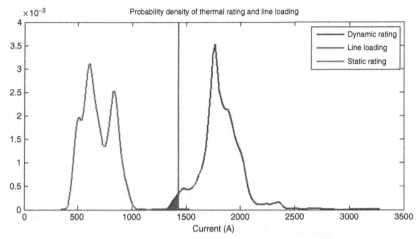

FIGURE 7.18 Probability density plot of the dynamic thermal rating and line loading of a transmission line in Malaysia.

load failure, and others. Overvoltages are transitory phenomena. It is defined as any transitory voltage between phase and ground or between phases with a crest value higher than the crest value of maximum system voltage [14]. These overvoltages are often expressed in per unit values, with maximum system voltage as per unit base.

For example, the phase-to-ground per unit overvoltage is

$$V_{pu} = \frac{V_{L-G}\sqrt{3}}{V_m\sqrt{2}}$$

where V_{pu} = per unit phase-to-ground overvoltage

V_{L-G} = crest value of phase-to-ground overvoltage

V_m = maximum rms system voltage

This condition can be transient or permanent depending on its duration. Understanding overvoltage behavior including power frequency phenomena as well as switching and lightning surges are important prerequisite conditions in determining insulation coordination in a power system network.

The overvoltage phenomena are classified into four major classes as follows [15]:

a. Power frequency overvoltages (also known as temporary overvoltages) that include ferranti effect, self-excitation generator, overvoltage of unfaulted phases due to single-line-to-ground fault and sudden load failure

b. Lower frequency harmonic resonant overvoltages that include lower order frequency resonance and local area resonant overvoltages

c. Switching surges that include breaker closing overvoltages, breaker tripping overvoltages, and switching surges by line switches

d. Overvoltages by lightning strikes direct to phase conductors (direct flashover), direct stroke to overhead shieldwire or to tower structure (back flashover) or induced flashovers

e. Overvoltages due to abnormal conditions such as interrupted ground fault of cable, overvoltages induced on cable sheath or touching of different kilovolts line, etc.

In this chapter, the emphasis is given on overvoltages due to lightning strikes and switching surges as well as temporary overvoltages that have great impact in the selection of insulation requirements for transmission lines.

7.8.1 Overvoltage Phenomena by Lightning Strikes

By definition, lightning is an electrical discharge. It is the high-current discharges of an electrostatic electricity accumulation between thundercloud and ground (cloud-to-ground discharges) or within the thundercloud (intracloud discharges). Other types of discharges such as cloud-to-cloud lightning and cloud-to-air lightning also exist but not very frequent. The thunderclouds contain ice crystals which are positively charged upper portion and water droplets that are negatively charged, usually located at lower portion of the thundercloud. The vertical circulation in terms of updrafts and downdrafts due to height and temperature effects caused disposition of charges in the thunderclouds.

The mechanism of a lightning stroke is typically explained as follows. As the negative charges build up in the cloud base, positive charges are induced on the ground. When the voltage gradient within the cloud build up to the order of $5-10$ kV/cm, the ionized air breaks down and creates an ionized path (leader) moving from thundercloud to ground. This initial leader progresses by series of jumps called a stepped leader. As this stepped leader (or downward leader) nears the ground, the upward leader (or return stroke) is initiated that meets the downward leader. This return stroke propagates upward from the ground to the thundercloud following the same path, at the speed of 10–30% that of light, which is visible to naked eye. The current involved may exceed 200 kA and lasting about 100 μs. The second leader called dart leader, may stroke following the same path taken by the first leader, about 40 μs later. The process of dart leader and return stroke can be repeated several times. The complete process of successive strokes are called lightning flash.

Lightning surge phenomena can be classified into three different stroke modes: (a) direct strike on phase conductors (direct flashover), (b) direct strike on shieldwire or tower structure, and (c) induced strokes.

In the case of direct strike, the lightning current path is directly from the thundercloud to the line. This will cause the voltage to rise at stroke termination point, either on one phase or more phase conductors. The stroke current propagates in the form of a travelling wave in both directions and raises potential of the line to the voltage of the downward leader. Overvoltage may exceed the

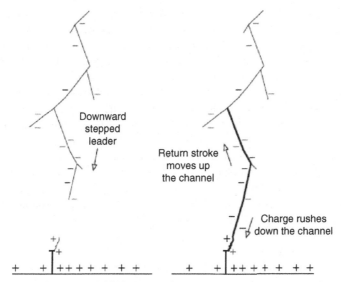

FIGURE 7.19   Cloud-to-ground lightning.

line-to-ground withstand voltage of line insulation and cause insulation failure, if not properly protected.

In the case of direct strike on shieldwire or tower structure, a lightning stroke directly strikes the shieldwire or tower structure. If lightning strikes a tower top, some of the current may flow through the shieldwires, and the remaining current flows through the tower to the earth. For stroke with average current magnitude and rate of rise, the current may flow into the ground provided the tower and its footing resistance are low. Otherwise, the lightning current will raise the tower to a high voltage above the ground, causing flashover from the tower over the line insulators to one or more phase conductors.

In induced strokes, lightning strokes terminate at some point on the Earth at a short distance from a transmission line. This can be considered as a virtual wire connected to a cloud and an earth point, and a surge current $I(t)$ flowing along the virtual wire. This wire has mutual capacitances and mutual inductances across the line conductors and the shieldwires of the neighborhood transmission lines. In this regard, the capacitive induced voltages as well as inductive induced voltage appear on the phase conductors as well as on the shieldwires [15].

The insulation for lines is composed of air and solid dielectric insulators. The geometry of the insulators and their insulation strengths are selected to ensure that if an insulation failure occurs, the failure will be a flashover in air. This flashover produces a low impedance path through which 50 Hz power current will flow. Generally, these arcs are not self-extinguishing. To interrupt the power fault a protective device (circuit breaker) operating to de-energize the circuit will be required. Four types of lightning-caused flashover can occur on transmission lines: back flashover, shielding failure, induced, or midspan.

A back flashover event can occur when lightning strikes a grounded conductor or structure. In this case, a flashover proceeds backward from tower metal to the insulated conductor. A lightning stroke, terminating on an overhead ground wire or shield wire, produces waves of current and voltage that travel along the shield wire. At the tower/pole, these waves are reflected back toward the struck point and are transmitted down the tower/pole toward the ground and outward onto the adjacent shield wires. Riding along with these surge voltages are other surge voltages coupled onto the phase conductors. These waves continue to be transmitted and reflected at all points of impedance discontinuity. The surge voltages are built up at the tower/pole, across the phase-ground insulation, across the air insulation between phase conductors, and along the span across the air insulation from the shield wire to the phase conductor. If this surge voltage exceeds the insulation strength, flashover occurs. The parameters that affect the line back flashover rate (BFR) are:

- Ground flash density
- Surge impedances of the shield wires and tower/pole
- Coupling factors between conductors
- Power frequency voltage
- Tower and line height
- Span length
- Insulation strength
- Footing resistance and soil composition

Sometimes, the design engineer can vary the shield wire surge impedance and the coupling factors, e.g., using two shield wires instead of one. Normally, only insulation and footing impedance can be varied to improve back flashover performance. Reducing the footing impedance directly reduces the voltage stress across the insulator for a given surge current down the tower.

A shielding failure is defined as a lightning stroke that terminates on a phase conductor. For an unshielded line, all strokes to the line are shielding failures. For a transmission line with overhead shield wires, most of the lightning strokes that terminate on the line hit the shield wire and are not considered shielding failures.

The calculated number of shielding failures for a particular transmission line model depends on a number of factors, including the model's electrogeometric parameters; the stroke current distribution; and natural shielding from trees, terrain, or buildings. Not all shielding failures will result in insulator flashover. The critical current is defined as the lightning stroke current that, injected into the conductor, will result in flashover. The critical current for a particular transmission line conductor is calculated by:

$$I_c = \frac{2 * (CFO)}{Z}$$

where

CFO is the lightning impulse negative polarity critical flashover voltage
Z is the conductor surge impedance

Severe transient overvoltage can be induced on overhead power lines by nearby lightning strikes. On lower voltage distribution power lines, indirect lightning strikes cause the majority of lightning-related flashovers. Estimation of indirect lightning effects is crucial for proper protection and insulation coordination of overhead lines. The problem of induced flashovers from nearby lightning strikes has received a great deal of scientific attention in the past 20 years, and the result has been the development of more accurate estimation models of lightning-induced overvoltages.

Important points to remember when dealing with induced flashovers from nearby lightning strokes include:

- Insulator CFO voltages above approximately 400 kV prevent nearly all induced flashovers.
- The presence of an effectively grounded overhead shield wire or neutral on the line will reduce insulator voltages by 30–40%, depending on the line configuration.
- Line surge arresters installed every few spans can improve induced flashover performance for distribution voltage lines (spacing line arresters in this manner will seldom improve direct stroke lightning performance, only induced flashovers, and it is not recommended for transmission lines).

Power line flashovers caused by lightning strokes near midspan are unusual for most line configurations. Midspan flashovers become more likely when midspan conductor spacing is small, such as on-distribution lines, or when span lengths are very long (304.8 m or more). The voltage on a conductor follows the equation presented for a shielding failure. If the voltage rises to approximately 610 kV/m in the air gap between conductors, a long, relatively slow breakdown process might occur that might take many microseconds to complete.

7.8.2 Switching Surges

Switching surges can occur during operation of circuit breaker and line switch opening (tripping) and closing at the same substation. In general, switching surges occur in the vicinity of non-self-restoring insulation equipment such as generators, transformers, breakers, cables, etc. Overvoltages caused by switching surges is a concern since they can damage insulation or cause insulation flashover. Damage and flashovers often lead to power system outage which is highly undesirable. The amplitude of a typical switching surge is about 1.5 pu with duration of about one power frequency cycle [16].

7.8.3 Temporary Overvoltage

A temporary overvoltage is an oscillatory phase-to-ground overvoltage that is of relatively long duration and is undamped or only weakly damped. These types of overvoltages usually originate from faults, sudden changes of load, Ferranti effect, linear resonance, ferroresonance, open conductors, induced resonance from couples circuits, and others (e.g., backfeeding, stuck pole). In general, temporary overvoltage has two important characteristics: they are determinable and they persist for many cycles. The magnitudes of these voltages may be calculated by steady-state analysis, assuming the physical conditions on the system are known.

7.9 METHODS OF CONTROLLING OVERVOLTAGES

There are several means for controlling the overvoltages that occur on transmission system. Overvoltages due to switching surges can be controlled by modifying the operation of a circuit breaker or other switching device in various ways. The overvoltage may be limited to an acceptable level with the installation of surge arresters at the specific locations to be protected. The transmission system may be changed to minimize the effect of switching operations.

It is clear that if the source of the overvoltage is a random event such as lightning or fault initiation, the only sure method of controlling overvoltage will be the use of surge arresters or similar protective devices. In lightning protection design, two aspects are important to consider: (i) diversion and shielding, mainly for structural protection that also provides reduction of electric and magnetic fields within the structure and (ii) limiting the currents and voltages on electronic, power, and communication systems through surge protection. For the latter, the protection must include the control of currents and voltages from direct strikes to the structure as well as from lightning-induced current and voltages surges propagating into the structure from aerial or buried conductors from outside. Four general types of current- and voltage-limiting technique are commonly used [17] which are:

a. Voltage crowbar devices such as gas-tube arresters, silicon controlled rectifiers (SCRs) and triacs that limit overvoltages to values smaller than operating voltages and short circuit the current to ground.
b. Voltage clamps such as metal oxide varistors (MOVs), Zener diodes and p-n junction transistors. These solid state nonlinear devices reflect and absorb energy while clamping the applied voltage across terminals to 30–50% above the system operating voltage.
c. Circuit filters that are linear circuits that both reflect and absorb the frequencies contained in the lightning transient pulses.
d. Isolating devices such as optical isolators and isolation transformers that suppress relatively large transients.

Reduction of basic impulse insulation levels of transmission equipment is possible with the use of surge arresters. Arresters are primarily used to protect the system insulation from the effects of lightning. Nowadays, modern arresters can also control many other system surges that may be caused by switching or faults. Arresters provide indispensable aid to insulation coordination in electrical power system. It has basically a nonlinear resistor so that its resistance decreases rapidly as the voltage across it rises. At the surge ends and the voltage across the arrester returns to the normal line-to-neutral voltage level, the resistance becomes high enough to limit the arc current to a value that can be quenched by the series gap of the arrester.

Surge arresters may be used to control temporary overvoltages, such as due to a fault. It may be able to protect the equipment for the short time it takes to clear the applicable breakers. Similarly, surge arresters may be used to control switching surge overvoltages along a transmission line and thus reduce the length of required insulator strings. For example, the use of metal oxide arresters at the receiving end of a 280-km line can reduce the switching overvoltage along the line from a maximum of 2.2 to 1.8 p.u.

7.10 INSULATION COORDINATION

Insulation coordination is the selection of the insulation strength [18]. The strength is selected on the basis of some quantitative or perceived degree of reliability. It is also important to know the stress placed on the insulation and methods to reduce the stress, be it through surge arresters or other means. There are three different voltage stresses to consider when determining insulation and electrical clearance requirements for the design of high-voltage transmission lines: (i) the power frequency voltage, (ii) lightning surge, and (iii) switching surge. In general, lightning surges have the highest value and the highest rates of voltage rise. Properly done insulation coordination provides assurance that the insulation used will withstand all normal discharge and a majority of abnormal discharges. Several terms used in insulation coordination are defined as follows.

The Basic Lightning Impulse Insulation Level (BIL), is defined as the electrical strength of insulation expressed in terms of the crest value of the "standard lightning impulse" [18]. The BILs are universally for dry conditions. It is determined by tests made using impulses of a 1.2/50 μs waveshape. Several tests are performed and the numbers of flashovers are noted.

The Basic Lightning Impulse Insulation Level (BIL) is the electrical strength of insulation expressed in terms of the crest value of a standard switching impulse. BILs are universally for wet conditions.

Withstand Voltage is the highest voltage that the insulation can withstand without failure or disruptive discharge, under specified test conditions. It is a common practice to assign a specific testing waveshape and duration of its application to each category of overvoltage.

There are two major categories of insulation: external insulation is the distances in open air or across the surfaces of solid insulation in contact with open air that is subjected to dielectric stress and to the effects of the atmosphere. Most transmission line insulation is external insulation. Internal insulation is the internal solid, liquid, or gaseous parts of the insulation of equipment that are protected by the equipment enclosures from the effects of the atmosphere such as in a transformer.

Insulation can also be divided into self-restoring and non-self-restoring categories. Self-restoring insulation completely recovers its insulating properties after a disruptive discharge. Non-self-restoring insulation loses insulation properties or does not recover completely after a disruptive discharge. Most external insulation, such as external air gap, is self-restoring insulation and most internal insulation, such as solid insulation in a transformer is non-self-restoring insulation. Insulation failure consequences are different for the different categories. Failure of non-self-restoring internal insulation is catastrophic while failure of external self-restoring insulation may only be a nuisance. Testing may be done after a disruptive discharge for self-restoring internal insulation that provides statistical data for examination.

Insulation coordination involves the following processes: (i) determination of line insulation, (ii) selection of BIL and insulation levels of other equipment, and (iii) selection of lightning arresters. To assist this process, standard insulation levels are recommended such as in IEC Publication 71.1 *"Insulation Coordination Part 1, Definitions, Principles and Rules,"* 1993–1912.

Knowledge of the overvoltages that they system will be exposed may be gained in two general ways: by measurement on the real system or by analysis on related models. Field measurement on a real system can only be done after the system is built. Analytical method can be used to determine the magnitude of overvoltages. The transient behavior of a complex transmission system must be studied through models either analog or digital. Transient network analyzer is an example on an analog model. For transmission lines below 345 kV, the line insulation is determined by lightning flashover rate. At 345 kV, the line insulation may be dictated either by switching surge or by lightning flashover rate. Above 345 kV, switching surges become a major factor in flashover considerations. The probability of flashover due to a switching surge is a function of the line characteristics and the magnitude of the surges expected. The number of insulators may be selected to keep the probability of flashovers due to switching surge very low.

In the design of substation, the maximum switching surge level used is either the maximum surge that can take place in the system or the protective level of arresters. In this case, the insulation coordination in a substation involves the selection of the minimum arrester rating applicable to withstand the power frequency voltage and the equipment insulation level to be protected by arresters.

REFERENCES

[1] World Electrical Power Generation, US Dept. of Energy, EIA (Energy Information Agency), May 2009.

[2] D.C. Lawry, Overhead line thermal rating calculation based on conductor replica method, 0-7803-81 10-6/03, IEEE, 2003.

[3] J.M. Hesterlee, Bare overhead transmission and distribution conductor desks overview, 0-7803-2043-3/95, IEEE, 1995.

[4] IEEE Standard, Standard for calculating the current-temperature relationship of bare overhead conductors, IEEE Std 738-2006, 2006.

[5] Cigre Working-group 22.12, The thermal behaviour of overhead conductors, Section 3: Mathematical model for evaluation of conductor temperature in the unsteady state, Electra no. 174, October 1997, pp. 59-69.

[6] W.Z. Black, R.L. Rehberg, Simplified model for steady state and real time ampacity of overhead conductors, IEEE/PES Winter Meeting, Paper No. 88 WM 236-5, 1985.

[7] R. Adapa, Increased power flow guidebook: Increasing power flow in transmission and substation circuits, Technical Report, EPRI, Palo Alto, CA, 2005.

[8] T.O. Seppa, Accurate ampacity determination: temperature-sag model for operational real time ratings, IEEE Trans. Power Deliv. 10(3) (1995).

[9] C. Mensah-Bonsu, U. Femindez, G.T. Heydt, Y. Hoverson, J. Schilleci, B. Agrawal, Application of the global positioning system to the measurement of overhead power transmission conductor sag, IEEE Trans. Power Deliv. (2000).

[10] S.D. Foss, Evaluation of an overhead line forecast rating algorithm, IEEE Trans. Power Deliv. 7(3) (1992).

[11] A.K. Pandey, Development of a real-time monitoring/dynamic rating system for overhead lines, Consultant Report, California Energy Commission, December 2003.

[12] Wikipedia Website on Selective Availability, http://en.wikipedia.org/wiki/Global_Positioning_System#Selective_availability (accessed on February 2, 2009).

[13] A.A. Rahim, Pilot Implementation of Dynamic Thermal Rating Technology in TNB Transmission Line, TNB Research Sdn Bhd, 2008.

[14] EPRI, Transmission Line Reference Book: 345 kV and Above, second ed., Electric Power Research Institute, Palo Alto, CA, 1982.

[15] Y. Hase, Handbook of Power System Engineering, John Wiley & Sons Ltd., New York, 2007.

[16] Power Standards Lab, Transient Overvoltages: Tutorial, 2004.

[17] V.A. Rakov, M.A. Uman, Lightning Physics and Effect, Cambridge University Press, New York, 2003.

[18] IEC Publication 71.1, Insulation Coordination Part 1, Definitions, Principles and Rules, 1993-12.

Index

Printed in the United States
By Bookmasters